Pressure
Vessel
Systems

Pressure Vessel Systems

A User's Guide to Safe Operations
and Maintenance

Anthony Lawrence Kohan

*Manager, Boiler and Machinery Technical
Specialists, Boiler and Machinery Department,
Royal Insurance Company; National Board
Commissioned Inspector (various state boiler
inspector commissions); Certified Safety
Professional; Member: American Society of
Mechanical Engineers, National Society of
Professional Engineers, American Welding
Society*

McGraw-Hill Book Company

New York St. Louis San Francisco Auckland Bogotá
Hamburg Johannesburg London Madrid Mexico
Milan Montreal New Delhi Panama
Paris São Paulo Singapore
Sydney Tokyo Toronto

Library of Congress Cataloging-in-Publication Data

Kohan, Anthony Lawrence.
    Pressure vessel systems.

    Bibliography: p.
    Includes index.
    1. Pressure vessels.   I. Title.
TS283.K74   1987   681'.76041   86-21475
ISBN 0-07-035238-0

1234567890   DOC/DOC   893210987

ISBN 0-07-035238-0

*The editors for this book were Betty Sun and Laura Givner,
the designer was Naomi Auerbach, and the production
supervisor was Thomas G. Kowalczyk. It was set in Century Schoolbook
by Rocappi/R. R. Donnelley.*

Printed and bound by R. R. Donnelley & Sons Company.

Contents

Preface ix
Abbreviations and Symbols xi

1. Pressure Vessel System Hazards and Risk Analysis 1

Effects of Failures • Pressure Part Failure • Pressure, Temperature, and Volume Relations • Pressures below Atmospheric • Temperature Scales • Characteristic Equation of a Perfect Gas • Molal Calculations • Effect of Heat on Pressure Vessel • Need for Safety Relief Valves • Hazards of Fire • Vapor-Cloud Explosions • Chemical Reactions • Exothermic • Endothermic • Safety Considerations • Runaway Reactions • Lethal Substance Confinement • ASME Code Requirements for Lethal Service • Codes and Standards • ASME, National Board Inspection Code, and Legal Requirements • ASME Code Promulgation • The National Board • Authorized Inspectors • Division 1 and Division 2 Vessels • Owner-User Specifications for Pressure Vessels • Need for Maintenance after Installation • Record Keeping • Automation Risks • Operator Training • Development Work • Computer Simulation Studies • Risk and Hazard Analysis • Fault-Tree Analysis • Management Concern with Risk • Hazard Evaluation Team • Engineering Evaluations of Computer-Controlled Processes • Possible Modes of Failure • Indirect Exposure • The Need for Hazard Analysis in Computer Control • Questions and Answers

2. Pressure Vessel Classification 33

Storage Vessels • Rotating Pressure Vessels • Heat Exchangers • Cooker-Type Vessels and Digesters • Chemical Reactors • Acetylation • Alkylation • Amidation • Amination • Cracking • Depolymerization • Diazotization • Distillation • Esterification • Extraction • Friedel-Crafts Reaction • Halogenation • Hydrogenation • Nitration • Oxidation • Polymerization • Reduction • Sulfonation • NB Pressure Vessel Accident Statistics • Common Installation Requirements • Stamped Allowable Pressure • Relief Valve Protection • Expansion and Contraction • Supports • Clearance • Pressure Gages • Vibration • Capacity of Relief Valves on Air Tanks • Flammable and Combustible Liquid Storage Tanks • Overpressure Protection • Venting • OSHA Standard • Emergency Relief Venting • Water Spray • Inert Gas Use • Cryogenic Storage Tanks • Material Considerations • Notch Toughness • ASME Code Requirements for Cryogenic Service • Repairs to Cryogenic Vessels • Rotating Pressure Vessel Dynamic Stresses • Slashers • Paper Machine Rolls • Yankee Dryer Roll • Head-to-Shell Attachments • Suction and Press Rolls • Ligament Cracks • Head-to-Shell Weld Cracks • Heat Exchangers • Thermal Performance • Thermal Duty Design Considerations • Parallel and Counterflow • Logarithmic Mean Temperature Difference • Mechanical Design Considerations • Shell- and Tube-Side Arrangements • Heat Exchanger Types • Shell-and-Tube • Spiral • Panel Coil • Plate Fin • Double Pipe • Reboiler • TEMA Classification • Severity of Service • Fixed Tubesheet • U-Tube • Floating Head • Kettle Reboiler • Outside Packed Lantern Ring • Bayonet Tube Heat

Exchanger • Plate Heat Exchanger • Double Tubesheet • Tube-to-Tube-Sheet
Joints • High-Pressure Feedwater Heaters • Deaerator Heaters • Cracking
Incidents • Evaporators • Chemical Plant Evaporators • Codes and Standards for
Heat Exchangers • Vulcanizers, Digesters, and Similar Pressure Vessels • Need for
Internal Inspection • Quick-Opening-Door Closures • Code Requirements •
Chemical and Petrochemical Plant Pressure Vessels • Catalyst Trays • Corrosion-
Resistant Lining • Telltale Holes • Jacketed Pressure Vessels • Conventional
Jacket • Dimple Jacket • Half-Pipe • Glass-Lined Vessels • Agitators •
Jacketed-Vessel Safety Considerations • Thick-Wall Pressure Vessels • Layered
Shells • Wickel Type • Continuous Strip • Multilayer • Hydrogen Service •
Fiberglass-Reinforced Plastic Pressure Vessels • Material Properties • Chemical
Attack • Questions and Answers

3. Pressure Vessel Materials, Joining By Welding, and Nondestructive Testing

**3. Pressure Vessel Materials, Joining By Welding, and Nondestructive
Testing**   99

Need for Knowledge of Properties of Materials • Crystalline Structure of Iron-Steel
Metals • Space Lattice • Slip Planes • Control by Chemical Means • Rate of
Cooling • Shaping Operations • Carbon-Steel Equilibrium Phases • Phase
Diagrams • Effects of Reheating • Welding Considerations • Mechanical Properties
of Materials • Elasticity, Endurance, and Ultimate Stress • Strength • Hardness •
Ductility • Brittleness • Charpy V-Notch Impact Test • Nil-Ductility Temperature •
Creep • Ferritic Materials • Cast-Iron Types • Code Restrictions • Tests to
Destruction • Manufacture of Steel • Effects of Adding Alloys • Impurities from
Steel Manufacture • Sulfur • Phosphorous • Oxygen • Hydrogen • Nitrogen •
Nonmetallic Inclusions • Heat Treatment of Steels • Annealing • Normalizing •
Hardening • Quenching • Common Steel Types • Rimmed Steel • Killed Steel •
Semikilled Steel • Low-Alloy Steels • Stainless Steels • Austenitic • Ferritic •
Martensitic • Duplex • Precipitation-Hardening • High-Alloy Austenitic • Handling
Stainless Steels • Welding Stainless Steels • Sigma Phase • Stainless-Steel Clad-
Plate Material • Code Considerations in Using Clad Material • Telltale Holes • Low-
Temperature Steel Application • ASME Code Material Requirements • Division 1 and
Division 2 • Joining by Welding • Welding Engineers • Types of Welding • Arc
Welding • Oxy-Fuel • Shielded-Metal-Arc • Gas-Tungsten-Arc • Submerged-
Arc • Plasma-Arc • Electron Beam • Heli-Arc • Electrodes • AWS
Classifications • Electrode and Wire Storage • Welding in Adverse Weather • Clean
Surfaces • ASME Welding Code • Qualification Responsibilities of Manufacturers
and Contractors • Quality Control • Welding Procedure Qualification • Qualification
of Welders • Welder Positions and Qualification • Test Plates • Tack Welds •
Welded-Joint Efficiencies • Welding Defects • Repairs • Welding Inspections •
Nondestructive Testing and Examination • Selection of NDT in Maintenance
Inspection • Visual Inspection • Radiography • Code Radiography Requirements
for Weld Inspections • Limitations of Radiography • Magnetic-Particle Testing •
Liquid-Penetrant (Dye) Inspection • Ultrasonic Testing • Eddy-Current Testing •
Acoustic Emission Testing • Destructive Tests • NDT Personnel Qualification •
Outside NDT Service • Questions and Answers

4. Imposed Stresses and Service Effects in Pressure Vessels

4. Imposed Stresses and Service Effects in Pressure Vessels   162

Importance of Defect Analysis • Loads on Pressure Vessel Systems • Resistance of
Materials to Load • Types of Stresses • Strain and Stress-Strain Diagrams •
Proportional Limit • Yield Point • Ultimate Strength • Modulus of Elasticity •
Elastic Limit • Longitudinal and Circumferential Stresses in Shells • Effect of
Temperature • Stress-Concentration Factors • Endurance Limits • Fatigue
Cracking • Fracture Mechanics • Corrosion Fatigue • Notches • Brittle
Fracture • CAT Curves • Hydrogen Embrittlement • Nelson Curves • Hydrogen
Embrittlement of Welds • Cracking from Welding • Creep Considerations • Effect
of Radiation • Corrosion • General Corrosion • Electrochemical Theory of

Corrosion • Atmospheric Corrosion • Coatings • Stainless Steel's Passivity •
Types of Corrosion • Uniform • Galvanic • Concentration Cell • Pitting •
Crevice • Fretting • Intergranular • Corrosion-Induced Fatigue • Methods to
Minimize Corrosion • Stress-Corrosion Cracking • Stainless-Steel Corrosion •
Austenitic Stainless • Sensitization • Chloride Stress Corrosion • Other Attacks on
Pressure Vessels • Oxidation • Caustic Gouging • Erosion Corrosion • Heat
Exchanger Service Effects • Galvanic Corrosion • Tube Erosion • Tube Stress-
Corrosion Cracking • Tube Fatigue • Flow-Induced Vibration • Tube Pitting •
Dezincification • Effects of Mercury and Ammonia • Venting of Heat Exchangers •
Tube Fouling • Alloy Development • Questions and Answers

5. Minimum Code Strength Calculations   207

Plant Level Use of Codes • Allowable Pressure • Allowable Stress • Safety
Factor • Analyzing Pressure Vessel Component Strength • Tubes • Effect of High-
Velocity Flow • Tube Manufacturing • Calculating Allowable Tube Pressure • Tube-
to-Tube-Sheet Joints • Stresses on Tube-to-Tube-Sheet Joints • Field Stress
Calculations • Code Shell Calculations • External Pressure • Spherical Shells •
Dished Heads • Hemispherical Heads • Ellipsoidal Heads • Torispherical
Heads • Conical Heads • Toriconical Heads • External Pressure on Dished
Heads • Shell-to-Head Juncture Stresses • Flatheads • Bracing and Staying •
Ligaments in Drums or Shells • Thick Cylinders and Spheres • Openings and
Reinforcements in Shells and Flat Plates • Bolted Head Covers • Proof Tests •
Bursting-Test Procedure • Stresses Other Than Those Caused by Pressure • Wind
Loads • Earthquake Loads • Structural Supports • Quick-Actuating Head
Closures • Questions and Answers

6. Operation, Controls, and Safety Devices   253

Integrating Maintenance and Inspection with Operating Results • Chemical Plant
Considerations • Operating Surveillance • Operator Training • Operating
Manuals • Training and Preparation for Emergencies • Start-up and Shutdown
Procedures • Clearance for Maintenance • Freeze-up Protection • Interaction of
Operators, Controls, and Computers • Pressure Vessel Instrumentation • Code
Requirements • Pressure and Pressure Gages • Bourdon Tubes • Bellows •
Diaphragms • U-Tube Manometer • Temperature Gages • Bimetallic
Thermometers • Filled-System Thermometers • Electronic Temperature Devices •
RDTs • Thermocouples • Thermistors • Infrared Sensors • Flowmeters •
Differential-Pressure Devices • Magnetic Flowmeters • Point-Velocity Devices •
Level Measuring Devices • Density Measurements • Combustible-Gas Indicator •
Strip Charts • Cathode-Ray Tubes • Piping and Instrument Diagrams • Outdoor
Instruments • Instrument Maintenance • Sensors and Controls • Manual,
Regulating, and Control Valves • Valve Selection • ASME Code Requirement •
Valve Material Selection • Valve Selection Specifications • Valve Pressure and
Temperature Limits • Types of Valves • Hand-Operated Valves • Block Valves •
Drain Valves • Bleed Valves • Dump Valves • Gate Valves • Globe Valves •
Needle Valves • Diaphragm Valves • Butterfly Valves • Check Valves • Wearable
Valve Parts • Packing and Seats • Valve Connections • Regulators • Control
Valves • Fail-Safe Concept • Electrical Equipment in Hazardous Areas • National
Electrical Code Definitions and Requirements • Transmitters • Electronic Control •
Safety Considerations on Electronic Controls • Regulatory Controls • Actuators •
Distributed Computer Control • Alarms • Interlocks • Hard Restraints • Safety
and Relief Valves • Restraint of Chemical Runaway Reactions • Pressure Relief
Valves • Pilot Operated • Safety Valve • Code Safety Valves • Liquid Relief
Valves • Pressure Relief Valves in Combination with Rupture Disks • Rupture
Disks • Materials • Rupture Disk Discharge • Relief Device Settings •
Capacity • Installation • Intervening Stop Valves • Test Levers • Seals •
Conversion of Steam Capacity • Steam Safety Valves • Safety-Valve Springs •
Blowback • Escape Pipes • Capacity of Steam Safety Valves • Popping-Point

Tolerance • Pilot-Operated Pressure Relief Valves • Back Pressure on Pilot-Operated Valves • Pressure and Vacuum Relief Valves • Liquid-Seal Valves • Volatile-Substance Storage Tanks • Breather Valves for Storage Tanks • Flare Stacks • Questions and Answers

7. Maintenance, Inspection, and Repair   322

Maintenance • Planned or Preventive Maintenance • Predictive Maintenance • Performance Monitoring • Contract Maintenance • Reasons for Maintenance • Loss in Performance • Establishing Benchmark Readings • Vendor Surveillance • Equipment Inventory • Work Order and Checklist System • Computerized Maintenance Programs • Maintenance Organizations • Preventive Maintenance in High-Pressure Plants • Safety in Maintenance, Inspection, and Testing • General Safety Rules • Welding and Cutting • Protective Clothing • Use of Masks • Use of Lifelines • Supervision of Maintenance • Pressure Testing and External Inspection • Block and Tag Controls for Rotating Vessels • Safety in Performing Internal Inspections • Vessel Entry • Purging and Internal Inspections • Interior Illumination • Inspection of Internal and External Physical Conditions • Corrosion • Erosion • Cyclic Cracks • Sharp Corners • Stress-Corrosion Cracking • Weld-Seam Deterioration • Creep Effects • Evidence of Distortion • Deciding on the Extent of Inspection Programs • Types of Inspections • Jurisdictional Inspections • Frequency of Inspections • Management Responsibility • Internal Inspections • Cleaning of Vessels • Interval of Internals • External Inspections • NDT and Inspection • Reports of Findings • Thickness Testing • Interpretation of Inspection and Test Results • Minimum Thickness and Wear Rates • NDT Inspections for Flaws • Leak Tests • Hydrostatic Tests • Precautions to Follow • Pneumatic Tests • Leak Tests • Vacuum Tests • Search Gas Tests • Air Tests • Service Fluid Tests • Inspection and Stress-Corrosion Cracking • Causes of Stress-Corrosion Cracking • Stainless-Steel Welds • Stainless-Steel Weld-Overlaid Vessels • Rotating Rolls • Low-Pressure Vessels • Deaerators • Heat Exchanger Maintenance and Inspection • Performance Problems • Fouling of Heat Exchanger Surfaces • Mechanical Problems • Tube Leakage • Corrosion and Erosion at Tube Inlets • Impingement Attack on Tubes • Hole Wire Drawing • Tube-Sheet Ligament Cracking • Crevice Corrosion on Tube Seats • Vibration Damage • Preventive Maintenance Inspections on Heat Exchangers • NDT • Hydrostatic • External • Performance • Tube Plugging • Repairs to Pressure Vessels • National Board R Stamp • Requirements for Stamp • Written Quality Control Program • Repairs and Alterations • Pressure Vessel Field Repairs • Determining the Cause of Defect • Welding Parameters • Unfavorable Environment • Slag Inclusion • Undercut, Overlap, Incomplete Joint Penetration • Porosity • Cracks in Welds • Pitting Repair • Wasted or General Corroded Areas • Weld Inserts • Crack Repairs • Weld Cracks • Lap Cracks • Heat Exchanger Repairs • Sleeving • Seal Welding • Stainless Steel Welding Repairs • Carbide Precipitation • Need for Metallurgical Review • Repairs Involving Dissimilar Metals • Stainless-Steel Weld Overlay • Management Support of Pressure Vessel System Safety • Questions and Answers

Appendix A   Metallurgical and Welding Definitions   388
Appendix B   Types of Stainless Steels   416
**Appendix C   Mechanical and Physical Properties of Some Stainless
              Steels**   426
Appendix D   Design Specifications for Pressure Vessels   429
Appendix E   Organizations Active in Pressure Vessel Rules and Codes 435

Selected Bibliography   437
Index   439

Preface

Pressure vessel systems are too often taken for granted once the design and installation to nationally recognized standards have been completed. Unfortunately, materials degrade with usage, hidden flaws from fabrication make their appearance in service, and plant upsets cause unforeseen problems to arise in the pressure vessel system. It is by periodic review of the condition of the vessels that wear and tear can be detected. Once this deterioration is detected, plant people who operate pressure vessel systems need to go further, namely, to determine (1) if the vessel can continue to be operated at the same pressure, (2) what repairs may be needed for the vessel system to be operated at the original pressure, and (3) whether pressure must be reduced in order to operate the system safely.

It is the aim of this book to stress safety and the avoidance of accidents in pressure vessel systems. The book is not one for the design of pressure vessels, nor for the complete interpretation of nationally recognized standards. However, of necessity, reference must be made to these standards and illustrative problems used in order to stress the need for operating people to use these standards as benchmark criteria to judge whether minimum safety standards are still being applied to their pressure vessel systems.

The intent of the author is to emphasize a total system approach to pressure vessel safety in order to consider the many risks and hazards, such as fire, chemical reactions, vapor-cloud explosions, lethal substance release, loss of production, adverse environmental effects, and possible legal and liability actions. Risk and hazard analysis is also stressed from a system perspective.

Thus, the purpose of this book is to provide practical information on minimizing accidents to pressure vessel systems for the many plant personnel who are responsible for (1) specifying this equipment to fabricators and (2) maintaining, inspecting, testing, repairing, and supervising the installation of the equipment.

Pressure vessel systems have an extremely broad range of application, from simple storage vessels to complicated chemical reactor systems. It is the author's intent to show some common features of these vessel systems in order to develop an organized approach in judging whether a system is reasonably safe to operate. This requires a review of some metals that are most often used in pressure vessel systems, in order to refresh the reader's knowledge of the properties of materials, how they are affected in manufacturing and use, and what the effect may be of welding repairs after service.

ASME code material selection is stressed, as are the many welding-procedure and operator-qualification requirements. Nondestructive testing is an important part of quality control in the pressure vessel industry; therefore, some material on the different types of NDT is presented in the text.

The basics concerning the strength of materials are reviewed in order to lay the groundwork for stress analysis and code equations that are applied to determine safe working pressures. Installation requirements to prevent overpressure and similar areas of protection are also included for study. Pressure vessels in service require periodic inspection, record keeping on deterioration, and possibly repair; therefore, a chapter is included on problems that may develop in this equipment as it is exposed to service factors of aging, cracking, rusting, thinning, pitting, weld deterioration, and similar use effects. Repair procedures per code and jurisdictional requirements are continuously emphasized to the reader.

The author is grateful for the assistance provided by many organizations and manufacturers. The latter have provided illustrations and pictures with product information. Information has also been provided by technical magazines such as *Power, Mechanical Engineering, TAPPI, Pulp & Paper,* and *Chemical Engineering.* Professional societies have provided useful information on their requirements for pressure vessels, and these should be referred to for specific requirements, because in some states or cities the professional society standards have been adopted as legal requirements; the most noteworthy of these is the nationally recognized *ASME Boiler and Pressure Vessel Code.* Other societies involved include the American Welding Society, the National Fire Protection Association, Industrial Risk Insurers, Factory Mutual Engineering, the National Board of Boiler and Pressure Vessel Inspectors, and state enforcing authorities, who are part of the state labor department in most jurisdictions. Each plant and its personnel should be cognizant of jurisdictional requirements regarding installation and maintenance so that they can comply with existing statutes, regulations of insurance companies that provide property insurance coverage, and rules of boiler and pressure vessel insurance companies.

The reader should note that due diligence and care were used in gathering and compiling all information in this book; however, no legal responsibility is assumed for the information and the accuracy of the contents herein, or possible consequences of the use thereof. The author, however, will appreciate being advised of any errors or omissions that may appear in the text or illustrations.

Anthony Lawrence Kohan

Abbreviations and Symbols

A or a	area
ASME	American Society of Mechanical Engineers
ASTM	American Society for Testing and Materials
AWS	American Welding Society
Bhn	Brinell hardness number
Btu	British thermal unit
C	carbon
Ca	calcium
°C	degree Celsius (centigrade)
cm^3	cubic centimeter
cm	centimeter
CO	carbon monoxide
CO_2	carbon dioxide
code, the	*ASME Boiler and Pressure Vessel Code*
Cu	copper
D	diameter of a shell or drum
E	Young's modulus of elasticity = unit stress divided by unit strain (29 million for steel)
°F	degree Fahrenheit
Fe	iron
FS	factor of safety
ft	foot
gal	gallon
gr/gal	grain per gallon (concentration)
gal/min	gallon per minute flow
H	hydrogen
H_2O	water
HAZ	heat-affected zone
ID	inside diameter
in	inch
k	a constant
kg	kilogram
kW	kilowatt
L	liter
l or L	length, in inches, unless otherwise specified
lb	pound

max	maximum
min	minimum
min	minute
mm	millimeter
Mn	manganese
NB	National Board of Boiler and Pressure Vessel Inspectors
N	nitrogen
NaOH	sodium hydroxide
NaSiO$_2$	sodium silicate
NDE	nondestructive examination
NDT	nondestructive testing
Ni	nickel
O	oxygen
OD	outside diameter
OH	hydroxide
oz	ounce
p	pitch, in inches, usually of a series of holes
P	pressure
Si	silicon
SMAW	shielded-metal-arc welding
SO$_4$	sulfate
std	standard
t	thickness, in inches, unless otherwise stated
temp	temperature
TS	tensile strength
V	vanadium
YP	yield point

Pressure
Vessel
Systems

Pressure Vessel System Hazards and Risk Analysis

Pressure vessel systems have contributed to the high standard of living existing today in the technically advanced countries. Their complexities are marvels of engineering capability as they spew out useful products. Because of the interconnections that exist in these fast-moving process streams, instrumentation, controls, and safety devices are needed far more than when operators were stationed near all the process equipment to watch, observe, hear, and smell developing abnormal operating conditions.

Effects of Pressure Vessel System Failure

Many pressure vessel systems can have great potential force stored within them, and when this potential energy is released, for example, as a result of material failure, there is the possibility for the destruction of life and property. Figure 1.1 shows the result in a textile mill of a cast-iron dryer roll's exploding owing to overpressure. The regulating valve and relief valve on the steam line to the dryer were found to have been tampered with in order to obtain higher operating temperatures.

The use of pressure vessels in high-technology process streams has created huge complexes of high-pressure equipment that must be periodically inspected, tested, and perhaps repaired as service conditions affect

1

Figure 1.1 Cast-iron dryer roll explosion after pressure regulator and relief valve were tampered with. (*Courtesy of Hartford Steam Boiler Inspection and Insurance Co.*)

the material. To do otherwise may result in explosions, fires, and equipment breakdown, with associated repair costs and business interruptions. The need for high equipment reliability is directly proportional to the effect such reliability may have on property, occupants of the surrounding area, the environment, jurisdictional considerations, efficiency of performance, quality of the product, potential cost to repair, and similar safety, economic, and legal considerations.

The effects of pressure vessel system failure can be broadly grouped into the following:

1. Product impairment

2. Interruption of production

3. Repair or replacement costs from pressure part failure

4. Repair or replacement costs and potential liability costs due to a lethal substance's escaping from a pressure vessel system into the surroundings

5. Repair or replacement costs and potential liability costs due to the

release of flammable substances with potential for combustion explosions and fire

There are many reasons for pressure part failures, which will be covered in later chapters. Besides overpressure from inoperative controls and safety devices, pressure parts can fail from cracking, thinning, escape of flammable gases that cause fires and explosion of the vessel, and runaway chemical reactions.

Since pressure vessel systems are required to confine safely all types of fluids and gases at various pressures, there is a need to review briefly the pressure, temperature, and volume relations that exist in a pressure vessel system, and how these may be affected by process disturbances.

Pressure, Temperature, and Volume Relations

Pressure, temperature, and volume relations are usually established by the fundamental gas law equations. Some examples of how to use these gas laws in relation to pressure vessel systems should refresh the reader's knowledge of some of the basic relationships of pressure, temperature, and volume changes and show how they relate to plant disturbance.

Pressure

Pressure is the unit force imposed on an unit area by a fluid or gas, this force acting on the confining walls of a vessel. In English units, this is expressed as pounds per square inch, psi. *Gage pressure* is that pressure indicated on gages that measure the internal pressure of pressure vessels. This is a pressure that is above the atmospheric pressure surrounding the pressure vessel on the outside. *Absolute pressure* is the sum of the gage pressure and atmospheric pressure. At sea level, the standard atmospheric pressure is 14.696 psi. It has also been defined as a 760-mm (29.921-in) height of a mercury column at 32°F. In laboratory work requiring close precision, a mercury barometer must be used to obtain the exact mercury height and weight at different temperatures, because the weight of the column of mercury in a barometer also gives the pressure per unit area of cross section. At 32°F, mercury weighs 0.491 lb/in³. Thus, each inch of mercury height will exert 0.491 psi.

Pressures *below* atmospheric pressure are defined as vacuum conditions and are usually measured in inches of mercury.

Example What is the absolute pressure on an air tank when the pressure gage reads 148 psi and a barometer indicates 29.45 inches of mercury?

answer Atmospheric pressure = 29.45(0.491) = 14.46 psi

Absolute pressure = 148 + 14.46 = 162.46 psia

Vacuum problems are solved the same way.

> **Example** If the vacuum gage on a condenser shows 28.20 in of mercury, and a barometer indicates 29.58 in of mercury, what is the absolute pressure in the condenser?
>
> **answer** Absolute pressure (in inches of mercury) = 29.58 − 28.20 = 1.38 in of mercury
>
> Therefore   Pressure = 1.38(0.491) = 0.678 psia

Temperature and volume

Temperature can be classified into three scales:

1. *Fahrenheit.*   On this scale the freezing point of water is 32°F, and the boiling point of water is 212°F, at atmospheric pressure. As can be seen, this gives a 180° spread on the scale.

2. *Centigrade.*   The metric system uses this scale, with 100° between the freezing and boiling points of water. Conversion from one scale to the other involves this difference in the spread between freezing and boiling of water, and can be calculated using:

$$°F = \frac{9}{5} °C + 32 \qquad °C = \frac{5}{9} (°F − 32)$$

3. *Absolute temperature.*   This temperature scale was derived from experiments that revealed that a perfect gas expands at a rate of $\frac{1}{491.7}$ of the gas's original volume for each change in temperature of 1°F. On the centigrade scale, the expansion rate is $\frac{1}{273}$ of the original volume per °C. By lowering the temperatures of various substances, it has been established that all molecular activity would cease at 491.7°F below the freezing point of water, or −459.7°F. This absolute temperature is called "absolute zero." Thus, to convert a Fahrenheit or centigrade temperature to absolute temperature, use the following equations:

Absolute temperature, $°F_a = °F + 459.7$

Absolute temperature, $°C_a = °C + 273$

> **Examples** (*a*) What is the absolute temperature of a substance at −10°F? (*b*) What will be the volume of a gas that is heated from 32°F to 60°F at atmospheric pressure if the original volume was 500 in³?
>
> **answers** (*a*) $°F_a = −10 + 459.7 = 449.7°F_a$
> (*b*) A perfect gas expands at a rate of $\frac{1}{491.7}$ of the original volume per °F. Therefore,
>
> $$\text{Expansion} = \frac{(60 − 32)\,500}{491.7} = 28.45 \text{ in}^3$$
>
> Total volume at 60 °F = 500 + 28.45 = 528.45 in³

In solving pressure, temperature, and volume problems according to the perfect gas laws, it is necessary to use absolute pressures and temperatures in the gas law equations. It is also common to express pressure in pounds per square foot and volume in cubic feet.

Characteristic equation of a perfect gas

Boyle's and Charles's law can be combined into the following equation:

$$\frac{P_1 V_1}{T_1} = \frac{P_2 V_2}{T_2}$$

where P = absolute pressure of the gas, lb/ft²
V = volume of M pounds of the gas, ft³, at pressure P and temperature T
M = weight of the gas, lb
T = absolute temperature of the gas, $°F_a$ = °F + 459.7

Example Air at 90 psi gage and 70°F occupies 28.11 ft³. What would be the temperature at 5 psi gage and a volume of 142 ft³? The air has been expanded under this lower pressure.

answer

$$\frac{P_1 V_1}{T_1} = \frac{P_2 V_2}{T_2}$$

where P_1 = 144(90 + 14.696) lb/ft²
V_1 = 28.11 ft³
T_1 = 70 + 459.7 = 529.7°F abs
P_2 = 144(5 + 14.696) lb/ft²
V_2 = 142 ft³

Substituting,

$$\frac{144(90 + 14.696)(28.11)}{529.7} = \frac{144(5 + 14.696)(142)}{T_2}$$

Clearing,

$$T_2 = \frac{2796.83}{2943}(529.7) = 503.2°F \text{ abs}$$

$$T_2 = 503.2 - 459.7 = 43.5°F$$

Note that, for P_1 and P_2, the figures were multiplied by 144 to convert in² to ft².

Dalton's law states that in a mixture of gases, the total pressure equals the sum of the pressures that each gas would exert if it were present alone in a tank at constant pressure.

The Pv/T term in the gas equations for a particular gas is called a "constant," termed R, which is expressed as foot pounds per pound per

°F absolute. With v = the specific volume of the gas in cubic feet per pound, the volume is expressed by

$$V = Mv$$

By substitution in the equation, $v = V/M$.

$$\frac{P_1 v_1}{T_1} = R \quad \text{or} \quad \frac{PV/M}{T_1} = R$$

Clearing, the following relationship of pressure, volume, and weight of the gas at a given pressure can be determined:

$$PV = MRT$$

Example Air has an $R = 53.34$. What would be the volume of 15 lb of air at 70°F, with a pressure of 100 psi in a tank?

answer Use $PV = MRT$

where $M = 15$ lb
$R = 53.34$ ft.lb/(lb)(°F abs)
$P = 144(100 + 14.696)$ lb/ft²
$T = 70 + 459.7$°F abs

Substituting,

$$V = \frac{15(53.34)(70 + 459.7)}{144(114.696)} = 25.72 \text{ ft}^3$$

Table 1.1 provides some common gas constants that can be used in pressure, temperature, and volume relationship calculations.

TABLE 1.1   Gas Constants Used in Thermodynamic Gas Equations

Gas	Molecular weight	R^*	c_p†	c_v‡	k§
Air	29.67	53.34	0.240	0.1715	1.40
Carbon monoxide	28.00	55.18	0.249	0.1789	1.40
Carbon dioxide	44.00	35.10	0.199	0.153	1.30
Helium	4.00	386.00	1.25	0.754	1.66
Hydrogen	2.018	766.4	3.445	2.460	1.40
Methane, CH_4	16.00	96.5	0.529	0.404	1.31
Nitrogen	28.016	55.15	0.249	0.178	1.40
Oxygen	32.00	48.28	0.218	0.156	1.40

$^*R$ = Rankine number
†c_p = Specific heat at constant pressure
‡c_v = Specific heat at constant volume
§k = Ratio of c_p/c_v

Molar calculations

Quite often it is necessary to calculate the volume or pressure of a gas of different molecular weight. For example, in chemical process work molar quantities are used to determine the pressure or volume that a certain gas may have at different temperatures. A *mole* is defined as the quantity of a substance whose weight in pounds, grams, ounces, etc., is numerically equal to its molecular weight in that weight measurement. For example, if the weight is expressed in pounds, the molar is expressed in pounds-mole.

Use is made of the hypothesis of Avogadro in molar calculations. This hypothesis states that equal volumes of two ideal gases at the same temperature and pressure contain the same number of molecules. Standard temperature and pressure are usually taken as 60°F and 14.7 psia in engineering work. At this pressure and temperature, the molar volume is approximately 379.5 cubic feet per pound-mole. In scientific work, standard temperature and pressure are 0°C and 760 mmHg. The molar volume at this standard is 259 cubic feet per pound-mole. An example should help in understanding molar calculations.

Example Calculate the volume occupied by 300 lb of methane gas for the above two standard conditions. (Methane = CH_4; C = 12; H = 1.)

answer (a) *At 0°C and 760 mmHg*

Molecular weight of methane is 12 + (4 × 1) = 16

Number of pound-moles of methane = $\frac{300}{16}$ = 18.75

Volume of these pound-moles = 18.75 × 359 = 6731.25 ft³

(b) *At 60°F and 14.7 psia*

Volume of pound-moles = 18.75 × 379.5 = 7115.63 ft³

At 60°F and 14.7 psia, the molar volume of any gas can be expressed as

$$V = 379.5 \frac{W}{M}$$

where V = volume, ft³
W = weight of gas, lb
M = molecular weight

A ratio can be used to determine the volume of any gas at other than 14.7 psia pressure.

$$V = 379.5 \left(\frac{W}{M}\right)\left(\frac{14.7}{P}\right)$$

Example Liquid propane weighs 4.22 lb/gal and has a molecular weight of 44. What would the volume of this gallon be as a saturated vapor at standard 60°F and 14.7 psia? What would the volume be at 2 psia?

answer Using the previous equation and substituting,

For standard conditions, $V = 379.5 \dfrac{W}{M} = 379.5 \left(\dfrac{4.22}{44}\right) = 36.4 \text{ ft}^3$

For 2 psia, $V = 36.4 \left(\dfrac{14.7}{2}\right) = 267.54 \text{ ft}^3$

Effect of Heat on Pressure Vessels

Metal expansion is usually neglected when calculating the potential pressure rise inside a tank caused by a local fire or other source of heat around a tank. The potential pressure can thus be calculated by the constant volume equation.

Example An air tank operates at 100 psig with surrounding air at 70°F. What would be the pressure in the tank if the tank were subjected to 500°F by a local fire and isolated from pressure relief devices?

answer This is a constant volume problem. Use

$$\frac{P_1}{T_1} = \frac{P_2}{T_2}$$

where $P_1 = 144(100 - 14.696)\ 16/\text{ft}^2$
$T_1 = 70 + 459.7 = 529.7°\text{F abs}$
$T_2 = 500 + 459.7 = 959.7°\text{F abs}$

Substituting,

$$P_2 = \frac{144(100 + 14.696)(959.7)}{529.7} = 29{,}923.8 \text{ lb/ft}^2$$

$$= \frac{29{,}923.8}{144} = 207.8 \text{ lb/in}^2 \text{ abs}$$

$$= 207.8 - 14.696 = 193.1 \text{ lb/in}^2$$

Need for safety relief valves

Later chapters will cover safety relief valves. Isolated tanks also need safety valves, as the above example of pressure rise due to a fire around the tank illustrates. It must always be assumed that a pressure tank with gases, or liquids that can vaporize with heat, may be isolated by valves to provide overpressure protection.

If a source of excessive heat develops near the tank, the safety valve will protect the vessel against overpressure. The venting to a safe point of discharge of any release from a safety valve must also be considered. This could involve personnel safety considerations, as well as property protection. For example, a flammable gas must be directed from the discharge opening of the safety valve to a scrubber, flare stack, or similar area *away* from the fire or location of the faulty equipment. The venting of vessels will be covered in later chapters.

Hazards of fire

When exposed to the heat of a fire, pressure vessels confining flammable gases can rupture and feed the fire. The heat will cause gaskets to start leaking, and as the vapors from these leaks ignite, the plate temperature will rise, which a review of material stress tables will show causes the yield and ultimate stress to drop drastically. Failure due to gross yielding or rupture of the tank will result under these conditions.

The release of flammable gases can result in the following outcomes:

1. A flame may appear where the leak occurs. This is most likely if the leak is small and of low pressure.

2. Large leaks under higher pressure may cause pockets of the gas to ignite with resultant fireballs engulfing the area until the gas burns off.

3. Gases can spread over a large area from a leak, and then be ignited, causing a "vapor-cloud explosion" that can destroy property and kill people in the explosion zone.

4. Leaks, of course, can disperse and never reach the flash point, or never become an explosive mixture.

The first recognized vapor-cloud explosion occurred in 1943 in Germany when a tank car failed and released an 80 percent butadiene and 20 percent butylene mixture into the air and surroundings. The resultant vapor-cloud explosion killed 57 people and injured 439. Several similar incidents soon made process engineers aware of the fact that if the substance confined is flammable, even a pinhole leak in a pressure vessel could set off in the surrounding free air a vapor-cloud explosion, even though the pressure vessel may not fail because of the commonly recognized risk of overpressure.

Chemical Reactions

Great exposure can exist when a pressure vessel system is subjected to an uncontrolled chemical reaction within a pressure vessel. Most chemical and physical changes to process stream substances are also accompanied by energy changes, and these may generate heat. The gas laws show that heat in a confined volume can raise the pressure of the confined gas or fluid. If the pressure rise is not controlled or limited by safety devices, pressure vessel deformation and explosions may result. Reactions in which heat is liberated are called "exothermic." These types of reactions require special attention to avoid the reactions' running out of control, and thus exposing property and personnel to a possible explosion. The opposite of exothermic reactions are "endothermic" reactions, in which heat or other energy is absorbed instead of being given off.

Safety considerations

In addition to the overpressure risk that exists when a chemical reaction goes out of control, consideration must be given to whether the escaping gas or fluids from a pressure vessel puncture are toxic or flammable. If toxic, relief devices must be able to discharge them to a safe place, such as an empty pressure vessel nearby, flare stacks, or scrubbers. Flammable gases that escape from pressure vessels always pose the threat of forming an explosive mixture outside the vessel. This can result in vapor-cloud explosions that can be very destructive to a plant and the workers that may be near the explosion. The threat of chemical reactions going out of control has caused safety engineers to encourage remote-control operation and isolation of process streams subject to this type of exposure.

Runaway reactions

In chemical process plants, the hazard of a runaway reaction must be continuously watched and guarded against. There are two types of reaction involved. One is caused by exothermic *decomposition* of material, generating an unwanted reaction. The second occurs when, because of process mishaps of flow, mixing, or similar controllable processes, a runaway reaction develops that is no longer controllable. The resultant thermal rise can cause an uncontrolled pressure rise, and if the safety devices are inadequate, or don't function, an overpressure explosion may result in the equipment. Operators trained for this type of situation may prevent this from happening. For example, the contents could be cooled, the vessel could be emptied, or a neutralizing agent could be pumped into the vessel in order to slow down the reaction.

Lethal Substance Confinement

Pressure vessel systems confining lethal substances require special consideration in design and fabrication; then, in service, they require periodic inspection of the protective devices and the physical conditions of the pressure vessel system. The spread of lethal substances to the surrounding area can create special dangers to human life, as occurred at Bhopal, India, in 1984.

Minimum ASME code requirements will help in avoiding leaks due to poor fabrication practices. For example, the following are required in Section VIII, Division 1, rules:

1. Full butt weld construction

2. Full radiographic examination of the butt welds

3. Postwelding heat treatment of the welded joints when fabricated of carbon or low-alloy steel

4. Avoidance of cast iron or cast ductile iron in lethal service

5. Stamping of the vessel with an "L" on the nameplate if it was designed for lethal service

The ASME code requires the user, or a designated agent, to advise a fabricator that a vessel for lethal service is required. The fabricator must then conform to code design and fabrication standards for the lethal service.

Lethal service is one in which a pressure vessel system confines a lethal substance. A lethal substance is defined as a poisonous gas or liquid of such a nature that even small amounts of the gas, or of the vapor of the liquid, if mixed with air or escaping into the air unmixed, would pose a danger to life when inhaled.

It is important to duplicate as much as possible original design and fabrication requirements when performing repairs to pressure vessel systems, especially those confining lethal substances.

Codes and Standards

There are many codes and standards that are applicable to pressure vessel system safety. Among these are:

1. NFPA (National Fire Protection Association) 30—flammable liquids

2. NFPA 58, 59—liquefied petroleum gases

3. NFPA 50, 50A, 50B—hydrogen

4. NFPA 68—explosion venting

5. NFPA 69—explosion prevention systems

6. American Petroleum Institute (API) guide for inspection of refinery equipment

7. Standards of Tubular Exchanger Manufacturers Association (TEMA)

8. Heat Exchange Institute *Standards for Closed Feedwater Heaters*

9. *ASME Boiler and Pressure Vessel Codes*
    a. Section VIII, Divisions 1 and 2—"Pressure Vessels"
    b. Section IX—"Welding and Brazing Qualifications"
    c. Section V—"Nondestructive Examination"

    d. Section II—"Material Specifications"
    e. Section X—"Fiberglass-Reinforced Plastic Pressure Vessels"

10. National Board of Boiler and Pressure Vessel Inspectors—*National Board Inspection Code*

ASME, National Board inspection code, and legal requirements

Most states have statutes for the installation and periodic inspection of boilers and pressure vessels. See Tables 1.2 and 1.3. The *ASME Boiler and Pressure Vessel Code,* depending on what section of the code is adopted as constituting legal requirements, and the *National Board of Boiler and Pressure Vessel Inspectors Inspection Code* are extensively used by jurisdictions that have adopted boiler and pressure vessel safety laws. The ASME codes are used for original design, fabrication, and installation requirements. The National Board code is an inspection and repair code, which however uses the ASME codes as reference material.

ASME code promulgation

In addition to maintaining active boiler and pressure vessel committees in order to keep the published codes up to date with developing technology, the ASME issues to qualified manufacturers, assemblers, material suppliers, and nuclear power plant owners code symbol stamps indicating that the manufacturer has received authorization from the ASME to build boilers and pressure vessels to the ASME code.

    A fundamental principle of the *ASME Boiler and Pressure Vessel Code* is that a boiler or pressure vessel, to be stamped ASME code–designed, must receive third-party authorized inspection during construction for compliance with the prevailing code requirements. Most third-party inspections are performed by authorized boiler and pressure vessel inspectors who have appropriate experience and have passed a written examination in a jurisdiction. They must be employed either by the state or by an insurance company licensed to write boiler and pressure vessel insurance in the jurisdiction where the boiler or pressure vessel is to be built, and in some cases the installation's location also must be considered. With uniform requirements for inspectors that have been promoted and implemented by the National Board of Boiler and Pressure Vessel Inspectors, a boiler or pressure vessel inspected by a properly credited National Board inspector will generally be accepted in all jurisdictions.

    The manufacturer or contractor who wishes to build or assemble boilers or pressure vessels under an ASME certificate of authorization must first agree with an authorized inspection agency that code inspections will be performed by the agency. This is usually arranged by both parties signing a contract, with the inspection work done on a fee basis.

TABLE 1.2   States Having Boiler and Pressure Vessel Reinspection Laws

State	Accepts insurance company reports (X = yes)	Requires inspection for		
		High-pressure boilers	Low-pressure boilers	Unfired pressure vessels
Alaska	X	X	X	X
Alabama	No law	—	—	—
Arizona	X	X	X	—
Arkansas	X	X	X	X
California	X	X	—	X
Colorado	X	X	X	—
Connecticut	X	X	X	—
Delaware	X	X	X	—
District of Columbia	X	X	X	—
Florida	No law	—	—	—
Georgia	X	X	X	X
Hawaii	X	X	X	X
Idaho	X	X	X	X
Illinois	X	X	X	—
Indiana	X	X	X	X
Iowa	X	X	X	X
Kansas	X	X	—	—
Kentucky	X	X	X	—
Louisiana	X	X	X	X
Maine	X	X	X	—
Maryland	X	X	X	X
Massachusetts	X	X	X	X
Michigan	X	X	X	—
Minnesota	X	X	X	X
Mississippi	X	X	X	X
Missouri	No law	—	—	—
Montana	No	X	X	—
Nebraska	X	X	X	X
Nevada	X	X	X	X
New Hampshire	X	X	X	X
New Jersey	X	X	X	X
New Mexico	No law	—	—	—
New York	X	X	X	—
North Carolina	X	X	X	X
North Dakota	X	X	X	—
Ohio	X	X	X	—
Oklahoma	X	X	—	—
Oregon	X	X	X	X
Pennsylvania	X	X	X	X
Rhode Island	X	X	X	—
South Carolina	No law	—	—	—
South Dakota	X	X	X	—
Tennessee	X	X	X	X
Texas	X	X	X	—
Utah	X	X	X	—
Vermont	X	X	X	X
Virginia	X	X	X	X
Washington	X	X	X	X
West Virginia	X	X	—	—
Wisconsin	X	X	X	X
Wyoming	No law	—	—	—

SOURCE: Harry M. Spring and Anthony L. Kohan, *Boiler Operator's Guide,* 2d ed., McGraw-Hill, New York, 1981.

TABLE 1.3   Cities and Counties Having Boiler and Pressure Vessel Reinspection Laws

| City or county | Accepts insurance company reports (X = yes) | Requires inspection for | | |
		High-pressure boilers	Low-pressure boilers	Unfired pressure vessels (UPV)
Albuquerque, N. Mex.	X	X	X	—
Buffalo, N.Y.	X	X	X	—
Chicago, Ill.	No	X	X	—
Dearborn, Mich.	X	X	X	X
Denver, Colo.	No	X	X	X
Des Moines, Iowa	X	X	X	—
Detroit, Mich.	UPV only	X	X	X
E. St. Louis, Mich.	No	X	X	X
Greensboro, N.C.	X	X	X	X
Kansas City, Mo.	X	X	X	X
Los Angeles, Calif.	X	X	X	X
Memphis, Tenn.	X	X	X	X
Miami, Fla.	X	X	X	X
Milwaukee, Wisc.	X	X	X	X
New Orleans, La.	X	X	X	X
New York City, N.Y.	X	X	X	—
Oklahoma City, Okla.	X	X	X	—
Omaha, Neb.	X	X	X	—
Phoenix, Ariz.	X	X	X	X
St. Louis, Mo.	X	X	X	X
San Francisco, Calif.	X	X	X	X
San Jose, Calif.	X	X	X	—
Seattle, Wash.	X	X	X	X
Spokane, Wash.	X	X	X	X
Tacoma, Wash.	X	X	X	X
Tampa, Fla.	X	X	X	X
Tucson, Ariz.	X	X	X	X
Tulsa, Okla.	No	X	X	X
University City, Mo.	No	X	X	—
White Plains, N.Y.	X	X	X	—
Arlington County, Va.	X	X	X	—
Dade County, Fla.	X	X	X	X
Fairfax County, Va.	X	X	X	X
Jefferson Parish, La.	X	X	X	X
St. Louis County, Mo.	X	X	X	X

SOURCE: Harry M. Spring and Anthony L. Kohan, *Boiler Operator's Guide*, 2d ed., McGraw-Hill, New York, 1981.

The National Board

The National Board of Boiler and Pressure Vessel Inspectors is composed of chief inspectors of states and municipalities in the United States and Canadian provinces. This organization has established criteria for experience requirements of boiler inspectors, the promotion and conductance of

•

uniform examinations, and testing that are used by the jurisdictions. The National Board issues commissions to inspectors passing an NB examination, which are accepted on a reciprocal basis by most jurisdictions, thus providing a "portability" feature to a credential.

Most areas of the United States and all jurisdictions in Canada require that high-pressure boilers be subjected to periodic inspection by an authorized inspector. In most jurisdictions, this consists of annual internal inspection of power boilers and biennial inspection of heating boilers and usually of pressure vessels for those states that have adopted laws on low-pressure boilers or unfired pressure vessels. If the results of an inspection prove satisfactory, the jurisdiction issues an inspection certificate, authorizing use of the vessel for a specific period.

Authorized inspectors

There are three types of inspectors who make the legal inspections and report to a jurisdiction that a boiler or pressure vessel is safe or unsafe to operate or that it requires repairs before it can be operated:

1. State, province, or city inspectors see that all provisions of the boiler and pressure vessel law, and all the rules and regulations of the jurisdiction, are observed. Any order of these inspectors must be complied with, unless the owner or operator petitions for (and is granted) relief or exception.

2. Insurance company inspectors are qualified to make ASME code inspections. If commissioned under the law of the jurisdiction where the unit is located, they can also make the required periodic reinspection. As commissioned inspectors, they require compliance with all the provisions of the law and rules and regulations of the authorities. In addition, they may recommend changes that will prolong the life of the boiler or pressure vessel.

3. Owner-user inspectors are employed by a company to inspect unfired pressure vessels for direct use and not for resale by such a company. They also must be qualified under the rules of any state or municipality that has adopted the code. Most states do not permit this group of inspectors to serve in lieu of state or insurance company inspectors.

Pressure vessels built to code standards can usually be recognized by the U stamp shown in Fig. 1.2. If the vessel is registered with the National Board of Boiler and Pressure Vessel Inspectors, an NB number will

Figure 1.2 The U symbol, shown on a pressure vessel, indicates conformity with the American Society of Mechanical Engineers (ASME) code in constructing the vessel.

also appear on the stamping. This signifies that the vessel was inspected during construction by a qualified NB inspector to make sure all provisions of the code were complied with. Data reports are completed on all registered NB vessels and are kept on file at the NB headquarters in Columbus, Ohio. These data reports can be obtained from the NB by citing the NB number stamped on the vessel. The NB address is 1055 Crupper Ave., Columbus, Ohio 43229.

Many plant people must decide on whether to order ASME code Section VIII, Division 1, vessels or Section VIII, Division 2, vessels.

Section VIII, Division 2, vessels

Section VIII, Division 2, vessel systems were developed for more advanced designs for large pressure vessels, as used in the chemical and petroleum industry. Division 2 vessels require more extensive design analysis, but permit using yield strength as the design basis instead of the ultimate strength divided by 4 as is the case for Division 1 vessels. Division 2 rules require extensive analysis of imposed stresses from which a combined stress, or stress intensity, is developed for the part under analysis, and this must be compared with permissible Division 2 stress intensities, which generally must be below two-thirds yield stress of the material, or below one-third ultimate stress of the material. The higher basic allowable stresses permitted in Division 2 will save on material costs during construction, but engineering design costs may be higher and quality control inspections more frequent.

Division 2 vessels may be attractive for the following reasons:

1. There is no upper pressure limit intended in Division 1 of the code, but it is clear that higher pressures require more stringent treatment from both the designer and the manufacturer than meeting only the minimum requirements of Division 1 vessels. A plant can thus expect better and more uniform design criteria to be followed when ordering Division 2 vessels for high-pressure service [i.e., when a plant is contemplating pressure vessels with design pressures around 3000 psig (20 MPa) and over].

For Division 2 vessels, the code requires that an engineering analysis be performed. This includes a formal design report, prepared by the manufacturer or its agent's registered engineer. Pressure vessels to be used in cyclic service must receive a fatigue analysis in compliance with Division 2 rules, which also provide methods for stress analysis and evaluation.

2. Costs for engineering service for a Division 2 vessel compared with such costs for a Division 1 vessel can be up to 25 percent higher. These costs, however, can be offset by requiring thinner, and thus less, material for the vessel.

3. Heavier vessels of about 20 tons, or multiple vessels built at the same time for high pressure, may show a material saving cost under Division 2 design.

4. Vessels built of high-quality material, instead of carbon steel as per Division 1 rules, may show a material cost saving by being built to Division 2 rules.

5. Vessels of large size, multiple-layered construction, and similar special designs generally will receive better design and construction under Division 2 rules.

6. Vessels to be exposed to highly fluctuating pressures and temperatures, or other cyclic service, will receive a fatigue analysis under Division 2 rules; this is left up to the discretion of the purchaser or manufacturer under Division 1 rules.

The following factors should be considered as *not favoring* a Division 2 pressure vessel:

1. When the purchaser or the purchaser's agent is not familiar with Division 2 requirements. Both divisions of the code require the purchaser to establish the design conditions. However, the requirements of Division 2 are more exacting, and the user's design specifications must be prepared under direct supervision of a registered professional engineer. This specification becomes a recorded document.

2. When accredited manufacturing facilities are not available. Separate accreditation is required before a manufacturer can construct and certify pressure vessels according to Division 2 rules. Sometimes, substantial savings are passed up in order to obtain an earlier delivery from a manufacturer limited to Division 1 construction, rather than wait until a Division 2 shop is free to do the job. Users should obtain quotations for both Division 1 and Division 2 construction when competitive bidding is restricted or when the potential savings are in doubt.

3. When the vessel is manufactured for stock for later sale, or when the vessel is likely to be resold as a commodity. This is because the intended use of the vessel must be specified under Division 2.

4. When the vessel is likely to be moved from one site to another, as for resale as surplus.

5. When service conditions are likely to be changed.

6. When, due to the nature of the service, modifications or alterations of the vessel may be desired in the future to accommodate changes.

User-Owner Responsibility in Specifying Pressure Vessels

It is not sufficient for plant people involved with ordering or specifying pressure vessel systems simply to indicate "Code construction." Many purchasers of process equipment do not realize that the ASME code places the responsibility of establishing design requirements for pressure vessels to be ordered from ASME-qualified shops on the user or user's designated agent as defined in the code. It is the responsibility of the user to notify the fabricator if the vessels are expected to be in lethal service, because this will determine the extent of nondestructive tests that will have to be applied to welds as specified by code rules.

The owner or user must specify the *corrosiveness* of the process, thus advising the fabricator of the need for extra thickness of the metal above code requirements so as to provide for above-normal corrosion wear.

The owner or user must also specify *postweld heat treatment* that may be necessary above code requirements. For example, low-carbon-steel P-No. 1 material, by code classification, does not require postweld heat treatment for materials up to $1\frac{1}{2}$ in thick, provided that material over $1\frac{1}{4}$ in in thickness is preheated. However, there may be some process applications in which postweld heat treatment should be specified in order to reduce residual heat-affected stresses from the welding process. The user must specify to the fabricator whether postweld treatment will be applied; otherwise, the fabricator will furnish a code vessel that does not require postweld heat treatment. Of course, the user must do all the calculations regarding process needs as to volume, materials necessary for resisting attack from unusual contents such as acids, process temperatures and pressures, heats of reaction, and the means of controlling the reaction, to name a few. These variables involving process design are considered to be within the user's expertise and responsibility, and not the code fabricator's.

There are fabricators that specialize in certain industries and thus have the staff to assist or advise an owner on what should be specified in the design of the process beyond normal code requirements. These specifications will include such optional and variable items as corrosion allowance, fouling factors in the design of heat exchangers, selection of material, NDT inspections to be applied, Division 1 versus Division 2 construction, and similar matters involving the process, costs, and quality of construction.

Extreme *confidentiality* is an important consideration in many processes that use pressure vessels; this must exist between the purchaser and the designer and fabricator because of obvious fears that competition may learn the secrets of the process.

Figure 1.3 is a questionnaire used by a company specializing in constructing distillation columns. The code does permit fabricators specializing in certain industries to act as agents for the user. Many of these firms advertise their capability as designers and fabricators of engineered pressure vessel systems in the following areas:

- Feasibility studies
- Economic evaluations
- Complete process line layout
- Total energy considerations
- Design, engineering, and procurement of equipment
- Fabrication to relevant codes
- Systems preassembly and testing when feasible
- Full instrumentation and controls
- Technical supervision of erection and startup
- Comprehensive personnel training

Operating, maintenance, and purchasing personnel of plant pressure vessel systems must review in-house capability in specifying pressure vessel systems against what is available from specialized companies, who can act as agents for the plant as designers, specifiers, and perhaps even'fabricators per code rules.

Need for Maintenance after Installation

The manufacturers of pressure vessel systems, with the aid of the ASME codes, have demonstrated by actual performance that by following proven design and fabrication techniques, if properly specified by the user, safe pressure vessel systems can be produced. After a vessel system has been in service, however, maintenance must be adopted in order to monitor whether the original design criteria may be affected by in-service factors of wear and tear, process upsets, and similar possibilities of degradation that may require reviewing permissible pressures or mode of operation.

The maintenance of vessels in service will require periodic inspections and vessel system monitoring. This requires of the maintenance personnel the following:

1. Knowledge, through inspection and record keeping, of the pressure vessel's condition through the life of the process stream.

2. Control of the process parameters so that they are within safe set points regarding pressure, temperature, autoignition of contents, etc. Controls should include alarms, pressure relief devices, and shutdown trips or procedures in order to provide protection against any abnormal operating conditions.

3. Knowledge of possible failure mechanisms and consultation among maintenance, inspection, and operating staff, so that the symptoms of

Date _____

Your ref. _____

Date quotation required _____
Type of quotation required

☐Magnitude of cost   ☐ Budget   ☐ Firm

Approx. installation date _____

Company	Phone
Address	Name
City State Zip	Position

PROCESS LIQUOR

Component(s)	A	B	C	D
Chemical name				
Molecular weight				
Weight % in feed				
Analysis of Top Product Stream	% wt/ppm	% wt/ppm	% wt/ppm	% wt/ppm
Bottom Product Stream	% wt/ppm	% wt/ppm	% wt/ppm	% wt/ppm
Sidestream 1	% wt/ppm	% wt/ppm	% wt/ppm	% wt/ppm
Sidestream 2	% wt/ppm	% wt/ppm	% wt/ppm	% wt/ppm
Sidestream 3	% wt/ppm	% wt/ppm	% wt/ppm	% wt/ppm

PHYSICAL PROPERTIES

	FEED		PURE B	PURE C	PURE D
Thermal conductivity (BTU/ft.°F hr)					
Specific gravity					
Viscosity: cps	@ °F	@ °F	@ °F	@ °F	@ °F
Specific heat (BTU/lb.°F)					
Is (1 gallon) sample feed available?	☐Yes ☐No	☐Yes ☐No	☐Yes ☐No	☐Yes ☐No	☐Yes ☐No
Is solution corrosive?	☐Yes ☐No	Suggested material of construction:			
Are undissolved solids present?	☐Yes ☐No	If yes, give percentage			%
Does material tend to foul heat transfer surfaces?	☐Yes ☐No				
Is material heat sensitive?	☐Yes ☐No	Maximum processing temperature:			°F

Figure 1.3 Questionnaire of a specialty company for distillation columns. (*Courtesy of APV Equipment, Inc.*)

abnormal conditions can be quickly identified before major damage results. For example, process upsets may greatly accelerate corrosion rates, or production at high rates may increase erosion effects on pressure vessel components. Changes in feedstock or product specification may change known and established deterioration rates on tubes or shell surfaces. Awareness of changes of this type and knowledge of their possible effects will enable maintenance groups to make timely checks of those areas potentially affected by the changes, and thus ensure the pressure vessel's integrity.

4. Good loss-prevention management, from the time of product development through plant design, construction, and operation. Automation has placed an even greater premium on reliability of vessels that are in

<table>
<tr><td rowspan="9" style="writing-mode:vertical-rl">PROCESS DUTY</td></tr>
</table>

PROCESS DUTY					
☐ Continuous		☐ Batch			
Feed rate:	Gal/hr	Lbs/hr	Gal/batch	Inlet feed temperature:	°F
Operating time			Hours/day		Days/year
Is turn-down ability required?		☐Yes ☐No	To what extent?		
Is provision for expansion required?		☐Yes ☐No	By how much?		
Is cooling of products required?		☐Yes ☐No			
If yes, cooling temperature:	Tops	Bottoms	Side 1	2	3
Any other special processing requirements:					

SERVICES					
Steam pressure at plant in psi:	Low	(Variation ±)		High	(Variation ±)
Cooling water:	Source		Inlet pressure:		psi
Maximum water inlet temperature:		°F	Is chilled water available?	☐Yes ☐No	°F
Are services limited?		☐Yes ☐No	Max. steam:	Lbs/hr	Max. water: GPM
Electricity:	3 phase—— V	Hz(c/s)	Single phase——	V	Hz (c/s)
Cost data:	Steam $/1000 lbs.	Cooling water	¢/1000 gal.	Electricity	¢/kwh
Products pumps head (ft.):	Tops	Bottoms	Side 1	2	3
Remarks:					

CONTROLS

Quotations include all necessary local temperature and pressure gauges to enable manual operation of the plant.   The following equipment can also be provided. Please indicate or delete as required.

Degree of automatic instrumentation:	☐Full	☐Partial	☐Microprocessor ☐None
Type of automation:	☐Electronic	☐Pneumatic	
Preferred manufacturer:	Instruments ——	Control valves——	
Remarks:			

SITE		
Location:	Altitude above sea level	ft.
Restrictions: Headroom——	Floor area	
☐ Indoor ☐Outdoor		
Remarks:		

Other requirements:

_____

_____

_____

Figure 1.3 Questionnaire of a specialty company for distillation columns. (*Courtesy of APV Equipment, Inc.*)

operation continuously between scheduled maintenance periods. "Fail-safe" designs aim to avoid operational failures. Continuous planned maintenance is required to avoid breakdowns of equipment, instrumentation, and safety devices.

Questions that have to be asked in establishing maintenance and testing programs include:

1. What hazards does the process stream present, and will flammable

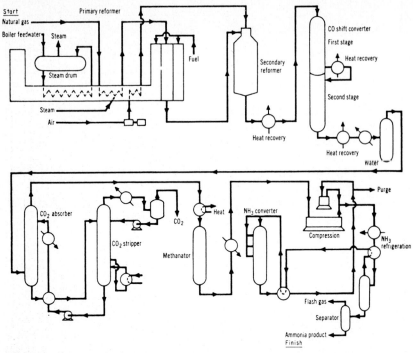

Figure 1.4   Ammonia single-train pressure vessel process flow. (*Courtesy of* Chemical Engineering.)

or possibly unstable chemical reactions result from an equipment breakdown?

2. Is the process stream of the single-train type, meaning that a failure of any major piece in the train may shut down the entire process while repairs are made? See Fig. 1.4.

3. Has automation reduced the working staff to the extent that operating personnel and maintenance staff may have difficulty in responding to emergencies such as fires or explosions?

4. Can spills of highly flammable substances create explosive vapor clouds outside the pressure vessel and thus cause extensive property damage?

5. Is the process stream or components of it possibly corrosive so that materials of construction will be affected? Corrosion has been defined in terms of degradation of material due to the chemical reaction between the material and the operating environment. Processing plants of all types must contend with corrosion, and it is considered the most frequent cause of equipment problems in process streams.

6. Is the process a new technology that requires demonstrating that design and research concepts are correct and that therefore no abnormal wear and tear on the equipment is occurring?

Management equates reliability of operation to productivity and profit. Similarly, it is necessary for designers, operating personnel, and maintenance people to equate their proposals for avoiding breakdowns to production and profitability goals if their recommendations are to be adopted.

Operator Training

The need for training operators of high-pressure, fast-moving process stream plants is apparent but still needs emphasis. Operators must be trained to start equipment very gradually to avoid pressure and thermal shocking of equipment, to open valves gradually for the same reason, and at all times to be cognizant of the effect on a process stream of equipment being brought on and off line. Training of operators should include formal classroom lectures on the process variables and on the equipment involved in the process. Classroom instruction should be supplemented by simulation training as well as by tours of plant facilities. On-the-job training should be under the supervision of a proven, skilled operator. All possible contingencies should be part of the training program. This should include malfunctions of control and safety devices, and the means to detect these by process readouts.

Development Work

Development work in the huge chemical, petrochemical, and nuclear industries will require that engineers, designers, and product developers create pressure vessel components to meet the special needs of these and other future industrial complexes. Emphasis in development work is on elevated temperature service, dynamic and seismic requirements on tall-structured pressure vessels, higher-pressure process requirements, flow-induced vibration on heat exchanger internal components, and the greater use of nondestructive testing to monitor a pressure vessel's materials under variable service conditions.

It must be stressed that pressure vessels represent the largest concentration of energy in chemical and allied industries. The constant striving for higher pressures and temperatures has prompted (1) a massive effort in research and development by pressure vessel manufacturers to develop new materials of higher strength and durability and new and improved methods in welding these materials, (2) establishment of better and more rigid quality control standards in fabrication practices, and (3) more extensive use of nondestructive testing techniques with the possibility of continuous monitoring of critical pressure vessels for crack growth by acoustic emission techniques.

Computer simulation of process dynamics is assisting designers and operators in evaluating possible process equipment malfunctions and the effects these may have on the increased complexity of modern processing plants.

Risk and Hazard Analysis

An essential objective in the design, fabrication, erection, operation, and maintenance of pressure vessel systems is to prevent incidents that might cause large property losses, injure people, and perhaps severely affect the environment. Past practices have depended on experience and precedent to provide safety and loss prevention. These included the reliance on codes and standards, checklist reviews, and audits of the installation. A variety of more rigorous reliability studies and a "what if" approach are now being used to more clearly identify hazards and what can be done to minimize them.

While pressure vessels may be built to code standards, the operation and safety analysis of a group of pressure vessels that have been assembled to perform a job in a process stream requires a complete system analysis approach to safety and health exposures, as well as to reasons for potential forced shutdowns due to process disturbances. The risk and hazard analysis should start with design proposals. For example, how can vapor-cloud explosion effects as a result of a leak be minimized? Design may consider implementing means of rapid depressurization of the leaking tank by directing the contents to a spare, empty tank, isolating the ruptured section by suitable valving, dyking the area around the tank, installing water curtains around the tank to limit the ignition temperature of the escaping fluid, and installing a tank with flammable or other dangerous contents away from other property and operating personnel.

Risk analysis versus hazard analysis

Risk analysis has been described as a method of estimating economic risk, such as the consequence of a failure producing either a probable maximum loss (PML) or, in the worst case, a maximum possible loss (MPL), as used by insurance underwriters. *Hazard analysis* tries to break down possible events that could occur that would lead to the creation of a *physical* hazard. For example, the loss of a cooling fluid during a pump failure could lead to an exothermic reaction in the vessels requiring cooling fluids. It can be seen that a study of possible operating mishaps is an important element of a hazard study. Another term used for hazard analysis is "event-tree analysis." This analyzes the sequence of events that can follow an initiating event, such as the failure or malfunction of a system component. The response to an initiating failure quite often is made by a safety control, such as a pressure switch, flow switch, or temperature alarm. In event-tree analysis, the safety device is considered to operate, but the consequences of its *not* operating are also reviewed. The event-tree analysis continues until an undesired condition is reached in the design or the physical or equipment limitations stop the event-tree from going any further, for example, a gas compressor's being shut down by low pressure, because ho more gas is available owing to a leak.

Fault-tree analysis traces back all reasons for a possible major, or "top event," failure. Eventually a point is reached where the failure can be reduced no further, or no more data are available.

Hazard analysis requires the following steps, which can be performed for batch operations as well as for continuous-flow processes:

1. Identify the possible incidents and their consequences to the operation.

2. Determine from a history search, insurance statistics, computations, or estimation how often the incident may occur.

3. Determine the consequences of a mishap not only for the plant but for employees and the public, and determine how these could cause state or federal legal actions or liability suits.

From the above analysis, it is usually possible to determine that the hazard should be further minimized by better controls, more frequent inspection and testing, better operator training and supervision, and possibly harder constraints, such as backup relief valves and automatic shutoff valves.

There are pitfalls in hazard analysis that must be recognized. One is the assumption of normal operator response. Situations arise in which an operator has to diagnose alarms and instrument readings. This diagnosis may not be correct. For this reason, codes generally favor hard restraints, such as temperature and pressure relief devices. On the other hand, full dependence on fully automatic systems means that reliance has been placed on the people who design, install, and test the automatic system, and they also make mistakes.

Some hazard analysts use a mathematical approach to analyze the chance of a serious equipment failure, such as 1 in 10,000 machine-years. If the failure that is of concern happens tomorrow, however, the statistical approach is no comfort to the plant experiencing the 1-in-10,000-machine-year failure. Nevertheless, it is necessary to establish an incident frequency and cost per incident, in order to determine if the additional cost of protection against the incident is justified. This is especially true if only property damage is involved, and no possibilities exist for personnel injuries.

One offshoot of hazard analysis can be increased reliability in maintaining production. This can be a very important item where a single failure could affect the product's quality, or seriously reduce daily output, which in single-train plants can involve large sums of money for each day of outage.

Management concern with risk

Management should be interested in risk evaluation, because a disastrous failure has an immediate effect on the financial status of an enterprise. In addition, society expects plants to be operated safely. There is always the threat of governmental restraints being imposed on a plant if safety rules and regulations have been found to have been violated. It is not uncommon, therefore, for large enterprises to have risk managers fully responsible for implementing a good loss-prevention effort so as to minimize risk, as well as being involved in assessing risk from a technical and financial-exposure point of view, such as determination of insurance coverage and placement. Smaller plants should assign risk assessments to plant engineers, crew leaders, and similar operating and maintenance supervisors who are familiar with the plant process. At times, a committee approach is needed to gain the expertise of workers, union delegates, technical people, and the person responsible for placing insurance coverage for the risks involved.

Hazard evaluation team

A hazard evaluation team must consist of various personnel who are knowledgeable of the process design, the method of operation, and the controls and safety devices that have been or will be installed in the pressure vessel system. The team should be provided with schematic diagrams, flowcharts, instrument and control logic diagrams, and similar information that shows the operation of the process from the beginning to the end. Included in the information should be details of operator attendance and similar personnel needs in order to determine and develop safety requirements.

A well-organized team will spot from experience and knowledge those areas of plant operation that will most likely present the most serious hazards and will work their way down from this area of the plant operation.

Engineering safety evaluations of computer-controlled processes

The variety of computer applications in process control—off-line, in-line, open loop, and closed loop—plus the functions and process boundary limits that have been assigned to computer control, should make it evident that detailing the type of computer, the hookup, its program of control, the variables it is considering, and so forth, would be an extremely time-consuming effort, and therefore, would be the wrong approach to use in effectively evaluating the computer control of a process so as to gauge how it functions in preventing accidents.

A more logical approach for fire and safety interests is to review a process in a fault-tree manner by looking at the process in detail and then evaluating the exposure possibilities in standard fashion. If this procedure is followed and, of course, if responsible engineering and operating people are consulted, the computer control function will become evident as to control limits assigned; restraints incorporated into the computer, whether this is by automatic action or by alarm system with the operator taking corrective actions; the response time; the adequacy of controls or controls not incorporated for a hazard in the computer; and so forth.

As an example, a chemical reactor vessel should be evaluated for possible breakdown failure, explosion failure, or fire possibilities. Assume that it is jacketed by cooling water because the contents react exothermically and that the vessel is of the closed type for pressure operation. Modes of possible failure are cracking, overpressure explosion, and overtemperature leading to ignition of contents.

Possible modes of failure of reactor vessels. Cracking of process vessels is usually due to stress concentration around sharp corners, in cyclic operation, or to thermal shocking. Direct damage control against cracking and thinning is usually by inspection—either by visual or by nondestructive tests such as x-ray or ultrasonic probes. An internal and external inspection will show potential crack sources and thinning. A study of the cycling pattern will suggest the frequency of nondestructive testing required. Computers are generally not programmed for detecting cracks or thinning. However, a record of on-off cycling could be incorporated into the computer so as to predict the time when the reactor will have to be checked nondestructively. This information can be secured by a record of the operation of the reactor. Thus, some form of computer information is already being found applicable for possible prevention of vessel failure.

Overpressure could occur because of a lack of sufficient cooling water, improper mixing of the ingredients going into the reactor, or a runaway exothermic reaction. Design pressure of the vessel must be checked and this pressure compared with the constraints provided on the vessel against overpressure. It may be found that an overpressure transducer has been installed on the computer hookup, but this only rings an alarm. In all cases of overpressure protection, hard constraints are absolutely essential. Inquiry on the chemical reaction taking place could show that a safety relief valve on the vessel is inadequate for overpressure protection and that a rupture disk is required for total relief in the event of a runaway reaction. The computer programming may have figures on the speed of the reaction.

The area of relief opening required to guard against overpressure can then be determined. Backup protection or feedforward concepts might also be included in the computer control to guard against loss of cooling water. A drop in cooling water flow might immediately reduce the flow of chemicals to the reactor by means of a proportioning valve. If not, a flow-failure safety switch is needed to immediately shut down the feed to the reactor in the event of loss of cooling water.

Overtemperature is another hazard to be considered. The degree of computer control might include a high-temperature alarm system or automatic feed reducer. Hard constraints required might be an internal CO_2-extinguisher system.

Indirect exposures on the reactor could be many, depending on the process. Cracking of the inner jacket could lead to cooling water's fouling the contents of the vessel. The reaction of water with the ingredients in the vessel could lead to a greater heat release or a slowing down of the process. These are the variables or disturbances that may be incorporated into the computer control of the process. Perhaps a chemical analyzer has been installed. Some means of detecting this "poisoning" should be considered. The immediate isolation of the reactor from downstream process equipment when this poisoning occurs should also be considered in the form of automatic or manually operated alarm valves, to prevent the poisoning from spreading to the downstream operation.

Batch or contents dumping or reclaiming under this poisoning action must also be considered, particularly if of hazardous or expensive proportions. Are pumping facilities provided to do this?

The need for hazard analysis in computer control.   The engineering approach to evaluating the degree of computer control of process and its adequacy still depends primarily on looking for possible exposures and then noting if these have been incorporated into the computer control. The countless variables in a process that can create uncontrolled disturbances are such that the designers of a computer control face the task of compromising even if the control design was given unlimited scope.

A careful safety evaluation in detailed manner will provide answers as to the degree of control by the computer in a far better manner than evaluating the computer control on an overall basis, which, at best, can be a very difficult and time-consuming effort.

Not to be forgotten in evaluating exposures in computer-controlled processes is computer malfunctioning itself. What backup system is installed if this occurs—can the plant return to standard automatic or semiautomatic control? Are there sufficiently skilled operators present to continue the process? For example, assume a flow valve under computer control develops electrical and mechanical trouble. Is a spare valve hooked into the system to divert the flow under computer control to this valve?

An area which needs careful attention concerning indirect equipment damage is the effect of a part of the plant experiencing an accident which can affect the entire computer programming. Will all the input data based on valve variables, flow variables, and so forth, designed for a specific modeling have to be reprogrammed to get the operation back on stream under computer control? How long would this take?

Certain conclusions are possible when evaluating computer controls:

1. The fact that a process is under some form of computer control is a variable. Its effectiveness should be analyzed by safety personnel using standard safety-evaluation and fault-tree analysis techniques to note potential failure paths in the process and the interaction of the computer with the process.

2. Where a process is known to be under computer control, the effect of computer outage, or of outage of the auxiliary equipment connecting the computer and the process, should be considered as it relates to accident and loss potential.

3. Most desirable in any contemplated computer control of a process is to have in the planning stage, including the mathematical modeling stage, well-rounded safety engineering appraisals of possible system disturbances and their effects on equipment and personnel, and estimates of what constraints should be incorporated into the computer to guard against possible accidents.

QUESTIONS AND ANSWERS

1-1 Whose responsibility is it to specify corrosion allowance and postweld heat treatment for pressure vessels if the ASME code has no specific requirements?

ANSWER: The purchaser, owner, or agent as defined in the code is required to specify the amount of corrosion allowance to be provided for the pressure vessel and whether the welds are to be postweld heat-treated. It is assumed under this guideline that the eventual user knows the process conditions and should consider the many factors involved in specifying the requirements.

1-2 What loading should be considered when specifying pressure vessels?

ANSWER: The following are among the items to be considered besides the usual internal and external pressures specified in the code:

1. Loads from impact forces, such as rapidly varying pressure or temperature
2. Weight of the vessel and contents, including the possible weights due to hydrostatic tests after erection or repairs
3. Superimposed loads on the pressure vessel, such as compressors mounted on tanks, the effect of connected piping's expanding and contracting with pressure and temperature, weight of insulation placed over the structure, and similar

stress producers additional to that emanating from the pressure within the tank

4. Wind and earthquake loads, depending on the location

5. Temperature and coincidental pressure fluctuations

6. Any other loads that can cause stress on the vessel

1-3 Define "lethal service."

ANSWER: Pressure vessel systems required to confine liquids or gases, or combinations thereof, that are considered to be dangerous to life if inhaled if they should escape even in very small amounts are considered to be in lethal service.

1-4 What code requirements exist for the pressure vessel welds that are to be in lethal service?

ANSWER: All welded butt joints must be fully radiographed and pass code requirements on permissible defects in the welds. Carbon- and low-alloy steel vessels must have all welds postweld heat-treated. This applies to repairs also. The code stamping for the vessel should include the letter "L" to indicate it was fabricated for lethal service and met all code requirements for this type of service.

1-5 What kind of iron cannot be used for lethal service?

ANSWER: Cast-iron and ductile cast-iron pressure vessels or parts cannot be used to confine lethal substances. Their crystalline structure is considered too porous.

1-6 What pressure and temperature limitations does the code specify for cast-iron pressure vessels?

ANSWER: The code specifies the following:

1. For vessels containing gases, steam, or other vapors, 160 psi is the maximum at temperatures not greater than 450°F.

2. For vessels containing liquids, a maximum of 160 psi is specified at temperatures not exceeding 375°F.

3. For those liquids at temperatures less than their boiling point per design pressure, 250 psi is specified, but the temperature cannot exceed 120°F.

4. The code permits cast-iron bolted heads, covers, or closures to be used up to 300 psi at temperatures not greater than 450°F, provided the closures do not form a major component of the pressure vessel.

1-7 An air tank has a volume of 30 ft^3 and contains 18 lb of air at 100 psi and 75°F. What would be the pressure in the tank, ignoring protective devices and tank expansion, if the air were heated to 600°F by a fire?

ANSWER: Use the constant volume equation

$$\frac{P_1}{T_1} = \frac{P_2}{T_2}$$

where $P_1 = 144(100 + 14.696)$ lb/ft^2
$\quad\quad T_1 = 75 + 459.7 = 534.7°F$ abs
$\quad\quad T_2 = 600 + 459.7 = 1059.7°F$ abs

$$P_2 = \frac{144(100 + 14.696)(1059.7)}{534.7} = 32{,}732.826 \text{ lb/ft}^2 = 227.3 \text{ psia}$$

$$P_2 = 227.3 - 14.696 = 212.6 \text{ psig}$$

1-8 Define the term "exothermic reaction."

ANSWER: A chemical reaction in which heat is liberated in the process is called an "exothermic reaction."

1-9 Name the three types of pressure vessel inspectors recognized by U.S. jurisdictions.

ANSWER: The three legally recognized pressure vessel inspectors are those who have passed the required jurisdictional examination and are:

1. Inspectors of the city, state, or province enforcing the pressure vessel laws of the jurisdiction

2. Qualified inspectors employed by recognized and authorized insurance company–qualified inspection agencies.

3. Owner-user inspectors if they are qualified under the jurisdiction and if they have an inspection program acceptable to the jurisdiction.

1-10 Define vapor pressure of a liquid.

ANSWER: Vapor pressure is a measure of the force that tends to vaporize any volatile liquid, such as petroleum. Molecular motion within the liquid is responsible for this force and is related to the composition of the liquid. Smaller molecules are more active; thus, vapor pressure increases as the proportion of these low-boiling (smaller-molecule) components increases. Higher temperatures also stimulate molecular motion and higher vapor pressure. The molecules that vaporize tend to disperse throughout the vapor space. At the same time, some of the molecules in the vapor space return to the liquid. Equilibrium is established when molecules leave and return to the liquid at the same rate. At any given storage pressure, the equilibrium percentage of hydrocarbon vapor in the vapor space, essentially, is directly proportional to the vapor pressure of the liquid.

1-11 What are upper and lower explosive limits?

ANSWER: Gases and vapors may form flammable mixtures in atmospheres other than air or oxygen, for example, hydrogen in chlorine. However, explosive limits are always expressed in terms of gases or vapors combining with air or oxygen. In the case of gases or vapors that form flammable mixtures with air or oxygen, there is a minimum concentration of vapor in air or oxygen below which propagation of flame does not occur on contact with a source of ignition. There is also a maximum proportion of vapor or gas in air above which propagation of flame does not occur. These boundary-line mixtures of vapor or gas with air, which if ignited will just propagate flame, are known as the "lower and upper flammable or explosive limits" and are usually expressed in terms of percentage by volume of gas or vapor in air.

In popular terms, a mixture below the lower flammable limit is too "lean" to burn or explode and a mixture above the upper flammable limit too "rich" to burn or explode.

There may be considerable variation in flammable limits at pressures or temperatures above or below normal. The general effect of increase of temperature or pressure is to lower the lower limit and raise the upper limit. Decrease of temperature or pressure has the opposite effect.

1-12   What is the purpose of a fault-tree or hazard analysis?

ANSWER: To determine the possible events that might occur singularly or in combination to cause a potentially unsafe condition to develop in a pressure vessel system. A probability is assigned to each potential failure, and the cumulative effect is evaluated to determine the ultimate risk that may develop. An analysis is made to determine the statistical basis for the possibility of the occurrence or occurrences.

Pressure Vessel
Classification

Pressure vessels can range from storage tanks in which a gas or liquid is stored under pressure to large chemical reactor vessels where with pressure, and perhaps temperature, chemical changes take place among the fluids or gases confined under pressure in the vessel system. A simple classification could be:

1. *Storage vessels.* These include air tanks, hot-water tanks, propane tanks, and similar tanks in which a gas or liquid is stored under pressure to be used when needed. Generally, design is to the ASME code, and with proper installation, maintenance on this type of pressure vessel is generally one of checking controls, such as pressure switches, overpressure protection settings, and capacity, and making periodic tests of the safety relief valves. Service considerations may include corrosion and vibration problems. Also, cryogenic, volatile, and combustible gas or liquid storage can present problems related to vapor pressure and materials of construction.

2. *Rotating pressure vessels.* These types of vessels usually use steam inside of them so that drying of a product will occur as it passes over the rotating roll's surface. They are used to dry clothing, paper, and plastic materials and may have names such as "Yankee dryer roll," "couch roll," and "calender."

3. *Heat exchangers.* This classification includes a large assortment of

component arrangements in which heating, or cooling, is performed on one side of the heat exchanger, with a heating or cooling medium on the other side.

4. *Cookers.*   Digesters, creosoting vessels, vulcanizers, and rendering tanks are included in this classification. These pressure vessels (cooking, rendering, etc., are performed under pressure) cause a physical change to take place to the contents and do not involve serious chemical reactions. Thus, in a digester, wood chips or rags are cooked with steam to make a pulp that is used in the manufacture of paper.

5. *Chemical reactors.*   These pressure vessels range from the steam-jacketed batch reactor to synthesizers such as those found in large chemical and petrochemical processes.

The reactions taking place in chemical reactors are quite numerous and may include some of the following:

Acetylation.   A process in which the basic chemical portion of acetic acid is combined with organic chemicals such as alcohols to produce substances such as vinyl acetate and acetone. Acetylation is a special process under the general heading of "esterification." The vessels in which the process is carried out are commonly referred to as "chemical digesters" or "chemical reactors."

Alkylation.   A process in which organic compounds are treated with combinations of various alcohols and acids, or of various alcohols and the halogens, to make medicines, dyes, explosives, plastics, solvents, butadiene for the manufacture of synthetic rubber, and even gasoline. The vessels in which the process is carried out are commonly referred to as "alkylators."

Amidation.   A process very similar to amination but in which the organic compounds formed are known as "amides." Amides are used in organic synthesis for the manufacture of commercial products. The process is carried out in chemical digesters or chemical reactors.

Amination.   A process in which an organic compound is treated with ammonia or one of its compounds to form what are known as "amines." Amines are used in the manufacture of dyes, and as insecticides, antiseptics, and medicines. The process is carried out in chemical digesters or chemical reactors.

Cracking.   A process which is sometimes referred to as destructive distillation, in which organic compounds are subjected to heat in such a manner that *different* organic compounds are formed, as in the cracking of certain petroleum oils to form gasoline. The process may be carried out in a chemical digester, or it may occur in a pipe coil commonly referred to as a "cracking coil."

Depolymerization.   A process whereby heat and sometimes pressure are applied to heavy hydrocarbons in such a manner as to change them to lightweight hydrocarbons; an example would be decomposition by heat of certain synthetic resins to recover more usable substances. The process is carried out in chemical digesters or chemical reactors.

Diazotization.   A process in which a selected type of organic compound is combined with nitrous acid, the products of which are used principally in the manufacture of dyes. The process is carried out in chemical reactors.

Distillation.   A process for the separation of different compounds from a mixture, such as alcohol from water, whereby the substance having the lower boiling point is vaporized and thereby driven off from the remainder of the mixture. It is usual practice to condense the vapor and thus retain the substance vaporized. The vessels in which the process is carried out are commonly referred to as "stills."

Esterification.   A process in which alcohols are combined with mineral and organic acids to form what are known as "esters," e.g., butyl acetate, ethyl acetate, and amyl acetate. They are used in the manufacture of lacquers, perfumes, plastics, and explosives. The process is carried out in chemical digesters or chemical reactors.

Extraction.   A process in which a substance is removed from a mixture by the use of a solvent in which the substance will dissolve, such as in the extraction of soybean oil by use of ethyl alcohol. It is the usual practice to reclaim the substance so dissolved by distillation or evaporation of the solvent. The vessels in which the process is carried out are commonly referred to as "extractors."

Friedel-Crafts reaction.   A process in which a comparatively active compound such as ethyl chloride or carbon tetrachloride is combined with an organic compound such as benzene in the presence of a certain type of catalyst such as aluminum chloride. The process is used in the manufacture of dyes, perfumes, synthetic resins, and plastics. The process is carried out in chemical digesters or chemical reactors.

Halogenation.   A process in which one of the halogens is combined with an organic compound to form substances used as cleaning fluids, refrigerants, or anesthetics and also for the manufacture of a great many other organic compounds. The process is carried out in chemical reactors.

Hydrogenation.   A process in which gaseous hydrogen is chemically combined with an organic compound. It is the process used in making shortening compounds, butter substitutes, and many other forms of "solidified oils." The vessels in which the process is carried out are commonly referred to as "hydrogenators" or "reactors."

Nitration.   A process in which a compound of nitrogen and oxygen, such as nitric acid or sodium nitrate, is chemically combined with organic compounds to form so-called nitro compounds, which are used principally in the manufacture of medicines, dyes, and explosives. The vessels in which the process is carried out are commonly referred to as "nitrators."

Oxidation.   A process in which oxygen is added to a substance or hydrogen is removed by the use of an "oxidizer," such as air, oxygen, potassium permanganate, or potassium chlorate. Some of the chemicals formed by the oxidation process are formaldehyde, acetic acid, camphor, and indigo. The process is carried out in chemical digesters or chemical reactors.

Polymerization.   A process in which heat and pressure are applied to lightweight hydrocarbons in such a manner as to change them to heavy hydrocarbons. It is a process used in the manufacture of synthetic rubber, synthetic resins, and gasoline. The process is carried out in chemical digesters or chemical reactors.

Reduction.   A process in which oxygen is removed in some degree from a compound by the use of a reagent such as hydrogen, carbon, or a hydrocarbon. Reduction is usually accomplished by the treatment of the substance with hydrochloric, sulfuric, or acetic acid and a metal such as zinc or iron in finely divided form. The process is carried out in chemical digesters or chemical reactors.

Sulfonation.   A process in which a compound of sulfur, oxygen, and hydrogen, such as sulfuric acid, is chemically combined with an organic compound to produce "sulfonic acids" or "sulfonated oils." The former are used in the manufacture of dyes, while the latter are used in the glue, paper, leather, textile, and soap industries. The vessels in which the process is carried out are commonly referred to as "sulfonators."

Chemical reactor pressure vessels can operate at very high pressures and therefore often require multilayer wall construction. These vessels may also require stainless-steel liners owing to the corrosiveness of the contents or, in the case of a gas such as hydrogen, because of the peril of carbon steel's becoming embrittled, and thus posing the potential for a disastrous cracking failure. The possibility of a runaway chemical reaction on these types of vessels also requires consideration.

All pressure vessels have their own peculiar hazards, points of wear and tear, control requirements, and overpressure and temperature safety devices that may differ on some points. These differences result from such factors as the contents being handled, environmental considerations, and lethal and flammability limits, to name a few.

The National Board publishes an annual accident incident report on boiler and pressure vessel accidents reported to them by various sources. Table 2.1 is a modification of the 1986 NB report summarizing the reported pressure vessel incidents. There were 477 pressure vessel accidents reported in the year, with 269 persons injured and 78 deaths.

More people were injured and killed from pressure vessel accidents than from boiler accidents. Shell failure occurred the most on pressure vessels, with corrosion and erosion of shells being named the most common cause for the failures (57 percent of total shell failures). In addition to loss of life and property, another result of pressure vessel accidents is the cost of downtime in terms of loss of production income. For many single-train plants, this can be a substantial daily amount, quite often not all recoverable from insurance because of deductible coverage programs.

Even ordinary storage vessels can fail violently unless they are built to minimum national standards such as the *ASME Boiler and Pressure Vessel Code,* installed properly, operated within their pressure and capacity ratings, and have installed on them proper controls and protective devices. Periodic inspection and maintenance are needed on this equip-

TABLE 2.1 Pressure Vessel Accidents for a One-Year Period

Initial part failure	Causes							Type of failure							Numbers		
	Low-water cutoff	Faulty design, fabrication, or installation	Corrosion or erosion	Operator error or poor maintenance	Burner failure	Pressure control failure	Other	Burning or overheating	Collapsing inward	Combination explosion	Cracking	Torn asunder (rupture)	Leakage	Other	Accidents	Injuries	Deaths
Shell	1	15	87	21	1	9	17	6	9	6	33	50	73	12	152	177	33
Head		7	55	12		5	3	2	4	1	10	18	2	40	36	14	8
Attachments		11	13	11		1	2			1	25	7	3	9	32	36	15
Piping	1	8	11	26	1	1	20			1	13	9	26	12	52	22	5
Safety valves		2	4	3	1		40				1	8	4	17	25		
Miscellaneous	3	10	53	11	2	3	65	10		2	57	18	73	33	180	20	17
															477	269	78

NOTE: The above report was compiled from data submitted by National Board jurisdictional authorities and authorized inspection (insurance) agencies. It also includes material submitted from several insurance companies that insure boilers but do not provide inspection services.

SOURCE: Adapted from the *National Board Bulletin,* April 1986.

ment also in order to ensure that operation is proper, that protective devices are functional, and that no deteriorating condition is developing. The weakening effects of corrosion and even erosion are ever a factor to consider in any loss-prevention maintenance; so is the effect of cycling service, which can initiate cracks on corners and transition sections owing to stress concentration. Later chapters will draw the reader's attention to specific code requirements and other problems of material failures.

Common Installation Requirements

There are some common installation practices that should be followed for practically all pressure vessels:

1. Pressure vessels should have an allowable working pressure stamped on them, as per ASME and NB requirements; this pressure should be above the operating pressure.

2. Pressure vessels should have operating controls, upper-limit controls, or alarms if possible, and should be protected against overpressure or overtemperature by safety relief valves or rupture disks.

3. Code construction directives should be treated as minimum requirements. Plant or safety engineers should supplement these requirements with plant safety rules and regulations.

4. Connected piping should be installed with due consideration of expansion and contraction effects on the vessels and piping. This will reduce nozzle connection stresses.

5. Pressure vessels should be properly supported so that the weight of the vessel and its contents is uniformly distributed around the vessel (see Fig. 2.1). This will prevent bearing stresses from being imposed on the vessel, in addition to the pressure stresses, and thus avoid local yielding or cracking. Hydrostatic tests require considering the weight and resultant stresses that may be imposed on the supports and the vessel walls. As Fig. 2.1 shows, the shell thickness should be substantial and the shell should be braced with stiffening rings to avoid high, localized, imposed stresses in the support area. For a cylinder that is to be mounted horizon-

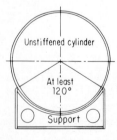

Figure 2.1 Pressure vessels should be supported so that the bearing stress caused by the weight of the vessel and its contents is uniformly distributed over a large enough bearing area to avoid overstressing the vessel wall.

tally with liquid contents, a 120° arc of contact is recommended, as shown in Fig. 2.1.

6. Adequate clearance around the pressure vessel should be provided for maintenance and inspection. Usually, at least 12 in is required. Clearance between the vessel and floor should also be at least 12 in so that corrosion from moisture is minimized.

7. Pressure vessels with tubes should have adequate clearance to permit tube replacement.

8. All access openings to pressure vessels, such as holes for entering the vessel, handholes, inspection plugs, and removable heads; pressure and temperature gages; and safety relief valves should be accessible at all times. Suitable permanent platforms for access will be appreciated by inspection and maintenance staff.

9. All welded seams should be accessible for external inspections and should not be placed against a wall where access would be difficult, if not impossible. The same applies to ASME and NB stampings. They should be accessible for determining the vessel's design ratings so that a comparison can be made with the control settings and with the overpressure protection that has been installed.

10. Pressure vessels should have provisions for safely draining the vessel of its contents.

11. Safety relief valves should have a pressure setting at or below the stamped allowable pressure and should prevent a rise in pressure of more than 10 percent above the allowable pressure.

12. Pressure gages should be provided on pressure vessels, graduated to at least $1\frac{1}{2}$ times the maximum allowable working pressure. Temperature gages for vessels with potential for overtemperature operation should also be provided as an aid in operation, inspection, and maintenance.

Vibration can present problems to common storage tanks, as well as to tubes inside heat exchangers that may develop vibration from the gas flows across the tubes. It is a consideration in installing pressure vessels. Air tanks have been known to fail violently from the weakening effect of corrosion. Many tanks have air compressors mounted on them, causing abnormal vibration on the assembly. Cracks have appeared where aprons were welded to the tank in order to mount the compressors on a flat surface. Cracks have also appeared on longitudinal and circumferential welded joints from the severe vibration. Tanks may have been designed only for confining the air under pressure, without consideration of the additional loading placed on the tank by a piece of rotating machinery that was mounted on the tank. After several explosions, one state now requires that air tanks and compressors be installed separately, unless the manufacturer offers proof that the stresses imposed by a compressor–air tank assembly have been thoroughly analyzed and the assembly strengthened accordingly.

Condensate in air tanks needs periodic removal; otherwise internal corrosion may occur. The compressed air coming from an air compressor can be quite hot. Usually aftercoolers are installed to cool this air. There have been cases of the *water to the air aftercoolers* being interrupted, with violent explosions resulting from residual oil in the discharge pipe and the hot air forming an explosive mixture that was ignited like fuel in a diesel engine. A flow switch or interlock on water flow could prevent this type of occurrence.

Capacity of relief valves on primary air tanks is directly related to the volume in cubic feet per minute that a connected compressor may deliver to the tank. (See Chap. 1 on pressure, temperature, and volume relationships for *secondary* air tanks.) The volume displaced by a compressor is the net area of the compressor piston swept during the compression stroke. For multistage compressors, only the low-pressure cylinder is considered. The capacity of a compressor is the actual volume of air compressed and delivered, usually expressed in cubic feet per minute at intake temperature and pressure. The displacement of a compressor does not equal the capacity found in practice, because of excessive clearance, friction during compression, and similar losses. This difference is expressed as the volumetric efficiency of the compressor E_v

$$E_v = \frac{\text{capacity}}{\text{displacement}}$$

However, applying an approximate E_v, it is possible to determine the capacity of a relief valve on an air tank by calculating compressor displacement.

Example What should be the minimum relieving capacity in cubic feet per minute for the relief valve of a primary air tank that is connected to a multistage compressor with pistons of 9 in, 7 in, and 4 in, and with a 9-in stroke? The compressor operates at 400 r/min and has a volumetric efficiency of 95 percent.

answer Only the low-pressure piston has to be considered, and this sweeps the following volume:

$$\frac{\text{Area of cylinder (in}^2) \times \text{stroke} \times \text{r/min}}{1728 \text{ in}^3/\text{ft}^3}$$

$$\text{Area of cylinder} = \pi\left(\frac{9}{2}\right)^2 = 63.6 \text{ in}^2$$

$$\text{Displacement} = \frac{63.6 \times 9 \times 400}{1728} = 132.5 \text{ ft}^3/\text{min}$$

Minimum relieving capacity required is

$$132.5 \times 0.95 = 125.9 \text{ ft}^3/\text{min}$$

Flammable and Combustible Liquid Storage Tanks

Storage tanks with flammable and combustible liquids pose the additional hazard of a fire erupting from escaping contents or of toxic and lethal gases or liquids escaping to the surroundings. Vapor pressure inside the tanks varies with ambient temperature. Pressure buildup is possible. A small leak can cause an external fire to erupt around the storage tanks. There is always the threat of a vapor cloud forming from the leak and then spreading the combustibles over a wide area. A source of ignition can cause a disastrous explosion affecting a wide area beyond the point of leakage.

Properly engineered installations can minimize the threat of leakage and consequent fire and explosion. Proper planning includes designing vessels to acceptable codes as a minimum, using appropriately labeled electrical equipment for the hazards involved (e.g., explosion-proof equipment), selecting a site with adequate clearance from adjoining properties, and separating equipment within the plant complex to avoid the spread or consequences of an explosion. Many installation requirements are now municipal, state, or federal requirements. One such standard is 1910.106, titled *Flammable and Combustible Liquids,* of the Occupational Safety and Health Administration. Table 2.2 lists some physical properties of hydrocarbon substances.

Overpressure protection

Overpressure protection is required for flammable and combustible liquid storage tanks. This is usually based on the potential pressure buildup from the surrounding heat, usually taken as 120°F. The corresponding vapor pressure at atmospheric pressure will determine the expected pressure buildup. For example, propane has a vapor pressure of 190 psia at 100°F (see Table 2.2). Propane tanks are usually constructed for pressure well above the expected vapor pressure.

A fire in the immediate vicinity of a propane tank creates the possibility of flame impingement on the tank. This can cause local shell overheating and rupture. The usual protection is to install sprinklers over such tanks in order to cool the shell surfaces and keep the vapor pressure inside the tank within the allowable pressure. Heat insulation over the tanks is another method used to prevent shell overheating and pressure rise within the tank.

Venting to prevent pressure buildup is also used, as is drawing off vapors, then compressing them so that the gas can be liquefied. The latter system is used for drawing vapors off in ammonia plants and in storing liquefied natural gas. This permits the storage tanks to operate within a designed vapor pressure.

TABLE 2.2 Some Physical Properties of Hydrocarbon Substances

Compound	Formula	Molecular weight	Boiling point, °F, 14.696 psia	Vapor pressure, 100°F, psia	Specific gravity, 60°F/60°F, liquid	Specific gravity Air = 1 Gas	Flammability limits, vol % in air mixture		Heat of vaporization, 14.696 psia at boiling point, Btu/lb
							Lower	Higher	
Methane	CH_4	16.043	−258.69	(5000)	0.3	0.5539	5.0	15.0	219.22
Ethane	C_2H_6	30.070	−127.48	(800)	0.3564	1.0382	2.9	13.0	210.41
Propane	C_3H_8	44.097	−43.67	190.	0.5077	1.5225	2.1	9.5	183.05
n-Butane	C_4H_{10}	58.124	31.10	51.6	0.5844	2.0068	1.8	8.4	165.65
Isobutane	C_4H_{10}	58.124	10.90	72.2	0.5631	2.0068	1.8	8.4	157.53
n-Pentane	C_5H_{12}	72.151	96.92	15.570	0.6310	2.4911	1.4	8.3	153.59
Isopentane	C_5H_{12}	72.151	82.12	20.44	0.6247	2.4911	1.4	(8.3)	147.13
Neopentane	C_5H_{12}	72.151	49.10	35.9	0.5967	2.4911	1.4	(8.3)	135.58
n-Hexane	C_6H_{14}	86.178	155.72	4.956	0.6640	2.9753	1.2	7.7	143.95
2-Methylpentane	C_6H_{14}	86.178	140.47	6.767	0.6579	2.9753	1.2	(7.7)	138.67
3-Methylpentane	C_6H_{14}	86.178	145.89	6.098	0.6689	2.9753	(1.2)	(7.7)	140.09
Neohexane	C_6H_{14}	86.178	121.52	9.856	0.6540	2.9753	1.2	(7.7)	131.24
2,3-Dimethylbutane	C_6H_{14}	86.178	136.36	7.404	0.6664	2.9753	(1.2)	(7.7)	136.08
n-Heptane	C_7H_{16}	100.205	209.17	1.620	0.6882	3.4596	1.0	7.0	136.01
2-Methylhexane	C_7H_{16}	100.205	194.09	2.271	0.6830	3.4596	(1.0)	(7.0)	131.59
3-Methylhexane	C_7H_{16}	100.205	197.32	2.130	0.6917	3.4596	(1.0)	(7.0)	132.11
3-Ethylpentane	C_7H_{16}	100.205	200.25	2.012	0.7028	3.4596	(1.0)	(7.0)	132.83
2,2-Dimethylpentane	C_7H_{16}	100.205	174.54	3.492	0.6782	3.4596	(1.0)	(7.0)	125.13
2,4-Dimethylpentane	C_7H_{16}	100.205	176.89	3.292	0.6773	3.4596	(1.0)	(7.0)	126.58
3,3-Dimethylpentane	C_7H_{16}	100.205	186.91	2.773	0.6976	3.4596	(1.0)	(7.0)	127.21
Triptane	C_7H_{16}	100.205	177.58	3.374	0.6946	3.4596	(1.0)	(7.0)	124.21

n-Octane	C₈H₁₈	114.232	258.22	0.537	0.7068	3.9439	0.96	—	129.53
Diisobutyl	C₈H₁₈	114.232	228.39	1.101	0.6979	3.9439	(0.98)	—	122.8
Isooctane	C₈H₁₈	114.232	210.63	1.708	0.6962	3.9439	1.0	—	116.71
n-Nonane	C₉H₂₀	128.259	303.47	0.179	0.7217	4.4282	0.87	2.9	123.76
n-Decane	C₁₀H₂₂	142.286	345.48	0.0597	0.7342	4.9125	0.78	2.6	118.68
Cyclopentane	C₅H₁₀	70.135	120.65	9.914	0.7504	2.4215	(1.4)	—	167.34
Methylcyclopentane	C₆H₁₂	84.162	161.25	4.503	0.7536	2.9057	(1.2)	8.35	147.83
Cyclohexane	C₆H₁₂	84.162	177.29	3.264	0.7834	2.9057	1.3	7.8	153.0
Methylcyclohexane	C₇H₁₄	98.189	213.68	1.609	0.7740	3.3900	1.2	—	136.3
Ethylene	C₂H₄	28.054	−154.62	—	—	0.9686	2.7	34.0	207.57
Propene	C₃H₆	42.081	−53.90	226.4	0.5220	1.4529	2.0	10.0	188.18
1-Butene	C₄H₈	56.108	20.75	63.05	0.6013	1.9372	1.6	9.3	167.94
cis-2-Butene	C₄H₈	56.108	38.69	45.54	0.6271	1.9372	(1.6)	—	178.91
trans-2-Butene	C₄H₈	56.108	33.58	49.80	0.6100	1.9372	(1.6)	—	174.39
Isobutene	C₄H₈	56.108	19.59	63.40	0.6004	1.9372	(1.6)	—	169.48
1-Pentene	C₅H₁₀	70.135	85.93	19.115	0.6457	2.4215	1.4	8.7	154.46
1,2-Butadiene	C₄H₆	54.092	51.53	(20.)	0.658	1.8676	(2.0)	(12.)	(181.)
1,3-Butadiene	C₄H₆	54.092	24.06	(60.)	0.6272	1.8676	2.0	11.5	(174.)
Isoprene	C₅H₈	68.119	93.30	16.672	0.6861	2.3519	(1.5)	—	(153.)
Acetylene	C₂H₂	26.038	−119	—	0.615	0.8990	2.5	80.	—
Benzene	C₆H₆	78.114	176.17	3.224	0.8844	2.6969	1.3	7.9	169.31
Toluene	C₇H₈	92.141	231.13	1.032	0.8718	3.1812	1.2	7.1	154.84
Ethylbenzene	C₈H₁₀	106.168	277.16	0.371	0.8718	3.6655	0.99	6.7	144.0
o-Xylene	C₈H₁₀	106.168	291.97	0.264	0.8848	3.6655	1.1	6.4	149.1
m-Xylene	C₈H₁₀	106.168	282.41	0.326	0.8687	3.6655	1.1	6.4	147.2
p-Xylene	C₈H₁₀	106.168	281.05	0.342	0.8657	3.6655	1.1	6.6	144.52
Styrene	C₈H₈	104.152	293.29	(0.24)	0.9110	3.5959	1.1	6.1	(151.)
Isopropylbenzene	C₉H₁₂	120.195	306.34	0.188	0.8663	4.1498	0.88	6.5	134.3

(continued)

TABLE 2.2 Some Physical Properties of Hydrocarbon Substances (continued)

Compound	Formula	Molecular weight	Boiling point, °F, 14.696 psia	Vapor pressure, 100°F, psia	Specific gravity, 60°F/60°F, liquid	Specific gravity Air = 1 Gas	Flammability limits, vol % in air mixture		Heat of vaporization, 14.696 psia at boiling point, Btu/lb
							Lower	Higher	
Methyl alcohol	CH_4O	32.042	148.1(2)	4.63(22)	0.796(3)	1.1063	6.72(5)	36.50	473.(2)
Ethyl alcohol	C_2H_6O	46.069	172.92(22)	2.3(7)	0.794(3)	1.5906	3.28(5)	18.95	367.(2)
Carbon monoxide	CO	28.010	−313.6(2)	—	0.801(8)	0.9671	12.50(5)	74.20	92.7(14)
Carbon dioxide	CO_2	44.010	−109.3(2)	—	0.827(6)	1.5195	—	—	238.2(14)
Hydrogen sulfide	H_2S	34.076	−76.6(24)	394.0(6)	0.79(6)	1.1765	4.30(5)	45.50	235.6(7)
Sulfur dioxide	SO_2	64.059	14.0(7)	88.(7)	1.397(14)	2.2117	—	—	166.7(14)
Ammonia	NH_3	17.031	−28.2(24)	212.(7)	0.6173(11)	0.5880	15.50(5)	27.00	587.2(14)
Air	N_2O_2	28.964	−317.6(2)	—	0.856(8)	1.0000	—	—	92.(3)
Hydrogen	H_2	2.016	−423.0(24)	—	0.07(3)	0.0696	4.00(5)	74.20	193.9(14)
Oxygen	O_2	31.999	−297.4(2)	—	1.14(3)	1.1048	—	—	91.6(14)
Nitrogen	N_2	28.013	−320.4(2)	—	0.808(3)	0.9672	—	—	87.8(14)
Chlorine	Cl_2	70.906	−29.3(24)	158.(7)	1.414(14)	2.4481	—	—	123.8(14)
Water	H_2O	18.015	212.0	0.9492(12)	1.000	0.6220	—	—	970.3(12)
Helium	He	4.003	—	—	—	—	—	—	—
Hydrogen chloride	HCl	36.461	−121(16)	925.(7)	0.8558(14)	1.2588	—	—	185.5(14)

SOURCE: Groth Equipment Corporation.
NOTE: Numbers in parentheses refer to Groth Equipment Corporation footnotes and should be referred to for further details.

Overpressure protection requirements

The U.S. Department of Labor, which administers the Occupational Safety and Health Act, has published overpressure protection requirements for storage tanks with flammable and combustible liquids. Some of these requirements are:

1. "Atmospheric tank" shall mean a storage tank that has been designed to operate at pressures from atmospheric pressure through 0.5 psig.

2. "Low-pressure tank" shall mean a storage tank that has been designed to operate at pressures above 0.5 psig but not more than 15 psig.

3. "Pressure vessel" shall mean a storage tank or vessel that has been designed to operate at pressures above 15 psig.

4. Low-pressure tanks and pressure vessels shall be adequately vented to prevent development of pressure or vacuum, as a result of filling or emptying and atmospheric temperature changes, exceeding the design pressure of the tank or vessel. Protection shall also be provided to prevent overpressure from any pump discharging into the tank or vessel when the pump discharge pressure can exceed the design pressure of the tank or vessel.

5. If any tank or pressure vessel has more than one fill or withdrawal connection and simultaneous filling or withdrawal can be made, the vent size shall be based on the maximum anticipated simultaneous flow.

6. Unless the vent is designed to limit the internal pressure to 2.5 psi or less, the outlet of vents and vent drains shall be arranged to discharge in such a manner as to prevent localized overheating of any part of the tank in the event vapors from such vents are ignited.

7. Tanks and pressure vessels storing Class IA liquids shall be equipped with venting devices that shall be normally closed except when venting to pressure or vacuum conditions. Tanks and pressure vessels storing Class IB and IC liquids shall be equipped with venting devices that shall be normally closed except when venting under pressure or vacuum conditions, or shall be equipped with approved flame arresters.

8. For normal venting for aboveground tanks atmospheric storage tanks shall be adequately vented to prevent the development of vacuum or pressure sufficient to distort the roof of a cone roof tank or exceeding the design pressure in the case of other atmospheric tanks, as a result of filling or emptying and atmospheric temperature changes.

9. Normal vents shall be sized in accordance with (*a*) the American Petroleum Institute Standard 2000, *Venting Atmospheric and Low-Pressure Storage Tanks* (1968), or (*b*) other accepted standards, or shall be at least as large as the filling or withdrawal connection, whichever is larger, but in no case less than $1\frac{1}{4}$-inch nominal inside diameter.

10. Where entire dependence for emergency relief is placed upon pres-

sure-relieving devices, the total venting capacity of both normal and emergency vents shall be enough to prevent rupture of the shell or bottom of the tank if vertical, or of the shell or heads if horizontal. If unstable liquids are stored, the effects of heat or gas resulting from polymerization, decomposition, condensation, or self-reactivity shall be taken into account. The total capacity of both normal and emergency venting devices shall be not less than that derived from Table 2.3, except as provided below for wetted surfaces over 2800 ft². Such devices may be a self-closing access cover, or one using long bolts that permit the cover to lift under internal pressure, or an additional or larger relief valve or valves. The wetted area of the tank shall be calculated on the basis of 55 percent of the total exposed area of a sphere or spheroid, 75 percent of the total exposed area of a horizontal tank, and the first 30 ft above grade of the exposed shell area of a vertical tank.

11. Every aboveground storage tank shall have some form of construction or device that will relieve excessive internal pressure caused by exposure to fires.

  a. In a vertical tank the construction of which complies with OSHA rules, it may take the form of a floating roof, lifter roof, a weak roof-to-shell seam, or other approved pressure-relieving construction. The weak roof-to-shell seam shall be constructed to fail preferentially to any other seam.

  b. For tanks and storage vessels designed for pressure over 1 psig, the total rate of venting shall be determined in accordance with Table 2.3, except that when the exposed wetted area of the surface is greater than 2800 ft², the total rate of venting shall be calculated by the following formula:

$$CFH = 1107A^{0.52}$$

where CFH = venting requirement, in cubic feet of free air per hour
A = Exposed wetted surface, in square feet

Note: The foregoing formula is based on $Q = 21{,}000A^{0.52}$.

The total emergency relief venting capacity for any specific stable liquid may be determined by the following formula:

$$\text{Cubic feet of free air per hour} = V\frac{1337}{L\sqrt{M}}$$

where V = cubic feet of free air per hour from Table 2.3
L = latent heat of vaporization of specific liquid, Btu per pound
M = molecular weight of specific liquid

TABLE 2.3   Wetted Area versus Cubic Feet Free Air per Hour (CFH)*

Square feet	CFH	Square feet	CFH	Square feet	CFH
20	21,100	200	211,000	1,000	524,000
30	31,600	250	230,000	1,200	557,000
40	42,100	300	265,000	1,400	587,000
50	52,700	350	288,000	1,600	614,000
60	63,200	400	312,000	1,800	639,000
70	73,700	500	354,000	2,000	662,000
80	84,200	600	392,000	2,400	704,000
90	94,800	700	428,000	2,800	742,000
100	105,000	800	462,000	and over	
120	126,000	900	493,000		
140	147,000	1,000	524,000		
160	168,000				
180	190,000				
200	211,000				

* 14.7 psia and 50°F.

SOURCE: U.S. Department of Labor, *Federal Register,* May 29, 1971.

Example   What is the free air venting in cubic feet per hour required for a tank holding liquid butane with a calculated 4070 ft^2 of wetted surface per API standards?

answer   The wetted surface is above the 2800-ft^2 criteria of Table 2.3. Use

$$V = 1107(4070)^{0.82} = 1,009,288 \text{ cu ft}^3$$

For butane

$$\text{Cubic feet per hour of air} = V \frac{1337}{L\sqrt{M}}$$

where $L = 165.65$ (from Table 2.2)
       $M = 58.124$ (from Table 2.2)

Substituting,

$$\text{Cubic feet per hour of air} = 1,009,288 \frac{1337}{165.65\sqrt{58.124}} = 1,068,508 \text{ ft}^3$$

The above example illustrates the method used to obtain data on relief valve capacity that may be needed on storage tanks containing flammable or explosive contents. The API has some excellent material on this subject, as do fire insurance groups, such as the Industrial Risk Insurers and the Factory Mutual Engineering group. Such material shows, for example, that venting requirements may be lowered if fire-resistant insulation is installed over the tank surfaces. The thicker the insulation, the lower the venting capacity may be.

   c. The required airflow rate, as determined by Table 2.3 and calculations, may be multiplied by the appropriate factor listed in the following schedule when protection is provided as indicated. Only one factor may be used for any one tank.

       0.5 for drainage in accordance with OSHA requirements, for tanks over 200 ft² of wetted area

       0.3 for approved water spray

       0.3 for approved insulation

       0.15 for approved water spray with approved insulation

   d. The outlet of all vents and vent drains on tanks equipped with emergency venting to permit pressures exceeding 2.5 psig shall be arranged to discharge in such a way as to prevent localized overheating of any part of the tank in the event vapors from such vents are ignited.

   e. Each commercial tank venting device shall have stamped on it the opening pressure, the pressure at which the valve reaches the full open position, and the flow capacity at the latter pressure, expressed in cubic feet per hour of air at 60°F and at a pressure of 14.7 psia.

The reader is referred to the Department of Labor OSHA consensus standard on the other safety requirements for storing flammable and combustible liquids, as published in the *Federal Register,* May 29, 1971.

Inert gas use

Another method used to minimize explosions in chemical process storage tanks is blanketing the top of the tank's contents with inert gas. The theory in using inert gas is to eliminate or avoid the formation of explosive mixtures within a tank or pipe. In combustible or flammable contents processes, inert gas may even be used to move the contents from one container to the other, thus preventing any combustible combinations from developing. Gases or gas mixtures that do not react with combustibles in an exothermic manner can be used as an inert gas. Any gas used for inerting must be kept free of air or other contaminants. A problem can arise if the process gas is at a higher pressure than the inert gas. To avoid contamination of the inert gas container or tank, double-block valves with a bleeder in between have been used, as well as double-check valves. This prevents process gas from flowing to the inert gas source tank.

Cryogenic Storage Tanks

The term "cryogenic" refers to a process that uses temperatures below −100°F. Special construction materials must be used, as many carbon-steel materials become brittle at temperatures below $-20°F$, and thus subject to possible sudden cracking failures (see Fig. 2.2a).

Among the most economically and technically important physical reactions using cryogenic temperatures are the cooling and liquefaction of gases and the distillation and fractional condensation of liquefied gas mixtures to yield pure-component streams. Cooling and liquefaction are the bases of every cryogenic process.

Gases with low boiling points are used as a working medium in cooling

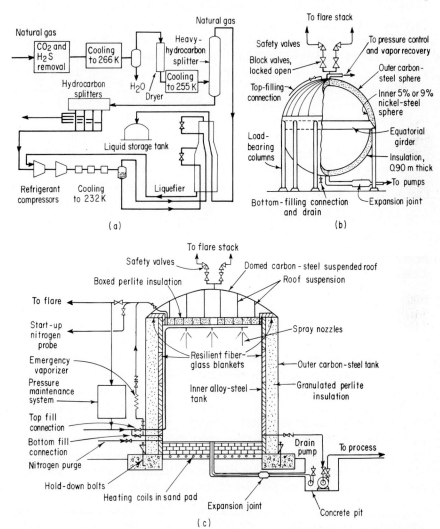

Figure 2.2 Cryogenic pressure vessels. (a) Liquefied natural gas (LNG) process flow diagram; (b) double-wall, sphere liquid ethylene storage tank with perlite insulation between the walls; (c) cylindrical ethylene storage tank, which also requires heavy insulation because ethylene has a boiling point of $-104°F$ at atmospheric pressure.

and liquefaction. For production of ultralow temperatures, helium is used, but below 1 K (-272°C, or -443.6°F), liquid helium has a very low vapor pressure so that adiabatic demagnetization must be employed to reach temperatures close to absolute zero. Economically significant cryogenic processes do not occur at such extremes.

Because of the contraction in volume of gases with decreasing temperatures, cryogenic facilities are smaller than standard process facilities and must be well insulated for energy conservation. This sometimes creates the impression that maintenance is more difficult than for warm plants. Mechanically moving parts can be designed and installed with ready access for easy maintenance. However, equipment behind the heavy insulation may require thawing out and purging of gases before repairs can be made.

Cryogenic liquid storage tanks are usually located outdoors, because indoor leakage of any gas may result in fire, explosions, asphyxiation, and even possible toxic effects. Cryogenic tanks are, thus, usually installed outdoors on reinforced concrete pads; they use low-voltage-powered pumps and may require a heating medium for liquid vaporization, as in the case of liquid natural gas's being vaporized before being piped into a natural gas line for distribution or plant use. Pressure regulation is usually by the use of the stored gas, and not by use of air or pneumatic regulators. This prevents instrument lines from freezing owing to moisture in the air.

Construction materials for cryogenic tanks

Low-temperature service requires careful consideration of the metals to be used for the storage tanks and piping connected to the system. Some metals become extremely brittle below a certain point called the "transition temperature" and can crack very abruptly from a shock load or through a stress concentration point. According to ASME code requirements, only notch-tough materials, including welding materials, should be used, especially with ferrous metals. This should be remembered when performing any repairs to vessels in cryogenic service. There is a problem in using nonferrous metals, such as copper and aluminum alloys. These may be less susceptible to brittle fracture but are more likely to fail in a fire, because of their lower strength at elevated temperatures in comparison with ferritic material. Some broad temperature limits and the corresponding metals that may become brittle are shown in Table 2.4

ASME code requirements

The ASME code should be consulted for allowable stresses at low temperatures and for the testing required of the material to ascertain if it is

TABLE 2.4   Transition Temperatures for Selected Metals

Material	Lowest permissible operating temperature, °F
Carbon steel	−28
Aluminum killed steel	−64
2% nickel steel	−82
3% nickel steel	−154
9% nickel steel	−373
18-8 stainless steel–monel	−422

suitable for low-temperature service. The ASME code stress tables designate $-20°F$ as the beginning of low-temperature service for ferritic material. Low temperature is usually taken as -20 to $-100°F$. Anything below that is considered cryogenic, which usually means that a gas can be liquefied or solidified at a temperature below $-100°F$.

Some Section VIII, Division 1, requirements of the ASME code are cited here to emphasize the need for consulting the code when ordering cryogenic equipment or when repairing a cryogenic vessel.

1. SA-36 and SA-283 carbon-steel material are not permitted for use at temperatures below $-20°F$.

2. Most carbon steels of the low-alloy type require Charpy V-notch impact tests on weldments and the materials for shells, heads, nozzles, etc., that may be subjected to temperatures below $-20°F$.

3. Longitudinal and circumferential joints must be of the butt type. Postweld heat treatment is required. All welded joints of the butt type must be x-ray-examined. Others require penetrant examination. See later chapters on NDT types.

4. The 5 to 9 percent nickel steels require notch ductility tests at the lowest temperature at which pressure will be applied.

5. The code permits waiving impact tests for certain code material, provided the vessel is designed for a pressure $2\frac{1}{2}$ times that required.

6. The code permits waiving impact tests for the high-alloy steels, such as types 304, 304L, and 347 stainless steels, and for 36 percent nickel steels, provided the operating temperature is not below $-425°F$. The code should be consulted on other permissible design and repair conditions that are permitted for high-alloy steels operating in cryogenic temperatures.

Figure 2.2 *b* and *c* is a schematic of a cryogenic tank with valving and piping. Note the inner and outer shell construction. For brevity, supports are not shown. Usually a vacuum is maintained between the two shells. The inner shell is the high-pressure shell and, for the gases shown, is usually made of nonembrittling alloy material, such as nickel steel, 304 stainless steel, or similar material. The outer shell is usually carbon steel and built for 15 psig. The space between the shells is filled with insulation, usually a form of perlite (expanded volcanic ash) and is maintained at a 10- to 100-mmHg vacuum.

Cryogenic tanks are usually protected against overpressure by a combination of rupture disks and safety valves. Safety relief valves should unload to a flare stack and not to the atmosphere if any flammable or combustible liquid is stored. Usually a reduction in pressure, as occurs when a relief valve functions, also lowers the gas or liquid temperature. Relief valve material should be of alloy steel suitable for this temperature, as should the connected piping to the relief devices.

Repairs to cryogenic storage tanks

Repairs to leaks on a cryogenic storage system can be quite involved, depending on what is stored. If welding is to be employed, great precautions are needed in emptying the contents, purging the tank and lines with an inert gas, and then finding the leak and making the area available for repairs. All lines to the area to be repaired should be physically sealed off by the use of line blinds or by breaking the line connections. Worker protection is also needed. The extremely cold cryogenic liquids will cause "cold burns" on the skin. Protective clothing in the form of gloves, coats, face shields, and similar gear should be worn around cryogenic equipment.

Rotating Pressure Vessels

Rotating pressure vessels are subject to possible failure not only from the stresses produced by internal pressure but also from repetitive-type stresses produced by machinery rotation. These include fatigue cracks from sharp corners where journals and heads meet, as shown in Fig. 2.3, and any similar stress concentration points of the rotating pressure roll. During inspections of rotating pressure rolls, the failure paths possible from the vessel's operating as a rotating machine should not be forgotten.

There are many industrial processes that require rotating pressure vessels, usually for the purpose of drying flat sheets of paper, cloth, plastics, and similar products. Such vessels are also used to remove moisture from pulp and also to press products to uniform thickness. Some of the designations of rotating pressure vessels may not indicate that they are rotating vessels.

Slashers

Drying cylinders are used on a *slasher*. This is a machine used in the textile industry to dry the size on the yarn before making the warp. The cylinders on the machine are driven mechanically and usually heated by low-pressure steam. The rolls may be as large as 84 in in diameter. There have been instances of these rolls exploding because the low-pressure steam was supplied by high-pressure boilers through reducing valves. If a reducing valve "hangs" up, there is the possibility of high-pressure steam's being introduced to these vessels, unless an adequately sized relief or safety valve is installed on the low-pressure side of the reducing valve. This must be set at a pressure not higher than the allowable pressure for the rotating rolls. Because of their thin construction, these vessels generally cannot resist a vacuum within the cylinder, such as may occur from the sudden interruption of steam. It is usual practice to install a vacuum breaker on each rotating roll to prevent this from happening. Much emphasis is also placed on preventing steam condensate from accumulating in these types of rotating vessels. If water accumulates inside the cylinders, it can act as a condenser for the steam, expediting the formation of low pressure in the cylinder, which can cause the cylinder to collapse inward. Water should not be allowed to accumulate in these types of vessels for another reason—thermal shocking on start-ups and the chance of accelerated internal corrosion due to the presence of the water.

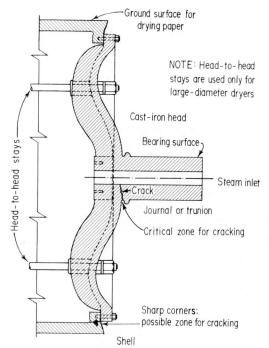

Figure 2.3 Cracks in rotating steam dryer roll were caused by sharp corner design where journal and head of the dryer meet. This area of rotating pressure vessels deserves periodic inspection for crack initiation, as does the head-to-shell juncture.

Paper machines

Rotating pressure vessel rolls are found on the three most commonly used papermaking machines.

Foudrinier papermaking machine. In the Foudrinier papermaking machine, the prepared pulp flows onto a horizontal traveling wire, or by a jet, the pulp is spread between twin moving wires. Water is removed from the pulp by normal drainage, assisted by suction. The pulp becomes a sheet and is further dried after leaving the wire and passing onto felts by being passed through rotating press and couch rolls with holes in them. A vacuum is created on some of these press and couch rolls to further remove the moisture from the traveling pulp sheet. Pressure is also applied on the press rolls to further squeeze the water out of the pulp. The pulp sheet is then passed through numerous steam-heated dryers to remove the moisture and basically create a fused sheet of dried pulp that is paper. To improve the finish, additives are used and the paper is passed through calender rolls.

Cylinder-type paper machine. In the cylinder-type paper machine, the pulp is passed over a cylinder with the wire and with the revolving cylinder operating partially submerged in a vat of thin paper stock. Suction is maintained on the inside of the cylinder with holes so that the stock adheres to the wire over the cylinder. The pulp sheet is removed from the cylinder by passing it on to a synthetic or woolen felt to continue the drying process through suction and couch and steam-heated dryer rolls as described for the Foudrinier machine.

Yankee dryer roll. A Yankee dryer roll is a large paper machine drying roll, generally 10 to 20 ft in diameter, having a true and highly polished finish and using a principle of application differing from ordinary paper drying rolls, inasmuch as the sheet of paper is pressed onto its heated surface while still in a wet and plastic state by means of a pressure roll. Actually, a Yankee dryer performs three functions that are normally performed by separate sections or groups of rolls in an ordinary paper machine. These functions are as follows:

1. Pressing to remove free water

2. Drying

3. Finishing or calendering

The Yankee dryer principle was originally developed in Germany and was introduced in this country just before the turn of the century. Most of the early Yankee machines and Yankee rolls were of foreign manufacture.

Yankee dryers are used in this country principally to produce tissue paper and other lightweight papers, although the Yankee dryer and its principle of handling paper can be incorporated into almost any paper machine, to manufacture almost any grade of paper. In Europe, Yankee dryers are used to produce papers ranging from cigarette paper to paperboard.

Most Yankee dryer rolls are made of cast iron because of its excellent heat transfer rate over the entire surface of the roll, which may have a length equal to the diameter of the roll. The large castings required for a Yankee roll, and the need for the castings to be machined and ground to a highly polished surface, limit the manufacturers of these rolls. At the present time, no American manufacturer is making these rolls. They are manufactured in Europe.

The principal difference between the Yankee roll and a conventional paper machine roll is the central shaft construction, which makes the journal a separate part from the head. The large-diameter heads are, of course, stayed by this type of construction. See Fig. 2.4.

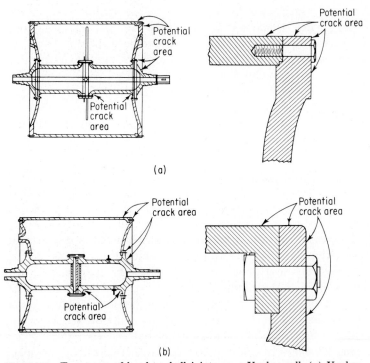

Figure 2.4 Two types of head-to-shell joints on a Yankee roll. (*a*) Yankee dryer roll features head attached to shaft and cylinder, with cylinder attached to shell by cap screws. This juncture is prone to cracking failure in early designs. (*b*) Head is bolted to shell with an internal flange to take the bending stress.

In this country, Yankee rolls are required to be constructed in accordance with the *ASME Unfired Pressure Vessel Code* in so far as the rules of that code are applicable. Working pressures have increased from 25 to 30 psi up to pressures above 100 psi, with corresponding increases in the required thickness of the cast shell. This increase in thickness has been greatly moderated by the introduction of higher grades of cast iron, which of course is a fortunate development for designers and manufacturers, since the rate of heat transfer becomes a serious problem as the thickness of shells is increased.

While stresses due solely to internal pressure are easily determined, thermal stresses and stresses due to dynamic loading are not readily determined and have been the center of much controversy. The method of attaching the heads to the shell has also been the subject of much controversy.

The explosion of a modern Yankee dryer roll because of its construction seems less probable than for a conventional roll. On the other hand, experience seems to indicate that Yankee rolls, particularly of certain designs, are prone to develop cracks in the shell. These cracks, from the standpoint of maintaining production of paper, are much more serious than head cracks, which are more prevalent in conventional paper machine rolls. Experience indicates that repairs are difficult, and even when temporarily successful, they are not apt to prove permanent.

The cast-iron construction of Yankee rolls requires care when warming up the roll as well as when shutting it down, in order to avoid any thermal stress differentials on the roll. Most cracking incidents have occurred in this period of operation, when time is required to equalize thermal conditions throughout the roll. Slow rolling is recommended, as is the use of low-pressure steam in bringing a roll up to operating conditions.

Problem areas in paper machines

Vacuum holes. Large *suction and press rolls* used on paper machines operate under a vacuum that requires holes to be drilled on their surfaces so that the moisture from the paper pulp passing over the rolls can be removed by suction. To expedite the water removal, the paper is also squeezed by press rolls that press the pulp against the cylinder with the holes. It can be seen that drilling the holes in a cylinder weakens the bending strength of the roll considerably. The holes are spaced quite closely. If a corrodent agent settles in these holes and enlarges them, the ligament strength between the holes is further reduced and cracking can result. Keeping the holes free of any caustic or other corrodent chemical is absolutely essential on these rolls.

Head-to-shell attachments. Earlier rolls built in this country used the so-called tap bolt construction, as shown in Fig. 2.4a, in which the head was attached to the thickened end portion of the shell by tap bolts. This type of construction has been partially superseded by the type of construction shown in Fig. 2.4b, in which the head is bolted to an internal, flush flange cast integrally with the shell. Experience with rolls using tap bolt construction indicates that they tend to develop cracks that start at the edge of the shell above a bolt hole and progress inward.

Head-to-shell welds.   Head-to-shell welds have also caused failures on rotating rolls. Figure 2.5 illustrates a head-to-shell attachment used in the 1950s on preheater rolls found on paper machines that corrugated sheets. The head-to-shell weld lacks full penetration in joining the two sections. The corners of the welds permit cracks to develop because of stress concentration. These rolls were manufactured to a 48-in diameter and up to 15 ft in length. Many are still in service; however, the insurance industry rechecked all their insured locations for these types of rolls and required a section of the weld to be cut out (trepanned) for examination for cracks. Those without cracks had slots cut across the weld to the minimum thickness required. If a crack should develop, leaking steam would show up in the slot, as with a telltale hole.

Heat Exchangers

Heat exchangers are an integral part of modern life and industry and range from shell and tube hot-water supply units to the petrochemical complexes, where they are extensively used to transfer heat from one substance to another. Proper utilization of heat exchangers results in energy conservation and, thus, more efficient use of available energy. Plant maintenance and inspection staffs have as their responsibility the maximizing of the effectiveness and service factor of the heat exchangers

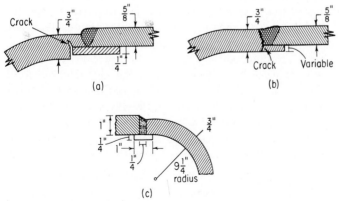

Figure 2.5 Cracks can develop in head-to-shell welds on rotating pressure vessel rolls when the weld does not fully penetrate. Figure parts (a) and (b) show such cracks. (c) Full-penetration head-to-shell weld.

under their control. The specifying and design of heat exchangers are related to the future performance and maintenance that will be needed.

Factors affecting thermal performance

The mechanical design of heat exchangers involves determining size, pressures, temperature, flows, toxicity, and flammable characteristics, so that the vessel is designed and built for these parameters in a safe and economical manner. The selection of material is similarly influenced and must take into account the corrosiveness of the heat exchanger substances.

The heat transfer design, as well as future service and maintenance, is influenced by the following:

1. The expected *flow rates* of the fluids or gases involved and their *physical and chemical properties,* such as specific heat, density, corrosiveness, viscosity, and similar considerations involving mass flow calculations. The flow rate must also be specified as to whether it is fixed or variable in nature.

2. The *molecular composition* of the exchanger substances being heat-transferred, and whether any content phase changes will occur in the exchanger.

3. The expected *in and out temperatures* of the fluids or gases. From this design consideration and the flow rate, required heating surfaces can be calculated.

4. *Fluid resistance to flow.*   This friction arises from the velocity of the moving streams and their viscosity.

5. *Fouling resistance.*   This involves scaling or deposits that may develop in service and thus affect performance. Designers use data from vessels in similar service. Sometimes an overdesign or safety factor is used to compensate for expected fouling resistance. In-use equipment will require more frequent internal cleaning if fouling is a persistent problem.

6. Expected *in and out pressures.* These are an important consideration in design, and in performance monitoring. For example, too high a pressure drop, in service, from the stipulated design conditions may be an indication that tubes are plugged.

7. Balancing of *heat mass flow rates* between the hot and cold fluids being handled. This involves the heat transfer equations to be applied and may affect output if not done properly. Today most of the heat transfer calculations are programmed for computer handling because of the high number of variables involved. The reader should consult books on heat transfer for more details.

A simple approximation of balanced flow can be made in terms of inlet and outlet temperatures, specific heats, and flow rates of the fluids involved.

$$Q = m_h c(T_i - T_o) = m_c c(T_o - T_i)$$

where $\quad Q$ = rate of heat transfer
m_h and m_c = the corresponding flow rates for specific heats and temperatures of the hot and cold fluids, respectively
c = specific heats of the fluids involved
T_i = inlet temperatures of fluids
T_o = outlet temperatures of fluids

Example A heat exchanger is to cool freon gas at constant pressure with c_p = 0.144. Water is to be used with an average inlet temperature of 60°F and an outlet temperature of 82°F. If the expected flow rate of the freon is 200 lb/h with an inlet temperature of 120°F and outlet temperature of 105°F, what flow of water will be required, assuming a constant c for water of 1 Btu/(lb)(°F)?

answer Substitute known quantities into the above equation: m_h = 200, and the corresponding c_h = 0.144; $T_{i,h}$ = 120°F; $T_{o,h}$ = 105°F. On the water side: m_c = unknown; the corresponding c_c = 1, $T_{i,c}$ = 60, $T_{o,c}$ = 82. Therefore

$$Q = 200(0.144)(120 - 105) = m_c\,(1.0)(82 - 60)$$

Solving for m_c

$$m_c = \frac{432}{22} = 19.6 \text{ lb/h of water flow needed}$$

Note that the heat required to be transferred is 432 Btu/h = Q.

Parallel flow and counterflow

Refer to Fig. 2.6. *Parallel* flow describes the situation in which water and a gas are flowing in the same direction from the inlet to the outlet point. *Counterflow* occurs when the water and gas travel through the heat exchanger in opposite directions as they go from inlet to outlet. In the counterflow unit, the hottest water and hottest gas will be at one end, while the colder portions of the fluids being handled will be at the other end. Counterflow is usually desired because (1) the exchange of heat may raise the temperature of the cooler medium more nearly to the initial temperature of the hotter medium than is possible with parallel flow and (2) for the same amount of heat transfer, it takes less heating surface to effect the transfer.

The *mean temperature difference* θ, as used in heat exchanger design, is not the arithmetic mean between inlet and outlet temperatures of the fluid, but rather is expressed as the *logarithmic mean temperature* difference. This temperature will give a true average value of the temperature of the medium as it travels through a heat exchanger, and is expressed by

$$\theta = \frac{\Delta T_l - \Delta T_s}{\log_e \Delta T_l - \log_e \Delta T_s}$$

where ΔT_l = larger temperature difference of fluid (see Fig. 2.6)
ΔT_s = smaller temperature difference of fluid

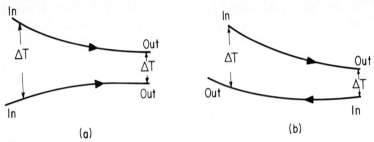

Figure 2.6   Types of flow possible in a heat exchanger. (*a*) Parallel flow; (*b*) counterflow. ΔT = temperature difference.

Example   Two fluids are to be used in a heat exchanger. The first is to operate between 140°F inlet and 100°F outlet temperatures, while the second will enter at 50°F and leave at 70°F. What are the (1) arithmetic and (2) logarithmic temperature differences for (*a*) parallel flow; (*b*) counterflow?

answer

1. *Arithmetic mean*
   a. Parallel flow: First fluid, $\theta = 140 - 50 = 90°F$
   Second fluid, $\theta = 100 - 70 = 30°F$
   b. Counterflow: First fluid, $\theta = 140 - 70 = 70°F$
   Second fluid, $\theta = 100 - 50 = 50°F$

2. *Logarithmic mean temperature difference*
   a. Parallel flow:

$$\theta = \frac{90 - 30}{\log_e \left(\frac{90}{30}\right)} = \frac{60}{\log_e 3} = \frac{60}{1.0986}$$

$$\theta = 53.6°F$$

   b. Counterflow

$$\theta = \frac{70 - 50}{\log_e \left(\frac{70}{50}\right)} = \frac{20}{\log_e 1.4} = \frac{20}{0.3365} = 59.4°F$$

Note that the end mean temperatures are used to obtain the logarithmic mean temperature differences.

Mechanical design

Plant personnel should consider some of the following factors influencing mechanical design, when ordering heat exchangers:

1. The type of exchanger and the expected use. For the pharmaceutical and food industries, the exchanger must have quick access to internal components for possibly frequent cleaning.

2. Size, pressure, and temperature requirements for the plant process.

3. The nature of the gases or fluids to be used. For toxic or flammable gases or fluids, welding requirements and the NDT required are more stringent.

4. The expected fouling characteristics of the fluids or gases to be encountered in the process. These will determine if closures are required for

frequent cleaning of internal parts of the exchanger. Many heat exchangers, if not specified otherwise, may be of welded head-and-shell construction. A tube rupture in this type of exchanger is far more expensive to repair than a similar rupture in an exchanger with a removable head closure. Cleaning of tubes because of unexpected fouling can also be done more expeditiously in the latter case.

5. The expected temperature difference from cold start to operating temperature. This can be quite severe in service, resulting in large expansion and contraction effects on the exchanger. For this reason, the plant may specify U-tube construction, or an open floating tube sheet.

6. The length of the heat exchangers. The plant must consider whether a long heat exchanger can cause tube vibration within the exchanger and thus abnormal tube wear at support sites. The fabricator should be advised to design the tube support to dampen any vibrational energy.

7. The corrosiveness of the fluids. Better material selection may help in reducing the wear from corrosion. Extra thickness or wear allowance can also be specified.

8. Expected load swings. Cyclic stresses may be imposed as well as expansion and contraction effects.

9. The possibility of coincidental maximum and minimum pressures and temperatures. The plant must specify these to the fabricator for design considerations.

10. The effect of a high-pressure tube failure in a low-pressure shell. It is absolutely essential to provide overpressure protection on the shell. The pressure setting of the safety device must be at or below the allowable pressure of the shell; the capacity should also be considered. Numerous failures occur in heat exchangers when tube failures at high pressure fill a low-pressure shell, and then spread to the connections to the shell, or expose the shell to pressure well above design.

The capacity of the overpressure protection device must be at least equal to the potential flow that is possible owing to a tube failure that floods the shell. Product impairment may result as the high pressure from the leaking tube spreads into the shell and connected piping, or the shell could fail from overpressure if the safety device capacity is too small.

Shell- and tube-side fluid arrangement

When a plant is required to order a heat exchanger for a process application, it is necessary to specify to the heat exchanger manufacturer whether the fluid to be cooled is to be on the shell side or tube side of the exchanger. Consultants in heat exchanger design usually consider the following:

1. Viscosity. Higher heat transfer rates are usually obtained by placing a viscous fluid on the shell side.

2. Presence of toxic and lethal fluids. Generally, the toxic fluid should be placed on the tube side, using a double tube sheet to minimize the possibility of leakage. The ASME code requirements for lethal service must be followed.

3. Flow rate. Placing the fluid having the lower flow rate on the shell side usually results in a more economical design. Turbulence exists on the shell side at much lower Reynolds numbers than on the tube side.

4. Corrosion. Fewer costly alloy or clad components are needed if the corrosive fluid is placed inside the tubes.

5. Fouling. Placing the fouling fluid inside the tubes minimizes fouling by permitting better fluid velocity control. Increased velocities tend to reduce fouling.

Straight tubes can be physically cleaned without removing the tube bundle. Chemical cleaning can usually be done better on the tube side. Finned tubes on square pitch are sometimes easier to clean physically. Chemical cleaning is usually not as effective on the shell side because of bypassing.

6. Temperature. For high-temperature services requiring expensive alloy materials, fewer alloy components are needed when the hot fluid is placed on the tube side.

7. Pressure. Placing a high-pressure stream in the tubes will require fewer (though more costly) high-pressure components.

8. Pressure drop. For the same pressure drop, higher heat transfer coefficients are obtained on the tube side. A fluid having a low allowable pressure drop should be placed there as well.

Heat Exchanger Types

The size of an exchanger is usually determined from some form of the general equation for heat transfer surface versus heat to be transferred: The required heat exchange surface area A is given by

$$A = \frac{Q}{U\Delta T}$$

where A = surface area required, ft^2
U = overall coefficient of heat transfer, Btu/(h)(ft^2)(°F)
Q = heat to be transferred, Btu/h
ΔT = average temperature difference between inlet and outlet of fluid, °F

The arrangement of heat transfer surfaces has led to numerous heat exchanger types (see Fig. 2.7). These include

1. Shell-and-tube type; further defined as straight and U-tube type

2. Spiral type

3. Panel coil type

4. Plate fin type

5. Double pipe type

6. Special reboiler type (e.g., deaerators and evaporators)

The shell-and-tube-type heat exchanger is the most prominent heat exchanger found in industry.

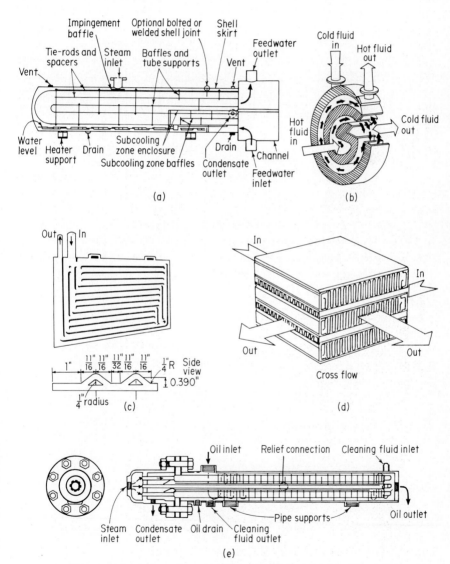

Figure 2.7 Heat exchangers can take many forms. (*a*) Shell and tube; (*b*) spiral type; (*c*) panel coil or dimpled; (*d*) plate fin; (*e*) double tube: steam tubes are rolled into one tube sheet (or welded), condensate return in another. (*Courtesy of Cleaver Brooks Co.*)

Shell-and-tube heat exchangers

Because of the variety of construction possible, the Tubular Exchange Manufacturers Association (TEMA) has classified these types by the shell and head arrangements, the method of tube attachment, and the number of passes. Figure 2.8 provides a sketch of some types of shell-and-tube arrangements and also lists some commonly used nomenclature for these types of pressure vessels.

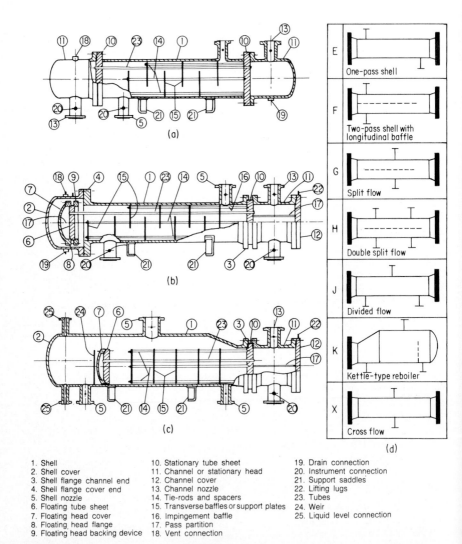

1. Shell	10. Stationary tube sheet
2. Shell cover	11. Channel or stationary head
3. Shell flange channel end	12. Channel cover
4. Shell flange cover end	13. Channel nozzle
5. Shell nozzle	14. Tie-rods and spacers
6. Floating tube sheet	15. Transverse baffles or support plates
7. Floating head cover	16. Impingement baffle
8. Floating head flange	17. Pass partition
9. Floating head backing device	18. Vent connection
19. Drain connection	
20. Instrument connection	
21. Support saddles	
22. Lifting lugs	
23. Tubes	
24. Weir	
25. Liquid level connection	

Figure 2.8 Shell-and-tube heat exchanger arrangements and nomenclature. (*a*) Fixed-tube-sheet, single-pass exchanger; (*b*) floating-tube-sheet, two-pass exchanger; (*c*) pull-through floating head kettle-type reboiler with removable channel and cover; (*d*) TEMA shell types.

TEMA shell types.   Refer to Fig. 2.8.

"E" designation is of a single-pass shell arrangement of heat transfer.

"F" designation is of a two-pass shell with longitudinal baffles.

"G" designation is of a split flow.

"H" designation is of a double split flow.

"J" designation is of a divided flow.

"K" designation is of kettle-type reboiler (see also Fig. 2.9).

TEMA classifies heads on shell-and-tube heat exchangers into front, or fixed, heads and rear heads, as is shown in Fig. 2.10.

Front-end heads

"A" designation is of a channel and removable covers.

"B" designation is of an integral or bonnet-type cover.

"C" designation is of a channel integral with tube sheet and removable cover.

"D" designation is of a special high-pressure closure.

Rear-end heads

"L" designation is of a fixed tube sheet and stationary head like "A."

"M" designation is of a fixed tube sheet and stationary head like "B."

"N" designation is of a fixed tube sheet and stationary head like "C."

"P" designation is of an outside packed floating head.

Figure 2.9   High-pressure tube-side kettle reboiler for a chemical plant. (*Courtesy of Struthers Wells Corp.*)

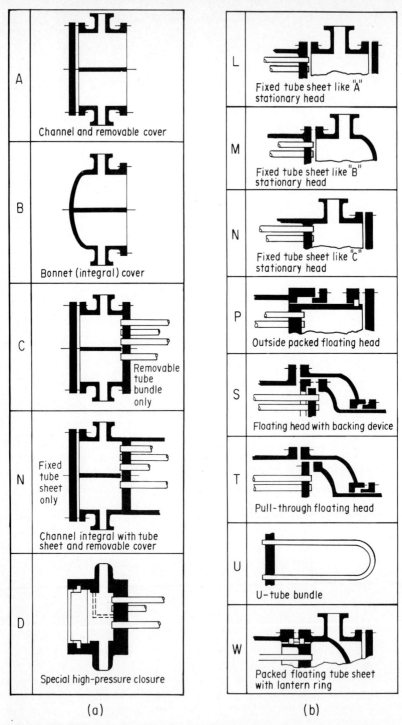

Figure 2.10 TEMA classification of shell-and-tube heat exchanger heads. (*a*) Front heads; (*b*) rear heads.

"S" designation is of a floating head with a backing device.

"T" designation is of a pull-through floating head.

"U" designation is of a U-tube bundle arrangement.

"W" designation is of a packed floating tube sheet with lantern ring.

Example   Describe a TEMA heat exchanger designated as "size 33-96 type AFM."

answer   This is a heat exchanger with a 33-in nominal inside diameter, always rounded to the nearest inch integer, with the number "96" indicating the length of the tube in inches. For straight-tube exchangers, this number is the tube length from end to end of the tube, while for U tubes, the length is from the end of the tube to the bend tangent. "A" represents a channel and removable cover head at the front; "F" represents a two-pass shell with longitudinal baffles; and "M" represents a fixed tube sheet and bonnet-type rear head.

Severity-of-service classification.   Another TEMA standard relates to construction requirements, based on severity of service:

TEMA Class R standards are intended for the severest service to be encountered in the petrochemical processing plants.

TEMA Class B standards are intended for chemical process services.

TEMA Class C standards are intended for the moderate service requirements of commercial and general process industries.

Classification by tube-removal arrangement.   Heat exchangers are usually classified by whether they have removable or nonremovable tube bundles as follows (see Figs. 2.8 and 2.10):

1. *Fixed tube sheet.*   This arrangement has the tubes secured at both ends to the tube sheet. If the heat exchanger is designed with removable channels or covers, tube leaks can be stopped by plugging or replacing the leaking tube. The shell side may not be accessible for mechanical cleaning; therefore, this type is suitable only for fluids having little fouling, corrosive, or similar adverse effect on the shell surfaces. Another disadvantage of fixed tube sheets is the possibility of high stresses developing on the tube joints because of differential thermal expansion. An expansion joint must be used to avoid this possibility.

2. *U-tube construction with one tube sheet.*   See Fig. 2.11. The U-bend section of the tube is free to expand in the shell, thus eliminating the need for an expansion joint, especially when thermal differences are great between cold-start and operating temperatures. The U bends are difficult to clean, however, thus requiring clean fluids on the tube side. Another problem is the replacement of individual tubes in case of leaks, except for

the outer row, which will not cause problems with adjoining tubes when being removed.

3. *Floating head exchangers.* These consist of straight tubes attached to tube sheets on both ends, but with one tube sheet free to move with any thermal expansion or contraction. There are four types of floating head exchangers: (*a*) floating head pull-through bundle, (*b*) floating head split backing ring, (*c*) floating head with outside packed lantern ring, and (*d*) floating head with outside packed stuffing box.

4. *The kettle reboiler.* This is a pool boiling unit used in evaporation and distillation processes. The unit has a tube bundle inserted into an enlarged shell that serves as a reservoir for column bottoms in the chemical industry and also acts as a disengaging space for vapor. An overflow weir keeps the liquid level above the top of the tubes. The tube bundle is usually a two-piece straight type, or U tubes may be used.

The *floating head* exchangers permit tube bundles to be removed for cleaning and tube replacement. In addition, tube-side headers, channel covers, gaskets, and similar parts are accessible for maintenance and repairs.

The *outside packed stuffing box unit* depends on packing compressed within a stuffing box to isolate tube fluids from the shell fluids. The packing permits the floating tube sheet to move back and forth with expansion and contraction. This type of construction is limited to shell-side pressures not over 600 psig and temperatures not over 600°F and is not permitted for handling hazardous or toxic substances because of the threat that a leak will flow to the outside environment.

The *outside packed lantern ring type* depends on shell-and-tube sealing on separate lantern rings or O rings with packing. The lantern ring is provided with a weep hole so that leakage through the packing can be

Figure 2.11   U-tube heat exchanger tube bundle. (*Courtesy of Struthers Wells Corp.*)

noted before it becomes serious. This design is generally used for pressures and temperatures under 150 psig and 500°F, respectively, and is not used where lethal, hazardous, or other conditions make any leakage or mixing of the fluids unacceptable.

The *pull-through bundle exchanger* has the bundle bolted directly to the floating tube sheet, with the assembly small enough to permit sliding it out of the shell. This permits better cleaning and maintenance of all parts of the exchanger. It is more expensive than other heat exchangers and cannot be used where mixing of shell and tube fluids cannot be tolerated.

The *inside split backing ring floating head* has the floating cover secured against the floating tube sheet by bolting to a secured split backing ring. The closure located beyond the shell is also enclosed by a shell cover of large diameter. In order to remove the tube bundle, the shell cover, split backing ring, and floating head cover must be removed. This design is usually considered to be the most expensive.

Feedwater heater. Figure 2.7a shows a low-pressure feedwater heater with 400-psi tube-side design pressure and 50 psi on the shell side. The heater has a 66-in diameter and is 37 ft long. Note the air vents, which are used to vent the heater on start-up. An important condition in operation is to have provisions on the feedwater heater for venting corrosive gases from the condensing zones of the heater; otherwise, serious internal component corrosion may occur. Venting these gases also improves heat transfer.

Bayonet tube heat exchanger

The bayonet tube heat exchanger permits the individual tubes to expand or contract. Tubes are individually replaceable. This is a modification of the U-tube design; therefore, tube bundles can be pulled for cleaning. This design is used for isothermal boiling or condensation service and generally is not suitable for multipass use. A major disadvantage is the expense of sealing the tube ends. The inner tube, called a "conveyor tube," requires a separate thin tube sheet, because the pressure on the conveyor tube is the same on the inside and outside. Bayonet tubes are also used in fired service in ammonia plants. They are usually of cast alloy steel construction.

Hairpin, or stacked pipe, heat exchanger

The hairpin heat exchanger is used for high-pressure service because small-diameter pipes can be used in a series-type flow pattern, thus not requiring a large shell diameter. The exchangers can be cleaned easily; therefore they are used for dirty fluid streams.

Plate heat exchanger

Figure 2.7*d* shows a plate heat exchanger. These units consist of a frame upon which are mounted a series of parallel, closely spaced metal plates that have been pressed with a patented trough or corrugation pattern. The plates and port edges are sealed with elastomer gaskets to prevent leakage. The product and heating or cooling fluids are directed between alternate plates for heat transfer between the plate surfaces. Models are available for working pressures to 300 psig and, by using special asbestos gaskets, to temperatures of 450°F. The plate heat transfer surface can be cleaned by removing the tie bars that clamp the plates together, thus exposing each plate's opening for cleaning. The units are available with total heating surfaces of over 16,000 ft^2 and with openings, or ports, to handle flows to 10,000 gal/min. One of the main advantages of plate heat exchangers is that they require less floor space than a shell-and-tube exchanger for the same duty. Other benefits include: (1) a higher overall heat transfer coefficient, (2) a higher mean temperature difference, (3) less expensive construction. Materials of construction will depend on expected service but include stainless steels, commercially pure titanium, Monel, Incolloy 825, and other similar metals that can be formed readily into plates.

Spiral heat exchanger

A modification of the plate exchanger is the spiral heat exchanger shown in Fig. 2.7*b*. Note that the hot fluid enters at the center of the unit and flows from the inside outward, while the cold fluid enters at the periphery and flows toward the center, thus achieving counterflow of the two fluids. Since the unit has only a single passage on each side of the fluids being handled, the spiral exchanger is usually cleaned with cleaning solutions without pulling the spiral plates out or dismantling the unit. The unit is claimed to have no interleakage, because the media do not intermix, being isolated by the welded closing on one side of each passage. Like the plate exchanger, this type of exchanger occupies less floor space than an equivalent shell-and-tube exchanger.

Double tube-sheet heat exchanger

Double tube-sheet heat exchangers are used when leakage and mixing of shell and tube-side fluids cannot be tolerated. This leakage usually comes from tube-to-tube-sheet joints or from tube-to-intermediate-tube-support joints. There are many process conditions in which mixing of fluids is not permitted or is dangerous:

1. Mixtures of the fluids could become explosive, flammable, toxic, or corrosive.

2. Health, hygienic, and similar legislation may prohibit any leakage of one fluid into another.

3. Extensive heat transfer fouling could result from the mixing.

4. There exists the chance of catalyst poisoning, which might stop the process.

5. Impure product could result downstream from the leakage.

6. A reduction in yield or efficiency could result in the process.

Figure 2.12 shows a double tube-sheet arrangement that will avoid the mixing of shell- and tube-side fluids. Note that two tube sheets are required at each of the tubes, unless one end is of the U-tube type. The adjacent tube sheets are joined to each other by the tubes' being rolled or welded to each tube sheet. The gap between the tube sheets can be sealed with a light shroud or plate, because there is no pressure in the gap under normal conditions. A drain or sight glass can be connected to this gap for detection of leaks. Some gaps are designed to be at a higher pressure than the shell or tubes, and the gap is filled with an inert substance. It is essential to align the adjoining tube sheets and also to consider the different rates of expansion between the tube sheets and the tubes in order to avoid tube-to-tubesheet joint leakage in service.

Tube-to-tube-sheet joints

Tubes can be roller-expanded into tube sheets, welded into the tube sheet, roller-expanded and then seal-welded, or roller-expanded and beaded or flared. In some applications, a ferrule is used to provide a packing or seal between the tube and the tube hole. It is common on high-pressure units, where axial force on the tubes may be substantial, to machine a groove in the tube hole, $\frac{1}{8}$ in wide by $\frac{1}{64}$ in deep, in order to provide additional holding power to the tube joint. See ASME, Section VIII, Division 1, for recommended tube-to-tube-sheet welds. Tubes are

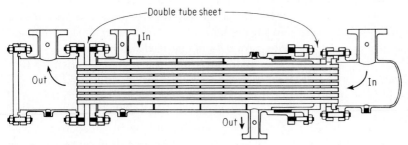

Figure 2.12 To avoid the intermixing of fluids, double-tube-sheet construction can be used. (*Courtesy of Patterson Kelley.*)

considered to act as a structural support or stay in resisting the forces imposed on the tube sheet by pressure. See later chapters for typical flathead and tube-sheet calculations on the strength required to resist pressure. Table 2.5 lists some of the advantages and disadvantages of the different types of tube-removal arrangements.

High-pressure feedwater heaters

High-pressure feedwater heaters in power plants have had tube failures due to vibration and due to inadequate venting of noncondensable gases. Since the feedwater is being heated by lower-pressure extraction steam from a turbine, a serious tube failure can also endanger the steam turbine by the high-pressure water flowing back into the turbine through the extraction lines. The feedwater heater has high-water alarms to warn the operators when this happens, but there have been incidents where the alarms did not function because of sludge buildup in the connections or because of noncondensable gases being trapped in the high-water alarm connections. Check valves in extraction lines have also failed at times in preventing water backflow to the turbine from a leaking feedwater heater tube.

Vibration and venting of feedwater heaters require special attention as a result of some of this adverse experience. The two most common flow-induced vibration phenomena are vortex-shedding and fluid-elastic excitation.

Most manufacturers have computer programs that are tailored to check, at the design stage, whether these conditions can occur. When the potential for flow-induced vibration is found to exist, the heater is redesigned to avoid it.

Proper vent system design is especially important where shell pressures are below atmospheric pressure, because the accumulation of noncondensable gases in these units seriously reduces heater effectiveness. Internal vent pipes are recommended by at least one manufacturer to release the large amount of corrosive gas that often collects in the center of the tube nest. Accumulation of noncondensables may also result in corrosion problems. Improper or inadequate venting may be detected only if it is severe enough to result in lower than normal feedwater outlet temperatures or increased corrosion products in the feedwater.

Currently, all major U.S. manufacturers use a central channel–pipe design for venting heaters. In some vertical heaters, and often in single-zone (condensing only) heaters, the air vents are located at the ends of the steam paths.

Continuous (operating) vent lines should not be cascaded to a lower-stage closed feedwater heater (CFWH). Cascading the vents to lower-pressure heaters leads to accumulation of gases in the lower-pressure heaters.

Deaerator heaters

Deaerator heaters are not of the shell-and-tube type but of the direct-contact-type water heater. The heating medium (steam) mixes with the water. See Fig. 2.13.

A deaerator has three main objectives. These are to remove oxygen and carbon dioxide gases, to serve as a storage vessel, and to preheat feedwater. All these objectives must be met for satisfactory operation.

The deaerating heater is a development of the open heater and increases its oxygen-removal function by operating at temperatures corresponding to pressures above atmospheric. Although for this reason it is no longer an "open" heater, it is nevertheless still a direct-contact heater. It is used with excellent results in moderate- to large-sized plants where a sufficient volume of low-pressure steam (5 to 50 psi) is available for the heating process.

There are three basic deaerator designs:

1. Simple tank in which condensate and makeup are heated

2. Spray-type deaerator

3. Tray deaerator

The basic physics behind deaerator operation is that oxygen and carbon dioxide gases are less soluble as temperature is raised (to a point). Hence the deaerator utilizes steam to heat the water to "strip it" of oxygen and carbon dioxide. This process is accelerated by large water contact surfaces. Hence the high efficiency deaerators use sprays and

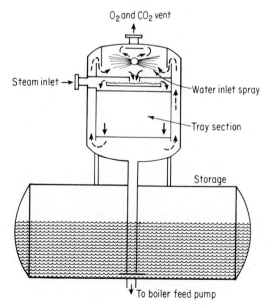

Figure 2.13 A combination spray- and tray-type deaerator heater is used to remove carbon dioxide and oxygen from boiler feedwater.

TABLE 2.5 Comparisons of Different Tube-Removal Designs on Shell-and-Tube Heat Exchangers

Construction	Advantages	Limitations	Selection tips
Nonremovable bundle, fixed tubesheet	1. Less costly than removable bundle heat exchangers. 2. Provides maximum heat transfer surface per given shell-and-tube size. 3. Provides multi-tube-pass arrangements.	1. Shell side can be cleaned only by chemical means. 2. No provision to correct for differential thermal expansion between the shell and tubes. (Exception—expansion joints available on shell.)	1. For lube oil and hydraulic oil coolers, put the oil through the shell side. 2. Corrosive or high-fouling fluids should be put through the tube side. 3. In general, put the coldest fluid through the tube side.
Removable bundle, packed floating tube sheet	1. Floating end allows for differential thermal expansion between the shell and the tubes. 2. Shell side can be steam-cleaned or mechanically cleaned. 3. Bundle can be easily repaired or replaced. 4. Less costly than full internal floating heat-type construction. 5. Maximum surface per given shell-and-tube size among removable bundle designs.	1. Shell-side fluids limited to nonvolatile and nontoxic fluids, i.e., lube oils, hydraulic oils. 2. Tube-side arrangements limited to one or two passes. 3. Tubes expand as a group, not individually (as in a U-tube unit); therefore, sudden thermal shocking should be avoided. 4. Packing limits design pressure and temperature.	1. For lube oil or hydraulic oil coolers put the oil through the shell side. 2. For air intercoolers and aftercoolers on compressors put air through the tube side. 3. In coolers with water through the tube side, with clean or jacket water, use $\frac{3}{8}$-in tubes; raw water, use $\frac{5}{8}$- or $\frac{3}{4}$-in tubes. 4. Put hot shell-side fluid through at stationary end (to keep temperature of packing as low as possible).
Removable bundle, pullthrough, bolted, internal floating head cover	1. Allows for differential thermal expansion between the shell and the tubes. 2. Bundle can be removed from shell for cleaning or repairing without removing the floating head cover.	1. For a given set of conditions, it is the most costly of all the basic types of heat exchanger designs. 2. Less surface per given shell and tube size than ring-type floating head.	1. If possible, put the fluid with the lowest heat transfer coefficient through the shell side. 2. If possible, put the fluid with the highest working pressure through the tube side. 3. If possible, put the high-fouling fluid through the tube side.

Type	Features	Recommendations
	3. Provides multi-tube-pass arrangements. 4. Provides large bundle entrance area. 5. Excellent for handling flammable and toxic fluids.	1. If possible, put the fluid with the lowest heat transfer coefficient through the shell side. 2. If possible, put the fluid with the highest working pressure through the tube side. 3. If possible, put the high-fouling fluid through the tube side.
Removable bundle, internal clamp ring-type floating head cover	1. Allows for differential thermal expansion between the shell and the tubes. 2. Excellent for handling flammable and toxic fluids. 3. Higher surface per given shell-and-tube size than pull-through units. 4. Provides multi-tube-pass arrangements. 1. Shell cover, clamp ring, and floating head cover must be removed before removing the bundle. Results in higher maintenance cost than for pull-through unit. 2. More costly than fixed tube-sheet or U-tube heat exchanger designs.	
Removable bundle, U tube	1. Less costly than floating head or packed floating tube-sheet designs. 2. Provides multi-tube-pass arrangements. 3. Allows for differential thermal expansion between the shell and the tubes, as well as between individual tubes. 4. High surface per given shell-and-tube size. 5. Capable of withstanding thermal shock. 1. Tube side can be cleaned only by chemical means. 2. Individual tube replacement is difficult, resulting in high maintenance costs. 3. Cannot be made single-pass on tube side; therefore, true counter-current flow is not possible. 4. Tube wall at U bend is thinner than at straight portion of tube. 5. Draining tube side difficult in vertical (head-up) position.	1. For oil heaters, wherever possible put steam through the tube side to obtain the most economical size.

SOURCE: American Standard.

trays to achieve large contact area. At full load, both spray and tray deaerators perform similarly. However, at part load, the spray is not as effective. Some manufacturers use continuous spray recirculation to maintain efficiency.

Cracking incidents on deaerator tanks have called attention to the fact that these relatively low-pressure vessels may fail violently if not inspected at least annually on the inside for corrosion and cracking due to the release of corrosive gases in operation.

Cracks are usually found in the welds or heat-affected zones (HAZ). All internal welds are suspect. Cracks may be located running parallel or transverse to the weld. Areas on the opposite surface of welds that attach internal or external supports, brackets, or accessories should also be examined.

Cracks usually start at weld defects or points of high residual stress. They are generally sharp and fine and appear to be filled with iron oxide. Areas at or below water level are most susceptible.

It has been generally determined that a wet, fluorescent, magnetic-particle examination of all internal welded surfaces is the most effective method to locate cracks.

Preheat and postweld heat treatments are recommended for any vessel that may be in corrosive service. Weld repairs are permitted, with the NB recommending the following if the crack penetration has pierced the minimum code-required thickness:

- Repairs should be performed in accordance with the *National Board Inspection Code*.
- Only qualified repair organizations should be used, i.e., ASME or National Board R (repair) stamp holders.
- Use weld and base metals that are compatible and of high ductility.
- Use only properly qualified welders and welding procedures.
- Inspect the welded repair for coarse ripples, ridges, valleys, undercut, excessive reinforcements, cracks, concavity, and overlaps.
- Ensure that there are no abrupt changes in surface contours.
- Use preheat and postweld heat treatment to minimize stresses.
- Reinspect, using the same nondestructive examination (NDE) procedure used to find the original crack.
- Document the repair on a National Board R-1 form to ensure continuation of the deaerator's ASME code integrity.

Evaporators

There is a need to distinguish the evaporators that are used in power plants from those used in process industries. The *power plant evaporator* is used to heat raw water to boiling, condense the resultant steam, and leave behind most of the objectionable solids and impurities. Continuous blowdown is necessary to prevent the buildup of impurities. Power plant evaporators have steam going through tubes that are submerged within the shell of the evaporator. An evaporator condenser is required to condense the vapor driven off the raw water in the shell of the evaporator.

Chemical plant evaporators are used to concentrate solutions that contain nonvolatile solutes and volatile solvents, without, however, fractionating the solvent. The aim is to vaporize the solvent out of the solution, but not to crystallize it. Thus, design of chemical plant evaporators must consider the properties of the solute and solvent in order to determine temperatures, flows, heat transfer surfaces, and similar variables so as to avoid the solvents' polymerizing or agglomerating in the evaporator (see Fig. 2.14).

Chemical plant evaporators are usually categorized according to the following three methods of operation:

1. Pool boiling, in which the bulk or pool of the liquid boils, as in kettle, internal-type, and natural-circulation thermosyphon reboilers typically used in distillation

2. Convection heating, as in forced-circulation evaporators

3. Film evaporation, in which a thin film of liquid is maintained on the heating surface

Chemical plant evaporators must be placed on a periodic inspection and maintenance program, depending, of course, on the contents being processed, in order to avoid some of the following:

1. Fouling or scale buildup

2. Control problems, leading to polymerization or plugging of the evaporator

3. Corrosion and erosion wear

4. Pressure drop and output loss from internal derangements

Material-of-construction choices can influence future performance and may be mandated by health regulations, the product being processed, and the potential corrosion and erosion that may be encountered in the process. The following summarizes some of the materials used in chemical plant evaporators:

Product	Materials of construction
Ammonium nitrate	Type 304 or 304L stainless steel
Ammonium sulfate	Type 316 stainless steel

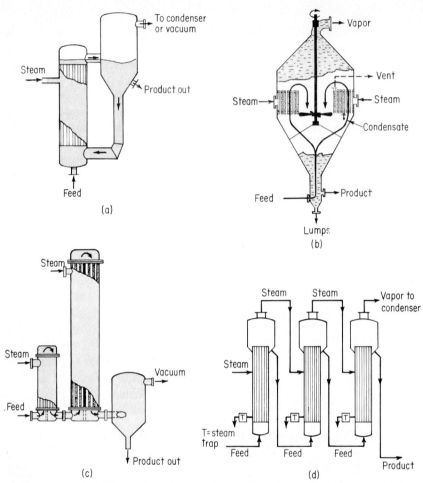

Figure 2.14 Types of chemical plant evaporators. (a) Natural-circulation pool boiling evaporator; (b) forced-circulation, calendria-type evaporator; (c) rising-falling film evaporator; (d) feed-forward, multiple-effect evaporator.

Product	Materials of construction
Caustic soda	Stress-relieved carbon steel, Monel, or nickel alloys, depending on concentration and contaminants
Pharmaceuticals	Type 304 or 316 stainless steel, Monel, or titanium
Phosphoric acid; dilute, pure urea	Type 316 stainless steel
Processed food, fruit juices, dairy products	Types 304 or 316 stainless steel
Salt (sodium chloride)	Type 316L stainless steel
Sulfuric acid	Rubber- or lead-lined carbon steel, lead, graphite, etc., depending on concentration

Codes and standards for heat exchangers

In the United States the most frequently used codes for heat exchangers are the following:

1. *American Society of Mechanical Engineers Boiler and Pressure Vessel Code,* Sections II, V, VIII, IX, and X, with Section X being for fiberglass pressure vessels. Section IX covers welding and brazing requirements. Section V is for nondestructive testing. Section II is a separate section and covers material specifications for all code-built boilers and pressure vessels.

2. TEMA standards, or *Standards of the Tubular Exchanger Manufacturers Association.*

3. American Petroleum Institute Standard, or API Standard 660, titled *Heat Exchangers for General Refinery Service.*

4. *Standards of the Heat Exchange Institute* (HEI). This pertains to the design of power plant feedwater heaters and condensers. This standard considers thermal and hydraulic factors in detail.

The TEMA code has some stiffer requirements for the internal components of heat exchangers, such as tube sheets, than the ASME code does. The ASME code is the most often adopted jurisdictional safety code. Some features of the TEMA code in relation to heat exchangers are as follows:

1. Heat exchanger nomenclature to be used for the different parts is clearly spelled out.

2. Flat-surface equations are more stringent than the ASME code in order to limit bending stress more and thus avoid tube leakage problems around tube sheets.

3. Internal welds must be ground flush to permit easy tube removal.

4. Precautions are provided to avoid the problems associated with tube rolling, ligaments in the tube sheets, and tube-sheet distortion.

An interesting feature of the API 660 code is that it prohibits the use of packed floating head construction in refineries in order to avoid packing being blown out and thus possibly releasing flammable fluids.

Vulcanizers, Digestors, and Similar Pressure Vessels

There are many pressure vessels in which solids in solutions are cooked by high-pressure steam to break down the solids into pulp or feedstock for further processing. These vessels require frequent inspections to detect wear due to product abrasiveness inside the vessel and to note the effects on the vessel of the chemicals that may be used in the process.

Vulcanizers

Vulcanizers are extensively used in the rubber industry to harden rubber in the presence of steam heat. The presence of sulfur in the rubber batches can cause corrosion problems if the shell material is not suitable for the application. Periodic internal inspections are required to determine if any shell deterioration is taking place. Such inspections furnish data for repairs needed to maintain the vessel in safe operating condition. Careful and close inspection and accurate recording and interpretation of data are necessary to prevent thinning below the required thickness and possible failure. Failure may cause explosions that result in loss of life and property. Only regular inspections by competent people, interpretation of the data, and prompt initiation of necessary repairs can keep the vessels in safe working condition.

Digesters

Digesters are used extensively in the paper industry and may be arranged as an array of many digesters programmed to cook wood chips by a batch system, or as one large, continuous-type digester, which may be 10 stories high. The continuous digester system is designed so that by the time the wood chips leave the bottom of the digester vessel, they are cooked or broken up enough to serve as the cellulose stock for making paper. See Fig. 2.15.

Corrosion and erosion of digesters are ever present problems in all paper mills. In sulfate, kraft, or soda digesters, nonmetallic linings are not commonly used, and the metal is exposed to the cooking liquors. The wear may result from corrosive action of the cooking liquors, corrosion or erosion due to loading practices, and erosion from the movement of the material being cooked. Changes in the cooking process, combined with the use of woods not used formerly, are believed to be contributing factors in the greatly increasing corrosion problems.

Both batch and continuous-type digesters require periodic internal inspections for cracks, weld deterioration, and metal thinning, to cite a few of the potential problems. There is a large amount of energy stored in these vessels that, if released, can be very destructive to property and can endanger human lives. The paper industry has developed inspection programs on batch and continuous digesters to minimize these types of occurrences. One involves periodic thickness testing of batch digesters, using a checkerboard layout system so that the same areas are checked each time for thinning. A recent experience on a continuous digester, in

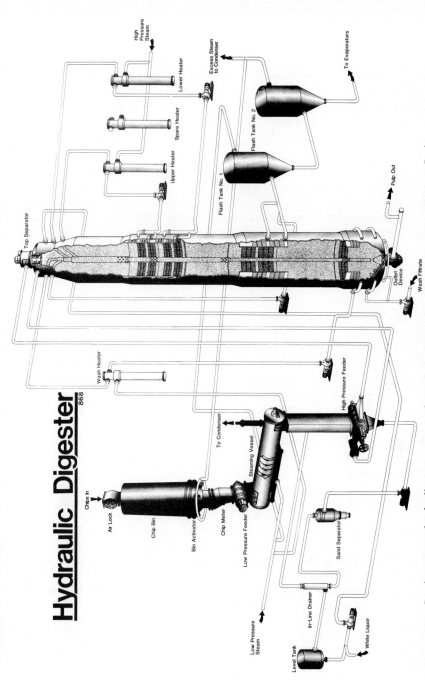

Hydraulic Digester
868

Chips In

Air Lock

Chip Bin

Bin Activator

Chip Meter

Low Pressure Feeder

Steaming Vessel

To Condenser

High Pressure Feeder

Sand Separator

In-Line Drainer

Low Pressure Steam

Level Tank

White Liquor

Wash Heater

Top Separator

Lower Heater

High Pressure Steam

Spare Heater

Upper Heater

Excess Steam to Condenser

Flash Tank No. 1

Flash Tank No. 2

To Evaporators

Pulp Out

Outlet Device

Wash Filtrate

Figure 2.15   Continuous wood pulp digester system makes pulp for a paper machine. (*Courtesy of Kamyr, Inc.*)

which the top of the vessel was blown off, has reemphasized the need for periodic weld examinations, by the wet, fluorescent, magnetic-particle method, of the heat-affected zone (HAZ) of the weld. It has also established the need for owners or purchasers to request postweld heat treatment of welds for this type of vessel, even though the ASME code may not require this, because of the class of material and because of the plate thickness's being under $1\frac{1}{2}$ in.

Locked-up stresses in the HAZ of the plate have permitted some stress corrosion to take place in many continuous digesters without postweld heat treatment. The notch formed by the corrosion has resulted in cracking. It is believed that any high caustic level in the digester can cause rapid failure due to caustic embrittlement. The paper industry has special task forces working on the cracking problem around welds on continuous digesters. These groups are reviewing liquor concentrations and makeup, acid cleaning of digesters, type of chip mix, cycling service, welding practices, NDT inspections, and inspection frequency. Later chapters will review inspection methods on digesters.

Quick-Opening-Door Pressure Vessels

These vessels are used to quickly load and unload the contents from a pressure vessel system. The types of vessels that use quick-opening-door mechanisms include vulcanizers, brick and concrete curing ovens, and similar vessels. The chief hazard in these types of pressure vessels is the chance of the door's being opened while there is still some pressure within the vessel. Another reason for failures is wear and tear on the door and on its seating surface on the vessel. One major insurer of pressure vessels for property coverage made a review of causes of failure. Its findings indicate the following:

> Structural failure is not the cause. The main factors are the design, maintenance, and operation. The types of designs are limited, and some require more maintenance than others. The most hazardous period is while the vessel is being brought up to pressure and temperature. This is indicated by the fact that the majority of failures have occurred prior to the vessel reaching its maximum operating pressure or at about 50 percent of the design pressure. Shifting of the door, movement of the shell or locking mechanism, and out-of-roundness appear to be the main contributing factors.

There are various types of interlocking devices advertised for preventing the door from opening while the vessel is under pressure. Among these are:

1. *Interlocking lug type.* The lugs on the door engage lugs on the locking ring or frame to hold the door against pressure in the vessel.

2. *Expanding or contracting ring type.*   Locking is by means of an expanding or contracting ring that engages the locking surfaces on the door and frame.

3. *Bar-locking type.*   The quick-opening door is locked by a series of radial arms or bars; the outer ends of the bars engage slots or circular holes in a shell ring.

4. *Swing-bolt type.*   Eye bolts are swung into slots on the cover and then fastened by nuts to hold the assembly tight.

The intent of the ASME code requirements on interlocking safety devices must be met in any design to prevent the door's being opened while the vessel is under pressure. A degree of redundancy is required, as the following paragraphs from Section VIII, Division 1, indicate:

> Closures other than the multibolted type designed to provide access to the contents space of a pressure vessel shall have the locking mechanism or locking device so designed that failure of any one locking element or component in the locking mechanism cannot result in the failure of all other locking elements and the release of the closure. Quick-actuating closures shall be so designed and installed that it may be determined by visual external observation that the holding elements are in good condition and that their locking elements, when the closure is in the closed position, are in full engagement.
>
> Quick-actuating closures that are held in position by positive locking devices and that are fully released by partial rotation or limited movement of the closure itself or the locking mechanism and any closure that is other than manually operated shall be so designed that when the vessel is installed the following conditions are met:
>
> 1. The closure and its holding elements are fully engaged in their intended operating position before pressure can be built up in the vessel.
> 2. Pressure tending to force the closure clear of the vessel will be released before the closure can be fully opened for access.
> 3. In the event that compliance with (1) and (2) is not inherent in the design of the closure and its holding elements, provision shall be made so that devices to accomplish this can be added when the vessel is installed.
>
> Any device or devices which will provide the safeguards broadly described in (1), (2), and (3) will meet the intent of the Code.
>
> Quick-actuating closures that are held in position by a locking device or mechanism that requires manual operation and are so designed that there will be leakage of the contents of the vessel prior to disengagement of the locking elements and release of closure need not satisfy Pars. (1), (2), and (3), but such closures shall be equipped with an audible and/or visible warning device that will serve to warn the operator if pressure is applied to the vessel before the closure and its holding elements are fully engaged in their intended position and further will serve to warn the operator if an attempt is made to operate the locking mechanism or device before the pressure within the vessel is released.

When installed, all vessels having quick-actuating closures shall be provided with a pressure-indicating device visible from the operating area.

Chemical and Petrochemical Plant Pressure Vessels

Petrochemical plants are huge complexes of pressure vessels, usually of sophisticated design and of the single-train type (see Fig. 2.16). The pressure vessels used in large chemical and petroleum plants can be multistoried in height and therefore require consideration of wind gust forces in addition to the forces on the vessels created by pressure and temperatures. These vessels may be built of high-alloy steel; they may be stainless-steel clad on the inside to a carbon-steel monowall; or they may be constructed of multilayers of carbon-steel plate with a stainless-steel liner.

Distillation column

Figure 2.17 shows a distillation column. Distillation columns are used to separate more volatile liquids from a liquid stream by means of different temperature zones in the vessel and condensation of the resulting vapors.

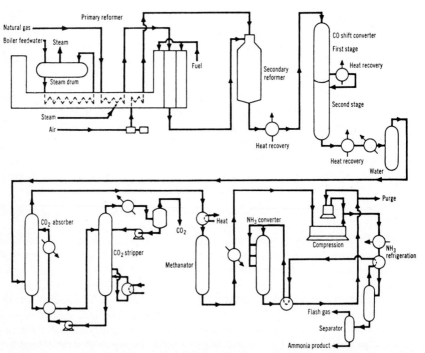

Figure 2.16   Single-train ammonia process flow features special pressure vessels typical of petrochemical plants.

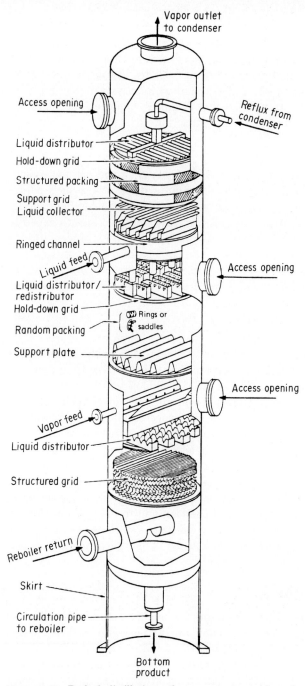

Vapor outlet
to condenser

Access opening

Reflux from
condenser

Liquid distributor
Hold-down grid
Structured packing
Support grid
Liquid collector

Ringed channel

Liquid feed

Access opening

Liquid distributor/
redistributor
Hold-down grid

Rings or
saddles

Random packing

Support plate

Access opening

Vapor feed

Liquid distributor

Structured grid

Reboiler return

Skirt

Circulation pipe
to reboiler

Bottom
product

Figure 2.17 Packed distillation column presents problems in making internal inspections. The unit shown is used in the petrochemical industry.

These types of vessels are usually designed by process specialists and constructed by fabricators specializing in the petrochemical field with associated capability for nondestructive testing, such as ultrasonic flaw detection, thickness measurement, dye penetrant testing, magnetic particle testing, x-ray and gamma ray testing, leak testing using mass spectrometers, halogen detection, and thermal conductivity detection. The fabricators also quite often have metallurgical facilities for chemical analysis of welds and materials of construction.

Catalyst trays

Catalyst trays are an important feature in many chemical process vessels. These can range from the expensive platinum metal trays such as are found in nitric acid plants to various nickel catalysts and can have values of over $500,000. Distillation columns use trays and packings to expedite the process. These internal components can be affected by process upsets, or the process may become exothermic on the loss of a catalyst. In a risk analysis of these types of vessels, failure of the internal components to perform according to design is an important consideration in providing controls and safety devices.

Corrosion-resistant lined vessels

Corrosion-resistant lined vessels are extensively used in the chemical industry. These involve lining a carbon-steel shell with a corrosion-resisting material such as stainless steel. The ASME code generally does not permit the thickness of material used for lining the vessel to be used in calculating the required thickness for the expected allowable pressure of the vessel. *Telltale holes* are used to indicate if the integrity of the lining has been pierced. These are $\frac{1}{16}$- to $\frac{3}{16}$-in holes drilled from the outside to the interface of the carbon-steel shell and the stainless-steel lining. Leakage through these holes will indicate if the lining or its weld is cracked or has pitted through or in other ways has lost its effectiveness. An internal inspection is usually required to find the leak and to make the necessary repairs. Postweld heat treating of lined pressure vessels is required when the base plate must be postweld heat-treated. This rule applies to repairs also.

Jacketed Pressure Vessels

These vessels are also called "reactors" when used to chemically and mechanically combine mixtures. They are extensively used for batch preparations in the food and pharmaceutical industries.

Jacketed pressure vessels are used to mix all types of products inside a tank or cylinder, which is either heated by steam, dowtherm, or some other fluid on the jacket portion of the vessel or supplied with a cooling fluid to *remove* heat due to a chemical reaction that may be taking place in the products being mixed, such as occurs in polymerization processes. There are three commonly used jacket types: (1) *conventional* jackets with inner and outer shells or heads (see Fig. 2.18), (2) *dimple* jackets, and (3) *half-pipe coil* jackets.

Overall heat transfer coefficients can vary with the type of reactor used and the materials employed in the mixing portion of the jacketed vessel. Jacketed vessels or reactors are extensively used in the pharmaceutical, cosmetics, and food industries. Product purity and the ability to clean the inside of the tanks are therefore paramount considerations, as is the fact that the metal or material used cannot have any effect on the product being mixed. As a result, most jacketed vessels are constructed of stainless steel or glass-lined carbon steel, or combinations thereof. Of interest is the fact that the overall heat transfer rate for a glass-lined vessel is 55 to 70 Btu/(h)(ft^2)(°F), while for a stainless-steel unit it may be as high as 125 Btu. The glass layer provides additional heat transfer resistance. The glass-lined unit helps to prevent fouling, but to improve heat transfer capability, the minimum thickness of glass possible is used. For cooling applications in the jacket, such as for processes generating exothermic reactions, chilled water is used to improve the heat transfer between the mixing chamber and the jacket. Another method of increasing heat transfer is to add supplementary or external heat exchangers that are connected to the jacket loop.

The mixing portion of the jacketed vessel can be open to the atmosphere, or may be of the closed type with pressure in it. The latter requires careful consideration concerning overpressure protection, especially when uncontrolled exothermic reactions are possible, as might occur if coolant were lost from the jacket. Rupture disks are generally used to prevent overpressure, sometimes in series with a safety relief valve. The relief valve opens first, and if pressure continues to rise, the rupture disk opens.

Coincidental pressures, or lack thereof, must also be considered. For example, the inner shell may have to be designed, and protected, both for inner pressure acting on the cylinder and for the fluid inside the jacket acting on the outer surface of the inner cylinder. The chief advantage of the dimple jacket is that it permits constructing the jacket from light-gage metals since the span for pressure resistance is considerably reduced. This also reduces manufacturing costs but does prevent easy cleaning of the jacket space. The half-pipe design is usually used for high-temperature heating service and is generally used with dowtherm-type fluids. Note that the half-pipe design permits multiple zones of heating and cooling to be applied because there are no limitations to the number of inlet and outlet pipe connections.

The dimple jacket design is treated as a welded, stayed construction by

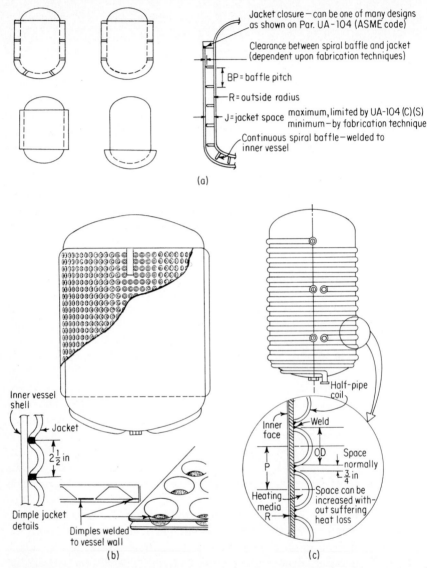

Figure 2.18 Types of jacketed pressure vessels. (*a*) Conventional jacketed pressure vessel arrangements; (*b*) dimple jacketed details; (*c*) half-pipe coil jacketed vessel. (*Courtesy of Chemical Engineering.*)

the ASME Section VIII, Division 1, code, specifically Par. UW-19(*b*), limiting allowable pressure to the dimple jacket to a maximum of 300 psi.

Glass-lined vessels

The chief advantages of the glass-lined reactors are their resistance to internal fouling and scale buildup and the ease with which glass can be cleaned for hygienic and other health considerations. Glass-lined reactors cost more than stainless-steel units. The practical size limitation is about an 18,000-gal capacity, and this may be lower if the pressure exceeds 200 psi. In the chemical industry, stainless-steel units are being used in sizes over the 18,000-gal capacity. Another advantage of glass-lined vessels is resistance to corrosion and acid attack.

Glass-lined vessel manufacturers caution against using glass-lined vessels with fluorides, which can cause rapid loss of polish and can cause glass pinholing with extended exposure. Thermal shock must also be avoided with glass-lined vessels, with one manufacturer specifying a maximum of 260°F at a graduated rise to the absolute maximum temperature limit in service of 450°F.

Glass-lined vessels can be spot-repaired on the glass portion by the use of patches or plugs made of tantalum. However, some chemicals, such as sulfur trioxide or oleum, may attack the tantalum, and it is recommended that the manufacturer of the vessel be consulted for repair details.

Agitators

Agitators are used in jacketed vessels to promote mixing and chemical combining as well as to promote heat transfer to or from the jacket. The agitators also help in maintaining certain chemical reactions in suspension, thus avoiding agglomeration. Figure 2.19 shows one use of an agitator in a jacketed vessel. Power requirements for driving the agitator are often estimated as being about 6 to 7 hp per 1000-gal tank capacity.

Jacketed-vessel safety considerations

There are several safety considerations that need to be reviewed on any jacketed-vessel installation:

1. Overpressure protection for the jacket, and for the mixing or reaction part of the vessel, is important. Many jackets are heated by steam supplied through a reducing valve. If the reducing valve becomes stuck, full boiler pressure may be imposed on the jacket; therefore, the allowable working pressure on the jacket must be compared with the steam-supplied pressure and suitable safety valves installed to protect the jacket.

Vacuum conditions inside the mixing chamber can cause the inner wall of jackets to collapse if the inner wall was not designed for the external pressure the wall must resist under vacuum conditions. This is especially true of thin-wall shells. A vacuum breaker may be necessary under these conditions. These usually occur when a heated vessel's vapor condenses rapidly inside the vessel. This could happen as the result of an interruption of steam to the vessel.

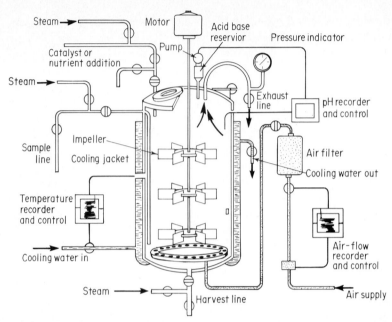

Figure 2.19 Agitators in jacketed vessels promote mixing and chemical combination and also promote heat transfer to and from the jacket. (*Courtesy of* Chemical Engineering.)

2. Overpressure protection for the mixing portion of a jacketed vessel also may involve determining the kinetics of a runaway reaction in the mixing operation. In some processes, the relief valve must handle not only vapor but also possible portions of the liquid or mass in the mixing chamber of the vessel. It is necessary, if possible, to determine the maximum amount of kinetic energy that may be developed and the expected rise in pressure and volume, in order to provide a safe relieving area with a pressure relief device set within ASME code limits established for allowable working pressure of the inner chamber of the vessel. Advances in the computer field now permit tracking reactions in these types of vessels. This permits using additional procedures to protect these vessels against a runaway reaction; for example, withdrawing or shutting off the heat source or adding cooling water to affect the catalyst, if the computer indicates that an above-normal pressure or temperature rise is taking place in the vessel.

3. Overtemperature is another consideration in reactor vessels. This may occur from a runaway chemical reaction or from the loss of a cooling fluid in the jacket that may be tempering the reaction.

4. The effects of a power outage or loss of water, heat, refrigeration, or similar services may create problems of product hardening and spoilage, in addition to the overpressure and overtemperature conditions previ-

ously described. If the exposure is severe, redundancy should be considered for supplying these services in an emergency situation.

Thick-Wall Pressure Vessels

Thick-wall pressure vessels are extensively used in the oil refining and chemical industries because of the need to use higher pressures and temperatures in the processes developed. In some applications, such as isostatic pressing, pressures of over 50,000 psi are being used. Solid-wall vessels of one thick shell present the possibility of a rapid failure from a crack that could reach the magnitude of an explosion. The cost of producing the shells for extrathick monowall construction can be considerable. The concept of *layered shells* being used in the place of one very thick shell gained great attention by designers when joining by welding appeared in the 1930s. There are *several layered-vessel designs:*

1. The wickel type, used in Europe, employs a corrugated metal tape or ribbon spiral-wound around an inner core cylinder. The inner cylinder has machined on its outer surface grooves to match the corrugations of the tape. The corrugated tape is wound on the vessel, a layer at a time, until the desired thickness is reached. Each succeeding layer mechanically locks the underlying layers together through the meshing of the corrugations in the tape or ribbon. It is claimed that the hoop stresses are borne by the ribbon's acting in tension and that the longitudinal stresses are resisted by the ribbon's acting in shear across the corrugations.

2. The Japanese have developed a layered-vessel concept in which a continuous strip of light-gage material or sheet is coiled around an inner cylinder until the proper wall thickness is reached. The ends are welded to complete the vessel wall.

3. In the United States, the most widely used design is the *multilayer* that was pioneered by the A. O. Smith Corporation (see Fig. 2.20). This design uses an inner shell over which successive cylinders are shrunk to the thickness required. The inner shell quite often is of a material that is resistant to the effects of the internal liquids or gases being handled. The layered, load-bearing shell material is usually carbon steel. For example, an ammonia converter operating at 5500 to 6000 psi may have 15 carbon-steel layers, each $\frac{1}{4}$ in thick with an inside stainless-steel liner, also about $\frac{1}{4}$ in thick. Vessels of this type may be over 9 in thick overall, have diameters to 16 ft, and be over 60 ft high. Replacement costs can be as high as $3 million.

Advantages of multiwalled pressure vessels

The advantages of multiwall over solid-wall construction, as claimed by fabricators and designers, are the following:

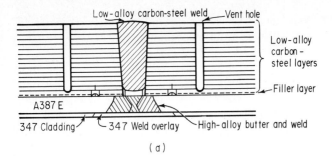

(a)

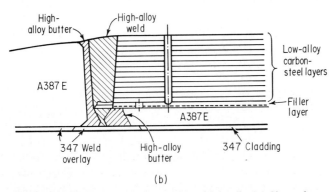

(b)

Figure 2.20 Multilayer pressure vessel weld details. (*a*) Circumferential weld detail for multiwall pressure vessel with inner layer of hydrogen-resistant material; (*b*) head-to-shell welded joint on a multilayer carbon-steel shell welded to a solid wall head of high-alloy steel material. Cladding and weld overlay protect the vessel in hydrogen service.

1. The multiple-layer shell provides a natural, safe design against risk of brittle fracture. A lower ductility transition temperature for the materials is more readily obtainable with the multiwall shell using individual plate thicknesses in a range of 25 mm (1 in) to 64 mm ($2\frac{1}{2}$ in). Any brittle fracture, if ever started, would not propagate through the total shell thickness and would be confined to one layer.

2. Materials in the thickness range required for the multiple-layer walls provide a much greater assurance for high quality than the heavy, thick plates used in monowall construction. Uniform quality of material and good metallurgical properties are obtained without requiring high-alloying elements and sometimes stringent heat treatments. Materials of different properties may be used for the various layers if desired.

3. There is no specific limit for the thickness and dimensions of a multiwall vessel. The size of vessel is restricted only to the physical capacity of the fabrication plants and to transportation limitations.

4. The multilayer shell is fabricated on conventional forming equipment at room temperatures using standard procedures. This assures quality, uniform workmanship, and integrity of product.

5. The multiwall venting system permits immediate detection of any defect on the inside surfaces of the weld or plate, long before the defect has propagated through the full thickness of the weld or the shell plate.

6. Multiwall construction does not involve any full-depth longitudinal welds, thereby minimizing problems associated with full-depth, heavy welds.

7. The temperature gradients through a multiwall shell create lower fiber stresses than in monowall shells because of the ability of individual multiwall courses to expand independently.

8. Multiwall construction permits 100 percent radiographic inspection of all welded longitudinal and girth seams.

9. Completed multiwall vessels may receive a full postwelded heat treatment.

10. The thermally shrunk cylinders of the multiwall shell place the inner layers in compression, which provides an added reserve strength not available in monowall shells.

Multilayer head closures are now used, thus eliminating large-end forgings. Details have been developed so that cylindrical sections of the multilayer can be field-assembled (see Fig. 2.20). Girth seams are usually staggered one from another layer, layer by layer, similar to longitudinal-layer welds. This avoids deep-groove welds and permits each weld to be NDT-tested more easily.

Hydrogen service

Multilayer vessels are extensively used for hydrogen storage or process stream applications because of the high pressures and temperatures (up to 5000 psi and 860°F, respectively) that may be required.

Hydrogen can embrittle carbon steels; therefore, the inner walls of pressure vessels are usually constructed of material resistant to hydrogen's attack on the crystalline structure of the material. Hydrogen combines with the carbon in steel to form methane. As the carbon becomes depleted, the steel becomes brittle. This reaction is dependent on the pressure and temperature of the hydrogen.

Section VIII, Division 1, of the ASME code now has some detailed requirements for multilayer vessels. For example, dummy layers cannot be considered as part of the required thickness of the vessel. Contact between layers must be smooth, with welds required to be ground smooth to assure contact between the layers. Longitudinal welded joints of the layer sections must be in an offset pattern, separated from each other by at least 5 times the layer thickness. The ASME code should be used as a reference for welded or other repairs.

Fiberglass-Reinforced Plastic Pressure Vessels

The fiberglass-reinforced plastic (FRP) pressure vessels were developed for highly corrosive service in the chemical industry. They are made of composite materials combining thermoset polymers with fiberglass in most applications. Thermoset polymers differ from thermoplastics. The latter have a defined melting point, while thermoset polymers have cross-linked molecules with no definite melting point. The strength of these materials is much lower in tension and bending than similar steel material. As a result, higher safety factors are used in design. The ASME has published a code for constructing these vessels in order to establish safe working pressures and temperatures. The strengths can vary by the composition of the reinforced plastic. Section X, titled *Fiberglass-Reinforced Plastic Pressure Vessels,* addresses this and other differences that these vessels have in comparison with steel pressure vessels.

Plastic materials have pressure and temperature limitations as Table 2.6 shows. It is therefore necessary for any user of these types of vessels to adhere to the specification limits for the vessel. These vessels are also affected by service conditions. Fiberglass-reinforced plastic pressure vessels may be affected by environmental exposures so that embrittlement may occur and any sudden shock load may cause cracks to develop. The material can also become crazed (develop a network of fine cracks), discolored, charred from heat, softened by chemical attack, or dissolved in spots, and in general may show evidence of deterioration in service. It is therefore essential to make periodic inspection of these types of pressure vessels.

Chemical attack on FRP is a complicated phenomenon that can occur in a number of different ways; it can be broadly classified as follows:

1. Resin attack. This includes (1) degradation and disintegration of a physical nature due to absorption, permeation, or solvent action, which, under mild conditions, may manifest itself as swelling, (2) oxidation, (3) hydrolysis, (4) radiation, including nuclear, (5) thermal attack, and (6) combinations of these.

2. Glass attack. Certain environments might not attack the resin itself, but upon absorption will attack the glass reinforcement. Fluorides, alkalis, hot water and its vapor, and hydrochloric acid under certain conditions are typical examples of glass-attacking agents.

TABLE 2.6   Properties of Fiberglass-Reinforced Plastic

Materials and construction	Average reinforcing constant, wt. %	Strength, 10³ psi		Modulus, 10⁶ psi	
		Tensile	Flexural	Tensile	Flexural
Mat-polyester (hand layup)	25	9	16	1.0	0.7
Mat and woven roving–polyester (hand layup)	40	15	22	1.5	1.0
Mat and roving–polyester (pultrusion)	65				
Longitudinal direction		30	30	2.5	1.6
Transverse direction	—	7	10	0.8	0.2

Maximum Recommended-Use Temperature (°F) in Acid and Other Applications

	Sulfuric acid			Nitric acid		Hydrochloric acid		Sodium hydroxide		Sodium hypochlorite, $5\frac{1}{4}\%$*	Ferric chloride, 40%*	Monochloro-benzene	Distilled water
	25%*	75%*	96%*	10%*	50%*	10%*	36%*	5%*	25%*				
Thermoset Polymer Composites (Fiberglass-Reinforced)													
Isophthalic polyester	180	NR†	NR	70	NR	180	125	120	NR	120	180	NR	160
Bisphenol A polyester	225	75	NR	150	NR	230	150	120	150	125	220	NR	200
Vinyl ester (standard)	210	120	NR	120	NR	210	150	150	120	150	210	NR	210
Vinyl ester (high-temperature)	220	100	NR	140	NR	220	150	150	120	150	220	75	220
Epoxy (special-cure)	NR	NR	NR	150	NR	150	NR	150	150	NR	205	200	225
Thermoplastic Polymers (Unreinforced)													
Polyvinyl chloride (PVC)	140	140	75	140	75	140	140	140	140	140	140	NR	140
Chlorinated PVC (CPVC)	180	180	120	180	75	180	180	180	180	180	200	NR	180
Polypropylene	150	NR	NR	180	NR	180	150	160	160	NR	150	NR	180
Polyvinylidene fluoride (PVDF)	240	175	120	240	120	240	240	NR	NR	NR	240	75	240
Ethlyene chlorotri-fluoroethylene (ECTFE)	240	120	NT‡	240	120	240	240	240	240	NT	240	NR	240
Fluorinated ethyl-enepropylene (FEP)	200	200	200	200	200	200	200	200	200	200	200	200	200

*Resin percentage
†NR = Not recommended at any temperature.
‡NT = Not tested.
SOURCE: *Chemical Engineering* magazine.

3. Glass-resin interface attack. The glass-resin interface can be attacked by many substances, even though the resin itself might be reasonably resistant to attack. Wicking (degradation of the reinforcement due to an inadequate resin cover) and delamination are typical effects of this type of attack.

ASME-stamped allowable pressures and temperatures should not be exceeded. Building and installing fiberglass-reinforced plastic pressure vessels according to the ASME code ensures that some minimum construction and testing requirements have been satisfied during fabrication. With properly installed protection against overpressure and overtemperature and by periodic inspections, these types of pressure vessels may offer an alternative to stainless-steel-type corrosion-resistant pressure vessels in some applications.

QUESTIONS AND ANSWERS

2-1 Define the term "hydrogenation."

ANSWER: This is a chemical reaction in which gaseous hydrogen is chemically combined with an organic compound to make shortening compounds such as butter substitutes. Reactor pressure vessels are used in this process.

2-2 What is the purpose of using inert gases in storage tanks?

ANSWER: To blanket the top of the tank's contents to prevent air or oxygen from combining with the contents and possibly forming explosive mixtures.

2-3 What can be the effect on some metals when exposed to cryogenic temperatures?

ANSWER: Some ferrous metals can become very brittle at low temperature and then unexpectedly crack in a rapid manner owing to a notch, stress raiser, or impact load.

2-4 Define the term "counterflow" as used in heat exchangers.

ANSWER: In the counterflow principle of heat exchanger design the two moving streams in the heat exchanger travel through the exchanger in opposite directions as they go from their respective inlet and outlet points. This permits the two streams to have their hottest parts at one end and their coldest parts at the other end. This design causes the final temperature of the stream being heated to be close to that of the heating stream.

2-5 What is the TEMA class designation for heat exchangers that are to be used for severe-type service?

ANSWER: TEMA Class R.

2-6 Name two types of tube and tube-sheet construction that permit thermal expansion and contraction of the tubes.

ANSWER: U-tube and floating head heat exchangers.

2-7 What type of tube-sheet construction is usually recommended to prevent the mixing of shell- and tube-side fluids owing to leakage?

ANSWER: The double tube-sheet construction is generally specified to avoid the two fluids' mixing.

2-8 What type of reactor costs more—the glass-lined or stainless-steel one?

ANSWER: The glass-lined reactor costs more for the same size, pressure, and temperature conditions.

2-9 What chemical can attack glass-lined vessels?

ANSWER: Fluorides can attack the glass polish and cause pinholes to develop in the glass lining.

2-10 What effect does hydrogen have on carbon steel at high pressure?

ANSWER: Hydrogen can combine with the carbon in the steel to form methane gas, causing enbrittlement. Proper steel selection is required to avoid this potential source of cracking failure.

2-11 Define a Yankee dryer roll.

ANSWER: This is a large-diameter, steam-heated, rotating roll, usually made of cast iron, that is used to dry thin paper stock while the stock is also being pressed against the smooth, cylindrical surface by a press roll. Thus, in one revolution of the stock on the Yankee dryer roll, dry paper can emerge from the paper machine.

2-12 What is the most pronounced cause of a cast-iron Yankee roll failure?

ANSWER: Usually cracking due to very fast thermal expansion and contraction.

2-13 Name three forms of chemical attack on fiberglass-reinforced plastic pressure vessels.

ANSWER: Chemical attacks from the contents, or surroundings, may include:

1. Attack by solvents, acids, and similar chemicals on the resin that holds the ingredients together. This will cause the vessel to swell in the area of attack.
2. Attack by fluorides on the glass component of the vessel. This usually results in pitting and roughness.
3. Attack on the glass-resin interface. This can cause delamination.

2-14 What is the minimum required diameter of vent holes in layered vessels?

ANSWER: $\frac{1}{4}$ in.

2-15 What is the purpose of vent holes in layered vessels?

ANSWER: To reveal leakage occurring through the inner shell of the vessel and to prevent buildup of pressure within the layers, which can cause layers to bulge.

2-16 What is the minimum thickness required of a shell in multilayered service?

ANSWER: The minimum thickness must be $\frac{1}{8}$ in.

Pressure Vessel Materials, Joining by Welding, and Nondestructive Testing

Need for Knowledge of Properties of Materials

The misapplication of pressure vessel materials, whether in new equipment or during subsequent repairs, may result in abnormal wear and tear, cracking, and similar accidental and unexpected occurrences, in most cases resulting in a loss for the plant due to business interruption. Certain process industries now require that maintenance and inspection personnel, as well as designers, have a more basic understanding of the properties of the materials most commonly applied to their plant. This understanding is necessary in order for them to analyze defects noted during inspections and to determine what repairs and materials are most appropriate for the condition that has been discovered.

Plant personnel should also be involved with equipment selection because of their intimate knowledge of the process, and of what may be required of the material in the process pressure vessel stream.

Factors to consider in choosing materials should include pressure, temperature, corrosion rates, cycling that may be imposed on the system, hygienic or pharmaceutical requirements, indoor versus outdoor installation, and similar variables of site selection and future use. There are ASME code requirements to be considered, as well as jurisdictional needs, including registration and installation permits. Pressure vessel manufacturers may also develop recommendations on materials to be used for their products, provided they are given sufficient process specifications regarding corrosiveness, wear allowance to be provided, weld-stress relief to be applied, and similar considerations that must be specified by the owner-user of the pressure system. Economic factors such as the cost of one material versus another must constantly be addressed during equipment selection and fabrication. The future repair needs and possible associated costs of repair must also be evaluated in material selection. Such evaluations must be made, for example, in deciding whether to use glass-lined reactors versus stainless-steel ones.

Large plants usually have a maintenance department that periodically inspects and repairs pressure vessel systems. Input on material selection from this group can be extremely valuable, because of their knowledge of how some materials stood up in service in comparison with others. The same knowledge is applicable to repair needs; also, the ease of making repairs is a paramount consideration in predicting the life of a pressure vessel system. Future cost considerations are also important.

For all these reasons, it is appropriate to review some of the properties of materials and the possible effect welding and other repairs or fabrication methods may have on pressure vessel systems.

Crystalline structure of iron-steel metals

Metallurgists carefully study the crystalline structure of new materials that are developed, in order to predict their properties; crystalline studies also provide information on the effects of service on materials.

If a piece of metal is carefully polished, immersed for a short time in an acid or other appropriate reagent, and then examined under a microscope, it will be found to be composed of small particles or crystals.

Materials that appear to be perfectly homogeneous in reality are composed of an aggregate of grains or crystals of distinctly different materials. A piece of the material will therefore have different properties at different points within the piece. Even pure metals are made up of an aggregate of crystals having different properties in different planes, depending on the space lattice structure.

One convenient method for explaining some of the similarities and differences in the characteristics of metals is to study the arrangement of atoms in the material.

Space lattice. A crystal of a given metal is composed of atoms arranged in a definite and regular geometric pattern. This pattern is known as a

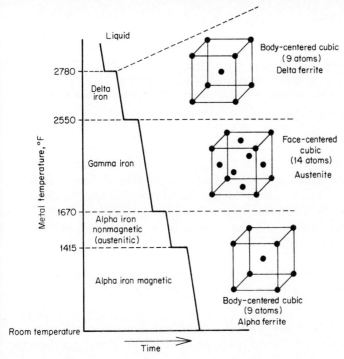

Figure 3.1   The crystal, or space lattice, structure of iron varies with temperature.

"space lattice" and is determined by x-ray studies. The atoms in iron at *room temperature,* for example, are arranged in a *body-centered cubic lattice*; that is, the atoms are located as at the corners of a cube, with one atom in the center. Iron at *high temperature* crystallizes as a *face-centered cubic lattice,* or with an atom at each corner of a cube and an atom in the center of each face. See Fig. 3.1. There are a total of 14 lattice systems.

A pure metal crystallizes when cooling to its freezing point; if a nucleus or seed is present at or below the freezing point, other atoms from the molten liquid start to attach to the nucleus. The atoms deposit along defined directions of the crystal, or space lattices.

A study of space-lattice or crystal formations that are created when metals are cooled to freezing, or solidification, temperatures assists in determining the effects of alloying, rate of cooling, and similar metallurgical considerations. Iron, as mentioned above, is transformed to various crystal types; the corresponding space-lattice arrangements for these types are as follows:

Type of iron	Type of crystal structure
Alpha,* delta, and ferrite iron†	Body-centered cubic (BCC) lattice
Gamma and austenite iron‡	Face-centered cubic (FCC) lattice

* Alpha iron reaches the BCC structure below 1670°F.
† Delta and ferrite iron with the BCC structure exist above 2534°F.
‡ Gamma and austenite iron with the FCC structure exist between 1670 and 2534°F.

Slip plane. A *slip plane* is a characteristic of a space lattice and is defined as the plane along which the atoms in a space lattice have the least resistance to relative displacement. In the slip plane, shearing displacement takes place more readily. Metallurgists can increase the shearing stress for a given crystal by introducing atomic displacements within the space lattice, thus disrupting the continuity or smoothness of the slip plane. This is accomplished by:

1. *Chemical means.* By this technique another material is introduced into the space lattice in order to disrupt the slip plane.

2. Control of the *rate of cooling* of the crystals. Alteration of the rate of cooling may change the crystal size and the chemical composition of individual crystals, thus changing the substance's ultimate properties.

3. Control of the *shaping operations.* This will give the material definite directional properties. Various types of shaping operations, such as forging, rolling, and drawing, have an effect in orienting crystals.

It is an established fact that a material composed of many small crystals will be stronger and harder than the same material would be if it were composed of a relatively few large crystals. In the fine-grained material, the individual slip planes are relatively small, and the amount of slip that may occur within the boundaries of any one crystal is limited by the restraint offered by the adjacent crystals having slip planes in other directions.

Carbon-steel equilibrium phases

In carbon steels, only three phases can be present in solid state under equilibrium conditions: austenite, ferrite, and cementite. The temperature interval within which austenite forms on heating, and also the temperature interval within which austenite disappears on cooling, is called the transformation-temperature range. The transformations on heating do not occur at the same temperatures as on cooling, except at almost infinitely slow rates of temperature change. The temperatures of transformation are dependent on carbon content.

Phase diagram.   The phase diagram for carbon steel (Fig. 3.2) shows the various solid substances existing at different temperatures for steels with varying carbon content. Ferrite is practically pure iron; cementite is Fe_3C, or iron carbide. Austenite is a high-temperature form of iron containing dissolved cementite.

To follow melting of 0.8 percent carbon steel, proceed along the 0.8 percent line. The room-temperature microstructure resembles the bottom of the three circular sketches presented. At about 1330°F, austenite begins to form and dissolves the cementite. By 1450°F, austenite is almost the only phase present. At about 2600°F, the austenite begins to melt; melting is complete around 2720°F. Slow cooling from the austenite region results in ferrite and cementite, once again. Fast cooling (quenching) to below 600°F changes austenite to martensite, a hard material supersaturated in carbon. Martensite makes carbon steel strong but brittle.

Effects of reheating.   In general, the effects of reheating a metal are the reverse of those obtained in slow cooling. For example, a steel containing 0.20 percent carbon and consisting of pearlite and ferrite is reconverted to austenite as the alpha iron is transformed to gamma iron in the critical range. However, for this steel, the temperatures of transformation are about 86°F, or 30°C, higher for heating than for slow cooling. Changes in the carbon content and the presence of other alloying elements affect the temperature differential.

Welding considerations.   Because they indicate the effects of different cooling rates, isothermal and continuous transformation diagrams are sometimes used by metallurgists in predicting the results of various weld-

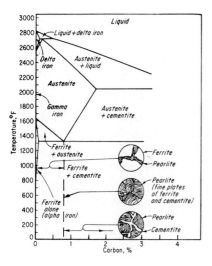

Figure 3.2   Carbon-steel phase diagram.

ing processes. In the arc welding of steel, the structure may range from ferrite to martensite, depending on welding conditions. In spot welding, large quantities of martensite are obtained, and a postweld treatment is essential if the carbon content is higher than approximately 0.25 percent. A transformation diagram will indicate, approximately, the type of microstructure in the zone adjacent to the weld. The type of microstructure desired will also determine the most suitable welding process to use.

Repair welding requires careful consideration of the effect that the welding process may have on the base material and the effect on the joint itself as it cools and solidifies after the welding operation. It is for this reason that welding engineers with metallurgical training are used to develop welding procedures and to control the welding done in fabrication and repair of many of the more sophisticated pressure vessel systems. This more precise control of the fabrication of metals and of welding has resulted in a high degree of uniformity in the production of materials and the assembly of this material through the welding process. This can be attributed to a better understanding of the scientific principles that are involved in metal property changes as metals are heated and cooled.

Physical and mechanical properties of materials

Properties of materials for engineering structures are generally described under the following headings:

1. *Physical properties.*   A material's composition, structure, homogeneity, specific weight, thermal conductivity, ability to expand and contract, and resistance to corrosion

2. *Formability.*   The manufacture of the material, such as fusibility (welding), forgeability, malleability, and ability to be shaped by bending or machining

3. *Mechanical properties.*   The ability of a material to resist applied loads (usually determined by tests)

The basic mechanical properties are elastic limits, moduli of elasticity, ultimate strengths, endurance limits, and hardness. Secondary mechanical characteristics determined from the basic ones or simultaneously with them are resilience, toughness, ductility, and brittleness.

Strength.   Strength depends on the type and nature of loading. The static strength of a material is expressed by the corresponding elastic limit stress. The impact strength is measured by the corresponding modulus of resilience. The endurance strength is expressed by the corresponding endurance limit. See later chapters on stress, pressures, and forces. Qualities other than strength are also important.

Hardness.   Hardness is a relative characteristic. There are several methods of measuring it, all of an arbitrary nature. The *Brinell hardness number* (Bhn) is obtained as follows: A hardened steel ball 10 mm in diameter is pressed under a certain load, F kg, into the smooth surface of the material to be tested; the diameter D of the indentation is measured in millimeters, and the depth h is calculated from it. The hardness number, Bhn, is then expressed as

$$\text{Bhn} = \frac{F}{10\pi h}$$

Another measure of hardness uses the *Shore scleroscope*. A small cylinder of steel with a hardened point is allowed to fall on the smooth surface of a material, and the height of the rebound of the cylinder is taken as the measure of hardness.

The hardness number obtained with the *Rockwell instrument* is based on the additional depth to which a test point is driven by a heavy load beyond the depth to which the same penetrator has been driven by a definite, lighter load.

Table 3.1 compares the different hardness numbers and also provides an approximate tensile strength of the material in comparison with the hardness numbers. This comparison is quite often used to decide if the welding process has possibly influenced the heat-affected zone (HAZ) of the weld.

Ductility.   A material is ductile if it is capable of undergoing a large, permanent deformation and yet offers great resistance to rupture. The measure of ductility is the percentage of elongation or the percentage of reduction of area during a tensile test carried to rupture; it is used as a relative measure. Ductility helps to relieve localized stress concentration through local yielding. It is a necessary characteristic of a material used to take live loads, especially where concentrated stresses may occur.

Brittleness.   Brittleness is a characteristic opposite to ductility and toughness. A material may be considered brittle if its elongation at rupture through tension is less than 5 percent in a specimen 2 in long. *Brittle materials* are those that are comparatively weak in tension. They will fail with very little give or stretching because of their lack of ductility. They have low toughness or low resistance to impact loading, sometimes called "lack of resiliency." As a result, impact testing is used as a measure of brittleness or toughness (described as the ability to absorb energy). When

TABLE 3.1 **Comparison of Hardness Indexes**

Hardness number determination permits approximate tensile-strength determination of a material as shown in the tensile-strength columns.

Rockwell hardness		Vickers hardness,* diamond pyramid	Brinell hardness†		Tensile strength, 1000 psi
C, 150-kg load, diamond	B, 100-kg load, $\frac{1}{18}$ ball		Tungsten-carbide ball	Steel ball	
67	...	918	820	...	...
66	...	884	796	...	...
65	...	852	774	...	...
64	...	822	753	...	...
63	...	793	732	...	...
62	...	765	711	...	...
61	...	740	693	...	...
60	...	717	675	...	...
59	...	694	657	...	...
58	...	672	639	...	...
57	...	650	621	...	...
56	...	630	604	...	...
55	...	611	588	...	...
54	...	592	571	...	...
53	...	573	554	...	283
52	...	556	538	...	273
51	...	539	523	500	264
50	...	523	508	488	256
49	...	508	494	476	246
48	...	493	479	464	237
47	...	479	465	453	231
46	...	465	452	442	221
45	...	452	440	430	215
44	...	440	427	419	208
43	...	428	415	408	201

Rockwell hardness		Vickers hardness,* diamond pyramid	Brinell hardness†		Tensile strength, 1000 psi
C, 150-kg load, diamond	B, 100-kg load, $\frac{1}{18}$ ball		Tungsten-carbide ball	Steel ball	
20	98.9	240	231	225	107
19†	98.1	235	226	220	106
18	97.5	231	222	215	103
17	96.9	227	218	210	102
16	96.2	223	214	206	100
15	95.5	219	210	201	99
14	94.9	215	206	197	97
13	94.1	211	202	193	95
12	93.4	207	199	190	93
11	92.6	203	195	186	91
10	91.8	199	191	183	90
9	91.2	196	187	180	89
8	90.3	192	184	177	88
7	89.7	189	180	174	87
6	89.0	186	177	171	85
5	88.3	183	174	168	84
4	87.5	179	171	165	83
3	87.0	177	169	162	82
2	86.0	173	165	160	81
1	85.5	171	163	158	80
0	84.5	167	159	154	78
...	83.2	162	153	150	76
...	82.0	157	148	145	74
...	80.5	153	144	140	72
...	79.0	149	140	136	70

42	…	417	405	398	194	…	77.5	143	134	131	70
41	…	406	394	387	188	…	76.0	139	130	127	68
40	…	396	385	377	181	…	74.0	135	126	122	66
39	…	386	375	367	176	…	72.0	129	120	117	64
38	…	376	365	357	170	…	70.0	125	116	113	62
37	…	367	356	347	165	…	68.0	120	111	108	60
36	…	357	346	337	160	…	66.0	116	107	104	58
35	…	348	337	327	155	…	64.0	112	104	100	56
34	…	339	329	318	149	…	61.0	108	100	96	54
33	…	330	319	309	147	…	58.0	104	95	92	52
32	…	321	310	301	142	…	55.0	99	91	87	50
31	…	312	302	294	139	…	51.0	95	86	83	48
30	…	304	293	286	136	…	47.0	91	83	79	46
29	…	296	286	279	132	…	44.0	88	80	76	44
28	…	288	278	272	129	…	39.0	84	76	72	42
27	…	281	271	265	126	…	35.0	80	72	68	40
26	…	274	264	259	123	…	30.0	76	67	64	38
25	…	267	258	253	120	…	24.0	72	64	60	36
24	…	261	252	247	118	…	20.0	69	61	57	34
23	…	255	246	241	115	…	11.0	65	57	53	32
22	100.2	250	241	235	112	…	0.0	62	54	50	30
21	99.5	245	236	230	110	…					28

* Vickers load: 50-kg load above 171 value; 30-kg, 95 to 170; and 10-kg, 62 to 94.
† Brinell load: 3000-kg load except for tungsten-carbide values 159 to 86, which are for a 1500-kg load, and 85 to 54, which are for a 500-kg load.
SOURCE: Courtesy of the American Welding Society.

a material has a notch and is brittle, failure can be unexpected and well below calculated allowable stresses.

Charpy V-notch test. Charpy V-notch tests are used to measure resistance to impact loading or brittleness. The test uses a pendulum-type apparatus to strike a specimen, usually notched, and the foot pounds of energy needed to cause a fracture are correlated with whether the material is considered brittle or not. The E-23 standard as developed by the American Society for Testing and Materials (ASTM) gives a complete description of notched-bar impact testing.

Figure 3.3 shows how a Charpy V-notch impact test is conducted. A V notch is cut out per standards on one face of a small specimen. When the heavy pendulumlike weight strikes the specimen's face opposite the notch, the foot pounds needed to cause fracture are calculated by weight-lever arm relationships.

Carbon steels with cold-weather resistance to impact failure are usually killed steels, steels that are deoxidized in manufacture so that no evolution of gas occurs during pouring and solidification. Nickel steels of 3.5 or 9 percent are considered superior impact-resistance steels, as are austenitic stainless steels.

Nil-ductility temperature. Low-carbon and low-alloy steels frequently exhibit a transition zone from ductile behavior to brittle failure over a small temperature range called the "nil-ductility temperature" (see Fig. 3.4). Any flaws in the material will aggravate the tendency of a material with a nil-ductility characteristic to act brittlely below the transition temperature.

Welding can also cause changes in a material that make it act brittlely in the weld metal or heat-affected zone. It is thus important to determine and consider nil-ductility transition temperatures of material to be used in boilers and pressure vessels in order to make sure this temperature is not experienced during service, and even perhaps under possible test or operating abnormalities. One such possible condition may occur during hydrostatic testing with water that will be below the nil-ductility temperature. Another is testing when the surrounding temperature may be too low, as may be the case in winter. Failures due to brittleness from not considering the effect of temperature transitions on the property of the material have occurred during hydrostatic tests of pressure equipment and other equipment such as water towers.

The effect of temperature's altering or changing the brittle nature of a material was dramatically displayed in World War II when merchant ships cracked unexpectedly. It was noted that this occurred in cold waters, but not in warm waters. It has been established since these failures that materials may exhibit ductile behavior in a normal environment but

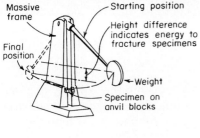

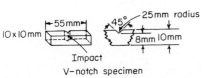

Figure 3.3 Charpy V-notch impact test is used to determine material toughness against brittle failure. (*Courtesy of* Power *magazine.*)

act in a brittle manner if the environment, such as temperature, is changed. Stress-corrosion cracking can also be of a brittle nature as a result of the combination of stress, susceptible material, and a corrosive surrounding.

Creep. A property of steel that is important in high-temperature service is creep. Above about 800°F there is a drop in the strength of steel, as the material ceases to be elastic and instead becomes partly plastic. With a constant load at the elevated temperature, permanent deformation takes place owing to the plastic nature of the material; this deformation is called "creep." Long-time exposure to conditions of creep will cause unacceptable deformations, and even rupture or tearing asunder of the mate-

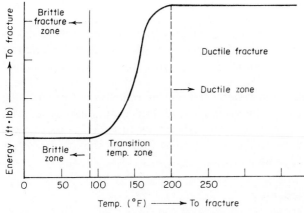

Figure 3.4 Nil-ductility temperature diagrams on certain steels are used to determine brittle-to-ductile transition temperatures.

rial. It is necessary to establish permissible times of creep and creep rates to avoid such occurrences. Manufacturers and code bodies have established allowable stresses, using long-term creep tests and stress-rupture tests at high temperatures.

Toughness. "Toughness" is a term used to denote the capacity of a material to resist failure under dynamic loading. The *modulus of toughness* is defined as the amount of energy per unit volume that a material can withstand or absorb without fracture occurring. The modulus of toughness is useful as an index for comparing the resistance of materials to dynamic loads, and it is especially applicable in the design of moving parts of machinery.

Ferritic Materials

Chemical control of ferritic materials starts in the manufacture of cast iron and steel, where the crystalline structure of the materials and their desired properties can be carefully controlled.

The first step in the production of cast iron and steel is the extraction of the iron from the ore. This is accomplished in the blast furnace. The blast furnace is a vertical, tubular steel chamber lined with refractory. The furnace may be 5 to 25 ft in diameter and up to 100 ft high. It is charged from the top with iron ore, flux, and coke. The usual flux is some form of limestone. When the "charge" is fired up, the coke burns at extremely high temperatures—up to 3600°F. The iron is melted out of the ore and flows to the bottom of the furnace. The molten iron is poured into molds, where it solidifies into "pigs." Pig iron contains a high percentage of carbon, which causes low ductility. Consequently, further refinement is necessary before the physical and chemical properties of pig iron are satisfactory for its use in boiler construction.

Cast iron is produced by remelting pig iron and pouring it into molds of the desired shape. The purpose of the melting is to reduce the amount of impurities and to secure a more uniform product than would be obtained by casting the pig iron directly as it came from the blast furnace. There are two types of furnaces in general use for the remelting of pig iron. In the production of most of the ordinary gray cast iron, a cupola is used, while an air furnace, known as a "reverberatory furnace," is more commonly used for the better grades of gray cast iron and the cast iron that is to receive heat treatment.

Cast-iron types and properties

Compared with steel, cast iron is decidedly inferior in malleability, strength, toughness, and ductility. The most important types of cast iron are the white and the gray cast irons.

White cast iron.   White cast iron is so known because of the silvery luster of its fracture. In this alloy, the carbon is present in combined form as iron carbide (Fe_3C), known metallographically as "cementite."

Gray cast iron.   Gray iron is the most widely used of cast metals. In this iron, the carbon is in the form of graphite flakes that form a multitude of notches and discontinuities in the iron matrix. The appearance of the fracture of this iron is gray because the graphite flakes are exposed. The strength of the iron increases as the graphite crystal size decreases and the amount of cementite increases. Gray cast iron is easily machinable because the graphite carbon acts as a lubricant for the cutting tool and also provides discontinuities that break the chips as they are cast. Gray iron, having a wide range of tensile strength, from 20,000 to 90,000 psi, can be made by alloying with nickel, chromium, molybdenum, vanadium, and copper. Table 3.2 shows permissible stresses specified by the ASME code, Section VIII, Division 1.

Physical defects.   Some of the most common physical defects which may be present in cast iron and which will weaken it are blowholes, cracks, segregation of the impurities, and coarse-grain structure.

Sulfur, which combines with the manganese or the iron to form a sulfide, makes the cast iron brittle (hot-short) at high temperatures. It also increases shrinkage. Hence, its amount is usually limited to less than 0.1 percent in specifications for cast iron.

Phosphorus, in amounts of more than 2 percent, makes the iron brittle and weakens it. However, it also has the effect of increasing the fluidity and decreasing the shrinkage, so it is desirable in making sharp castings for ornamental parts where strength is unimportant.

Section VIII, Division 1, code restrictions

Cast-iron pressure vessels cannot be used for the containment of lethal gases or liquids if a pressure above the limitation may be generated in such a closed vessel. Pressure limitations are (*a*) 160 psi at a temperature

TABLE 3.2   Permissible Stresses

Cast-iron class	Ultimate tensile strength, kilopounds per square inch (kips/in^2)	Allowable stress in tensile strength, kips/in^2
20	20.0	2.0
25	25.0	2.5
30	30.0	3.0
35	35.0	3.5
40	40.0	4.0

not greater than 450°F for pressure vessels containing gases, steam, or other vapors, (b) 160 psi at temperatures not to exceed 375°F for vessels containing liquids, and (c) 250 psi for liquids at temperatures below their boiling point, provided this is below 120°F. The code does permit parts such as bolted heads, covers, or closures to be of cast iron with a pressure limitation of 300 psi and temperature of 450°F.

If a design is based on calculation, cast-iron pressure vessels must be built with a safety factor of 10, as seen in Table 3.2, showing ultimate tensile strength and allowable tensile stress.

The code requires liberal radii to be used at corners and other abrupt changes in cast sections, so as to avoid stress-concentration points.

Hydrostatic tests on pressure vessels built of cast iron must be at a pressure twice the allowable working pressure. For steel, the hydrostatic test is $1\frac{1}{2}$ times the stamped allowable pressure.

Where calculations are not possible, tests to destruction of complicated shapes are used to establish the allowable working pressure of cast pressure vessels. The code stipulates that the allowable pressure for such a cast-iron (CI) vessel, and others of same shape, material, etc., shall be given by

$$\text{Allowable working pressure} = \frac{\text{hydrostatic pressure at failure}}{6.67}$$
$$\times \frac{\text{specified minimum tensile strength for the class of CI}}{\text{average tensile strength of a test specimen}}$$

All temperature restrictions for the intended service, as well as the previously listed maximum pressures for the service, must still be followed.

Manufacture of steel

The production of steel from pig iron involves the removal of as much impurity as is practicable, the adjusting of the carbon content to the desired value, and the addition of such alloying as may be required to alter the properties.

The manufacture of steel starts with pig iron. Pig iron is transformed into steel by the oxidation of the impurities with air, oxygen, or iron oxide, since the impurities combine more readily with oxygen than does iron. Pig iron may be refined by oxidation alone in the acid process or by oxidation in conjunction with a strong basic slag in the basic process. Carbon, silicone, and manganese are removed by both processes, but phosphorus and some of the sulfur in the pig iron can be removed only in the basic process. Phosphorus stays persistently with the iron if the slag is strongly acidic or when it is high in silica. The principal methods of manufacturing steel for subsequent rolling or forging are the basic-oxygen process, the basic-open-hearth process, and the basic-electric-furnace process. Cast steel can be made by the above methods and also by the acid-electric-furnace and acid-open-hearth methods. The furnace charge is steel scrap and pig iron in both the basic-oxygen and basic-open-hearth processes and selected steel scrap in the electric-furnace process. Prereduced iron pellets may also be used in these processes as part of the charge.

Effects of adding alloys.   The purpose of adding alloying elements to carbon steel is to impart to the finished product desirable physical or chemical properties that are not available in carbon-steel parts fabricated by standard procedures. These properties could involve desirable electrical, magnetic, or thermal characteristics, as well as engineering considerations such as (1) high tensile strength or hardness without brittleness, (2) resistance to corrosion, (3) high tensile strength or high creep limit at elevated or subzero temperatures, and (4) other desirable physical features required to resist special loading.

Carbon.   Up to 1.2 percent carbon in iron increases the strength and ductility of steel. When the carbon content is above 2 percent, graphite formation is promoted, which lowers both the strength and the ductility of steel. Carbon content above 5 or 6 percent causes the metal to be brittle with very low strength for any load-resistance application.

Manganese.   Because it combines with sulfur, manganese prevents the formation of iron sulfide at the grain boundaries. This minimizes surface ruptures at steel-rolling temperatures (red-shortness) and thus results in significant improvement in surface quality after rolling.

Nickel.   Nickel increases toughness, or resistance to impact. It is the most effective of all the common alloying elements in improving low-temperature toughness.

Chromium.   Chromium contributes to corrosion resistance and heat resistance in alloy steels. A strong carbide former, chromium is frequently used in carburizing grades and in high-carbon-bearing steels for superior wear resistance. An 18 percent chromium and 8 percent nickel (18-8 chrome-nickel) alloy is widely used as a high-strength stainless steel in pressure equipment requiring strengths and corrosion resistance.

Molybdenum.   Like nickel, molybdenum does not oxidize in the steel-making process, a feature that facilitates precise control of hardenability. It markedly improves high-temperature tensile and creep strength and reduces a steel's susceptibility to temper brittleness.

Boron.   Since boron does not form a carbide or strengthen ferrite, a particular level of hardenability can be achieved without an adverse effect on machinability and cold formability, such as may occur with the other common alloying elements.

Aluminum. In amounts of 0.95 to 1.30 percent, aluminum is used in nitriding steels because of its strong tendency to form aluminum nitride, which contributes to high surface hardness and superior wear resistance.

Silicon. Silicon combines with carbon to form hard carbides that, when properly distributed throughout the alloy, have the effect of increasing the elastic strength without loss of ductility.

Tungsten. Tungsten forms hard, stable carbides when it is added to steel. It raises the critical temperature, thus increasing the strength of the alloy at high temperatures.

Vanadium. Vanadium acts as a deoxidizing agent, as aluminum does, on molten steel. It forms very hard carbides, thus increasing the elastic and tensile strength of low-carbon and medium-carbon steels.

Other elements may be added to improve the rolled strength and toughness of steels in the high-strength, low-alloy category.

Impurities from steel manufacture. The strength, ductility, and related properties of the iron-carbon alloys can be affected by the presence of harmful elements such as sulfur, phosphorus, oxygen, hydrogen, and nitrogen.

Sulfur. Sulfur has the same effect on steel as on cast iron, making the metal hot-short, or brittle at high temperatures. As a result, it may be harmful in steel that is to be used at elevated temperatures or, more particularly, may cause difficulty during hot-rolling or other shaping operations. Most specifications for steel limit sulfur content to less than 0.05 percent.

Phosphorus. Phosphorus makes steel cold-short, or brittle at low temperatures; so it is undesirable in parts that are subjected to impact loading when cold. However, it has the beneficial effect of increasing both fluidity, which tends to make hot-rolling easier, and the sharpness of castings. Since cast iron is brittle anyway, phosphorus is sometimes added to make the castings clean-cut. Most specifications for structural steel limit the phosphorus content to less than 0.05 percent.

Oxygen. When iron is in the liquid state, any free oxygen combines readily with it to form iron oxide. In the finished steel or iron, the iron oxide usually appears in the form of tiny inclusions distributed throughout the metal. These inclusions introduce points of weakness and increased brittleness and stress concentration points from voids, which are undesirable, for they promote the formation of cracks that may result in progressive fracture.

Hydrogen. Hydrogen, such as is produced when steel is immersed in sulfuric acid to remove mill scale before cold drawing, makes steel brittle. The hydrogen may be removed by heating the steel for a few hours, or it will gradually work out of the steel at ordinary temperatures. When present, hydrogen increases the hardenability of the steel.

Nitrogen. Nitrogen has a hardening and embrittling effect on steel. This may be objectionable, or it may be desirable in producing a hard surface on the steel, under controlled manufacturing conditions. For example, in the nitriding process, the steel is exposed to ammonia gas at about 1112°F (600°C) to produce a hard, wear-resisting surface without lowering the ductility of the center of the piece.

Nonmetallic inclusions. Nonmetallic inclusions occur during the plate-rolling operation in steelmaking. Quite often, these inclusions cause laminations, or planes in the plate where metal separation or voids exist.

The quality of steel plate can be significantly affected by nonmetallic inclusions. The presence of inclusions, such as sulfides and oxides, primarily affects the ductile behavior of the steel. The quality of a particular grade of steel can be improved by eliminating or minimizing inclusions. Nondestructive testing is used extensively for critical-pressure applications, such as nuclear reactors, in order to find inclusions.

Heat treatment of steels. When steel is heated to certain temperatures and then rapidly or slowly cooled, its physical properties, such as elastic limit, ultimate strength, and hardness, can be changed.

Heat treatments fall into two general categories: those which increase strength, hardness, and toughness by quenching and tempering and those which decrease hardness and promote uniformity by slow cooling from above the transformation range or by prolonged heating within or below the transformation range, followed by slow cooling.

Annealing consists of heating steel to a certain temperature and cooling it by a relatively slow process. Annealing may be used to remove stresses such as are produced in forgings and castings, to refine the crystalline structure of steel, or to alter the ductility or toughness of steel.

Stress-relief annealing involves heating to a temperature approaching the transformation range, holding the temperature for a sufficient time to achieve temperature uniformity throughout the part, and then cooling to atmospheric temperature. The purpose of this treatment is to relieve residual stresses induced by normalizing, machining, straightening, or cold deformation of any kind. Some softening and improvement of ductility may be experienced, depending on the temperature and time involved.

Normalizing involves heating to a uniform temperature about 100 to 150°F above the transformation range, followed by cooling in still air.

Hardening consists of heating steel to above its transformation temperature range and cooling it suddenly by quenching in water, oil, or some other cooling medium that absorbs heat rapidly.

Quenching is defined as a process of rapid cooling from an elevated temperature, by contact with liquids, gases, or solids. Quenching increases the hardness of steel if its carbon content is 0.20 percent or higher. It also raises the elastic limit and ultimate strength and reduces the ductility; however, it induces internal stresses, and the metal is apt to become brittle.

Common steel types

"Plain" carbon steel.   Plain carbon steel is a broad range of steels representing about 90 percent of all steel produced. The term "plain" denotes a steel whose properties depend primarily on carbon content, whereas alloy steels owe their properties primarily to the alloys added. Carbon content in plain carbon steels ranges from 0.10 to 0.60 percent for some grades of carbon-steel pipe.

Rimmed steel.   Rimmed steel is a carbon steel with less than 0.20 percent carbon. The name "rimmed" was derived from the soft rim that forms on the steel ingot during steel manufacture. When rimmed steel is heated to its melting point, as in heli-arc welding, the iron oxide in solution reacts with the carbon in solution and forms carbon monoxide and carbon dioxide gas bubbles. Thus rimmed steel that is welded may have porosity in the weld from these escaping bubbles. To prevent this, oxyacetylene welding may be used, because this welding process permits the bubbles to escape and not cause voids, or gas pockets, in the metal. Covered electrodes with silicon or other deoxidizers also help to reduce the bubbling by combining with the oxygen in the rimmed steel.

Killed steel.   Killed steels contain enough deoxidizers to remove all the dissolved oxygen in the steel and thus prevent gases from forming. The sulfur and phosphorus contents are usually 0.05 and 0.04 percent, respectively, with the silicon content being a minimum of 0.1 percent. Killed steel is a fine-grained steel that has a high yield strength and is used in pipes and tubes. Grades A106 and A192 are used for boiler tubes.

Semikilled steel.   Semikilled steels are partially deoxidized and are intermediate between rimmed and killed steels, with less than 0.25 percent carbon and less than 0.90 percent manganese, with low sulfur and phosphorus content. Porosity is not likely to be encountered in welding, since this steel has a silicon range from 0.05 to 0.09 percent.

Low-alloy steel. Low-alloy steels are always killed steels with suitable alloys added. For example, the chromium-molybdenum grades A200 and A235 of ASTM specification are frequently used in high-temperature steam and oil-refinery applications. These steels are usually preheated before welding, and some form of postweld heat treatment is also required to remove residual stresses from the welding.

Stainless Steels

Stainless steels are extensively used in industry for corrosion resistance and for increased hygienic standards in cleaning the vessels.

Stainless steels are high-alloy killed steels that were developed to resist atmospheric and high-temperature oxidation and also to resist the corrosive effects of many salts and acids. The corrosion-resistance property is derived primarily from the principal alloys used—chromium and nickel. Thus, an 18-8 stainless steel has 18 percent chromium and 8 percent nickel.

Stainless steels are defined as alloys of iron that contain at least 11 percent chromium, because this percentage of chromium prevents the formation of rust. All stainless steels achieve corrosion resistance by forming a protective, or passive, layer. The chromium in stainless steel is the element most responsible for stainless steel's ability to form a protective layer. When chromium cannot be used, because of the chemicals being processed, nickel is used to increase the corrosion resistance of the stainless steel. Nickel establishes the austenitic structure of the steel. Molybdenum is also used to add to the strength of the passive layer and thus also adds corrosion resistance in stainless steels. It is especially effective in decreasing the possibility of pitting in the presence of halogen ions.

Stainless-steel classification

Classification of stainless steels can be extensive. Several hundred stainless steels are produced, but they can be classified into the following types:

Austenitic. Austenitic steels can be hardened by cold working but not by heat treatment. When annealed, all are nonmagnetic. They have excellent corrosion-resistance properties, and most are readily weldable. Type

304 is a typical grade and is an 18-8 stainless steel. Type 316 is also an 18-8 stainless steel but has 2 percent molybdenum added for improved pitting and crevice-corrosion resistance.

Ferritic. Ferritic steels are straight chromium-type stainless steels with 11 to 30 percent chromium content. They are good resistors of chloride stress-corrosion cracking. Ferritic stainless steels cannot be hardened by heat treatment but can be moderately hardened by cold working. They are a magnetic steel with good ductility. However, ferritic steels may lose their ductility at long exposures between 750 and 1000°F. The code limits these steels to 700°F maximum temperature as a result.

Martensitic. Martensitic stainless steels are austenitic at elevated temperatures and can be hardened by suitable cooling to room temperature. They usually have 11 to 18 percent chromium, with type 410 stainless steel a typical martensitic steel. These stainless steels do not have the corrosion resistance of the ferritic or austenitic types. The boiler code limits these steels to 700°F, since they have a tendency to oxidize at high temperatures.

Duplex. Duplex stainless steels are a mixture of austenitic and ferritic. They are resistant to chloride stress-corrosion cracking and are used in tube sheets of condensers and heat exchangers.

Precipitation-hardening. Precipitation-hardening steels are chromium-nickel stainless steels containing other alloying elements such as copper or aluminum. The addition of these elements results in the formation of precipitates during processing of the steel. These steels are used primarily in gears and aircraft parts.

High-alloy austenitic steels

As pressure levels increase in the chemical and petrochemical fields, and plants become more complex, with service conditions quite severe, more corrosion-resistant alloys are continuously being developed. High-nickel-chrome alloys contain more than 50 percent nickel and chromium; therefore, they are not technically considered steels. These are being used in industry and have many of the qualities of austenitic stainless steels.

Handling stainless steels

To retain their corrosion resistance, it is important to keep stainless steels free of notches and dirt, paints, and similar contaminants. Small crevices on the surface can act as a place for corrosion to begin, especially in any form of acidic service environment. Scratches, burrs, imbedded nonalloy particles, weld spatter, paint, etc., should be removed during vessel fabrication, and also in service. Care in handling these steels, so as to prevent damage, will prevent future expensive repairs. This is especially important in the welding of stainless steels. Weld spatter should be removed carefully by grinding. Tools used preferably should also be of stainless steel to prevent contamination from nonalloy steel material. The bottom of any stainless-steel tank should be covered with protecting material during fabrication or repair to prevent contaminants from adhering to the stainless-steel surfaces.

Welding stainless steels

When welding some stainless steels, carbide precipitation may occur. When unstabilized stainless steels are heated to 800 to 1500°F, the chromium in the steel combines with the carbon to form chromium carbides; this tends to form along the grain boundaries of the metal, thus the term "carbide precipitation." Carbide precipitation lowers the dissolved chromium in the grain boundary and makes the affected area much less resistant to corrosion. The corrosion that does take place is intergranular. Another term for carbide precipitation is "sensitization"—the migration of chromium in solution to form chromium carbides, thus leaving the chromium-depleted region "sensitive" to local corrosion where the depletion has taken place.

In stainless steels there are three well-known methods for minimizing carbide precipitation due to welding:

1. Employing postweld heat treatment. This requires temperatures of 1850 to 2050°F, which are difficult to attain in the field.

2. Keeping the carbon content of the stainless steel low so that very little is available to combine with the chromium.

3. Substituting or adding titanium and columbium so they will combine with the carbon and thus leave the chromium in solution to resist corrosion. The stainless steel is then classified as being "stabilized" under these conditions. AISI 347 and 321 are considered stabilized stainless steels. AISI 316 and 317 also contain molybdenum, and these steels are used for corrosive service and for steam-power-plant applications for temperatures above 1100°F.

Sigma phase

Ferrite in stainless steels can cause a brittle condition to develop at elevated temperatures, as during welding. Austenitic alloys containing delta ferrite can at temperatures between 900 and 1775°F develop a brittle phase called "sigma." The change to brittleness can be very rapid at about 1550°F. The sigma formation in the microstructure causes the corrosion resistance of the stainless steel to be reduced, and ductility is lowered; the hardness is increased, and notch toughness is decreased. For services above 1000°F, careful choice of alloys is necessary to limit the amount of delta ferrite. The sigma phase can be eliminated by annealing a weld at temperatures above 1800°F.

Stainless-steel clad-plate material

Function of stainless-steel cladding. Many pressure vessel process streams require the use of materials that are highly resistant to corrosion. This is especially applicable to the chemical and petroleum industries that use large-diameter vessels operating at high pressures. Both of these factors (large diameters and high pressures) require that the pressure vessels be extra thick in order to resist the large hoop stresses. To resist corrosion from contents and avoid making the whole vessel with stainless steel, carbon steel is used to resist pressure, and a stainless-steel inner surface is used to resist corrosion.

One choice in providing corrosion resistance is clad-steel plate (versus solid stainless steel). A *clad-steel plate* is a composite plate made by uniformly and inseparably bonding a layer of a corrosion-resistant material or metal to a carbon- or low-alloy steel backing plate. Cladding thickness can vary from 15 to 30 percent of the baseplate. Steel companies and suppliers have dollars-per-pound figures available for determining when to use cladding in comparison with solid stainless-steel walls. In general, cladding will decrease the total wall thickness required for a vessel, using high-strength baseplate steels to resist pressure and the cladding to resist corrosion.

Code considerations in using clad material. The ASME code places the responsibility on the owners and users for determining cladding material that will be suitable for the intended service. "Suitability" includes adequate corrosion resistance and the retention of expected mechanical properties for the planned service life.

In general, the code does not permit using the clad plate as additional thickness to resist pressure but rather treats it only as a corrosion allowance. An exception is for clad material made of SA-263, SA-264, and SA-265 steels. SA-263 is a corrosion-resistant chromium-steel clad plate, SA-264 is a corrosion-resistant chromium-nickel steel plate, and SA-265 is nickel and nickel-base alloy clad-steel plate. From the thickness of the cladding, it is still necessary to deduct the allowance for corrosion. All welds on clad vessels using the combined-thickness approach per code rules must have at least spot-radiography examinations. An example will illustrate how the strength of the cladding material and the strength of the baseplate determine what combined thickness is required under code rules.

Example $t_t = t_b + (S_c/S_b)t_c$

where t_t = permitted total thickness of clad plate plus baseplate, in
t_b = nominal thickness of baseplate, in
S_c = maximum allowable stress of cladding at design temperature, lb/in²
S_b = maximum allowable stress of baseplate at design temperature, lb/in²
t_c = nominal thickness of cladding *less* corrosion allowance, in

Given that baseplate thickness is 0.75 in, clad-plate thickness is 0.25 in, allowable stress for the baseplate is 17,500 lb/in², and allowable stress for the clad material is 11,200 lb/in², what is the allowable combined thickness t_t that can be used for pressure calculations if the corrosion allowance on the cladding is 0.050 in?

answer Substituting,

$t_t = 0.75 + (11,200/17,500)(0.25 - 0.050) = 0.878$ in

Telltale holes are required on clad-material pressure vessels in order to warn the user when the corrosion-resistant cladding has been penetrated. Such leakage could be due to cracks in the clad metal or in the welds in the clad plate, or due to a bulged-out clad plate that has separated from the base material. Immediate shutdown of the pressure vessel system is required if leakage is noted from the telltale holes, so that an inspection can be made of the internal components to determine the cause and institute repairs to the clad material. If this is not done, serious corrosion and other effects can occur to the baseplate material with possible disastrous results to the vessel and its surroundings.

Welds that are used on cladding material must have a corrosion resistance at least equal to that of the cladding. The bond strength between the cladding and the backing steel must meet shear strength requirements of the code to prevent separation of the cladding from the baseplate. This shear strength must be a minimum of 20,000 lb/in².

Steel material for special applications

Many process plant people must order material for special applications.

Low-temperature service.   Many chemical processes involve the use of liquefied gases that must be stored at low or cryogenic temperatures. In addition, columns and other components used in the extraction of liquefied gases often operate at cryogenic temperatures. Nickel-containing steels are often used for service temperatures below −100°F. Several of the commonly used steels are ASTM A203 grades D and E, 3.5 percent nickel steels; ASTM A645, a new 5 percent nickel steel approved for use to temperatures as low as −320°F in ASME pressure vessels; and ASTM

A353 and A553, 8 percent and 9 percent nickel steels that have been used for many years to temperatures as low as $-320°F$.

Crude distillation columns.   Vaporizing and condensing under heat conditions characterize the operations in a crude unit atmospheric and vacuum tower. Since corrosive conditions are present in part of or in all the processes, clad steels are often used for crude units.

While the materials for the cladding are usually selected because of the corrosive environment and vary by refinery, typical crude units include Monel clad in the top of the atmospheric unit, 316L or 410-S clad in the bottom section, and possibly carbon steel in the middle portion. The vacuum tower could include unclad carbon steel in combination with a 316L or 410-S clad material for the bottom portion. Usually the carbon-steel unclad areas have considerable corrosion allowance added to their thickness.

The choice of the backing steels includes carbon- and low-alloy steel. If the vessels are located in an area where ambient temperature could be low, some designers prefer a notch-tough backing steel.

Steel companies and specialty steel suppliers are quite willing to assist owners and users in selecting material for special process applications.

ASME Code Material Requirements

Specific code requirements on materials should be followed by users through the life of a pressure vessel system. This will ensure that the original material selected and the design remain appropriate. On the other hand, materials sometimes may be adversely affected by the environment, and therefore, a material not originally specified, but better, may need to be selected. It should still be a code-approved material.

Certain procedures must be followed by a fabricator or repairer in order to make sure only code-specified material is used in vessel construction or repair. It also is part of the responsibility of the authorized inspector to help implement a quality control procedure in order to make sure code material is used. These controls may include the following:

1. The material to be used for a pressure vessel must be one specified in the section of the code under which the pressure vessel is built.

2. The fabricator of the pressure vessel generally orders permissible code material from the steel mills. The steel mill is responsible for making the necessary tests to the specifications given in Section II of the *ASME Boiler and Pressure Vessel Code.*

3. Section II test requirements that the steel manufacturer may have to perform include the following:

a. Chemical analysis of the steel to determine if it is within code limits for the specification.
b. Tests to determine if the metallurgical grain structure is within code-specified limits.
c. Inspection of plate or tube to note if defects such as blowholes, slag, laminations, and any other imperfections are present and whether these are within code-permissible tolerance.
d. Tension and bend tests as stipulated in the code to note if these properties are within code specifications.
e. Notch-toughness tests to check on fatigue-failure strength.
f. Mill tests, showing that the material complies with code specifications; this must be certified by a responsible person of the material-testing laboratory of the steel manufacturer.

The code lists allowable stresses for the different permitted materials, and these should not be exceeded. See later chapters for stress calculations.

The *ASME Boiler and Pressure Vessel Code* consists of 22 sections. For plant people involved primarily with pressure vessels, not all volumes may be needed. The following may be sufficient, depending on what is in a plant or is going to be installed:

1. Section VIII, Division 1: *Pressure Vessels.*

2. Section VIII, Division 2: *Pressure Vessels, Alternative Rules.* This volume is intended for high-pressure design as well as for large vessels in which more sophisticated analysis is needed of the stresses, for example, to make the vessel lighter than a Division 1 vessel.

3. Section II: *Material Specifications, Part A—Ferrous Material.*

4. Section II: *Material Specifications, Part B—Nonferrous Materials* (if the plant has nonferrous-type pressure vessels or components thereof, such as nonferrous tubes in heat exchangers).

5. Section II: *Material Specifications, Part C—Welding Rods.*

6. Section V: *Nondestructive Examination.*

7. Section IX: *Welding and Brazing Qualifications.*

8. Section X: *Fiberglass-Reinforced Plastic Pressure Vessels* (for those plants having or contemplating fiberglass-reinforced plastic pressure vessels).

This book is concentrating on pressure vessels covered under Section VIII, Division 1.

Vessels designed for pressures over 3000 psi are not covered by the code entirely, because the code body believes that additional requirements that have to be met for these vessels at the present time fall into good engineering practice. However, the code does permit stamping the vessel with the applicable code symbol if the additional good engineering practices are followed and if the vessel complies with all the present requirements of Section VIII, Division 1, rules, including inspection, quality control, NDT examinations, material certification, and completion of certified data reports prior to stamping of the vessel.

Section VIII, Division 1, design restrictions on allowable stress due to pressure are based on reviewing the principal membrane stresses that pressure may produce. The allowable stresses from the code tables are reduced if, in welded construction, only spot radiography or no radiography is used to check the integrity of the weld. In general, the code requires heavier or thicker construction throughout the vessel when full radiography is not used on welded joints. See later chapters on stress calculations involving welded joints.

One fact that deserves mention is that Section VIII, Division 2, rules require that a fatigue analysis be made (design is by analytical methods), whereas Division 1 rules leave this in the area of good engineering practice.

Minor repairs may be made (such as grinding protrusions smooth) if they do not reduce in any way the basic suitability of the material for its intended service. Material that is defective, and cannot be repaired, must be rejected for use on a code pressure vessel. Defects would include cracks, excess blowholes, valleys and grooves, material too thin for the design, and similar problems. The code requires material identification markings to be used by material manufacturers and repairers, and these must be plainly visible, as specified in code rules.

Plates, heads, and similar parts to be joined may be cut to shape by machining, shear cutting, grinding, or oxygen or arc cutting. After arc or oxygen cutting is employed, all slag and discolored, heat-affected metal must be removed by mechanical means before further fabrication or repairs are conducted.

Joining by Welding

Modern pressure vessel systems are fabricated into their usable shapes by welding. Plant maintenance and repair of pressure vessel systems rely heavily on welding in such areas as restoration of worn sections of plate material, grinding out of cracks and rewelding, and similar plant operations on equipment that has seen service. Plant personnel involved with pressure vessel operation, inspection, and maintenance should recognize the need for tight control of the many variables involved in making a good weld. The ASME and the National Board have been instrumental in developing welding requirements for code-constructed vessels in order to assure sound welds. These are, however, minimum requirements; plant personnel should specify additional requirements on new construction for their plant, or on repairs, if deemed appropriate. For example, postweld heat treatment of welds might be specified, even though the code may not require it.

The ASME has recognized the joining of pressure vessel parts by welding since about 1931. Riveting was the chief method employed up to that time, and there are still some pressure vessels in service with riveted joints. Since the introduction of welding as an acceptable method of fabricating and repairing pressure vessels, great attention has been given to drawing up requirements that will assure sound welds on a uniform and consistent basis. The basic elements of a welding program include developing a welding procedure for the material; determining the type of welding to be used, the filler metal, and similar variables; and then testing the weld to make sure that the proposed procedure will produce a weld that has the strength, ductility, and similar material considerations for the service intended. Welding operator qualification follows a similar route of preparing specimens of the proposed material and welding process and then testing the specimens to make sure the operator can produce a sound weld to minimum required specifications.

Welding engineers

In large construction or fabricating shops, welding engineers are employed to ensure that welding is done according to the current level or state of the art. They are the liaison between production, design, quality assurance, and customers in all matters pertaining to welding and control of weld quality. This includes making sure that welders are qualified and knowledgeable in the welding techniques that may be required for a particular welding job. For example, they must ensure that welders know the difference between welding carbon steel and welding stainless steel. The welding engineer generally must have a knowledge of welding chemistry and metallurgy, electrical physics, welding processes and procedures, the welding equipment available, types and properties of welding consumables, and the methods needed to establish a good welding quality control program for an organization.

Types of welding

Welding is a localized coalescence (fusing together) or consolidation of metal in which joining is produced by heating to fusion temperatures, with or without the application of pressure and with or without the use of a filler metal. The filler metal (when used) has a melting point of approximately that of the pieces (base metal) joined together. The weld is that portion that has been melted during welding, and the welded joint is the union of two or more members produced by the welding process.

The most common method of welding pressure parts is by *fusion* (melting) of the metal, the heat being supplied in one of several different ways. In fusion welding, no pressure is applied between the pieces being welded. Arc welding, gas welding, and Thermit welding are classified as fusion welding, but arc welding is the most common.

Arc welding is a localized, progressive melting and flowing together of adjacent edges of the base-metal parts, caused by heat produced by an electric arc between a metal electrode, or rod, and the base metal. Both the welding material (welding rod or electrode) and the adjacent base metal are melted. On cooling they solidify, thus joining the two pieces with continuous material. See Fig. 3.5.

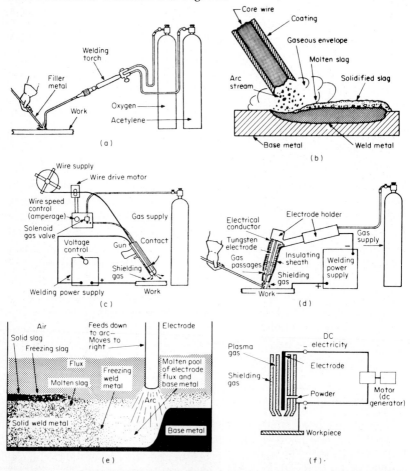

Figure 3.5 Some types of weld fusion processes. (*a*) Oxyacetylene welding; (*b*) shielded-metal-arc welding; (*c*) gas-metal-arc welding (MIG); (*d*) gas-tungsten-arc welding (TIG); (*e*) submerged-arc welding; (*f*) plasma-arc welding. (*Courtesy of Arcos Corp.*)

Following are some types of arc welding:

1. The *oxy-fuel gas* method offers the capacity to deposit smooth, precise, and extremely high quality welds. Low base-metal dilution is particularly important where filler and base metals differ considerably, as with cobalt-based filler metals applied to steels.

2. The *shielded-metal-arc* welding method uses a covered electrode whose coating during welding decomposes to form a slag and shielding gas.

3. The *gas-metal-arc welding* (MIG) process produces weld metal by fusion in an arc between the ends of a continuously fed bare electrode and the work; this progressively melts the electrode and the work. The welding current is carried to the electrode through a gun. The arc is protected by an externally supplied shielding gas, which is usually argon or helium. Some flux-cored wires do not require separate shielding. This process permits fully automatic welding, with the gun traveling on a mechanized carriage.

4. In the *gas-tungsten-arc* (TIG) process, the electrode is not consumed to form the weld. It carries the current between the tungsten electrode and the work with virtually no change in the tip of the tungsten. The filler metal is separately fed to the arc, which is shielded by argon, helium, or a mixture of the two. This process can also be fully automatic.

5. The *submerged-arc* process is particularly suitable for automatic welding because it permits continuous feeding of the filler wire and at relatively high input currents. The arc is completely shielded by a layer of loose granules of flux.

If composite wires are used, this process permits the application of high-alloyed metals. The metal-cored wires have the main alloy in the sheath with all other alloys in the core. A special flux shields the arc from contamination from the air during deposition and also provides a protective slag blanket during solidification. The flux can also provide alloying ingredients to produce the desired weld composition.

6. The *plasma-arc weld* is a true welding process (not a metal spray or coating process). To accomplish this weld, one employs the transferred plasma-arc process shown in Fig. 3.5. It is particularly suitable for highly alloyed hard surfacing. The deposit is formed from powdered alloys that are conveyed into the arc by the gas stream. Plasma-arc weld surfacing fits especially well into mechanized production applications requiring thin weld overlays, but heavy deposits can also be made at relatively fast speeds. Its high deposition rates, along with smooth deposits that reduce material usage and require less finishing, lower costs significantly over gas-tungsten-arc and oxy-fuel-gas methods.

7. In *heli-arc welding,* an electric arc is struck between a tungsten electrode within the heli-arc torch and the workpiece. The arc provides

the heat to melt metal, while simultaneously a stream of shielding gas covers the electrode and the weld zone, protecting both from contaminants in the atmosphere. An inert gas such as helium is used—hence the term "heli-arc." The inert gas protects the weld puddle by preventing oxygen and nitrogen from the air from entering the puddle and thus avoids weak and possibly porous welds.

Electron-beam welding was introduced in the 1960s. This permits high depth-to-width ratios for the welding groove with low distortion and better weld bead purity.

There are many types of welding being used in conjunction with new developments still being researched, such as laser welding, robotic welding devices, and welding using sophisticated electronic controls. Some welding methods are suitable only for certain applications.

Electrodes

In the selection of electrodes, or filler metals, the deposited weld metal should do more than join the base metal. It should also match its mechanical and chemical properties as closely as possible. This will assure a uniform structure capable of withstanding the service environment.

Both undermatching and overmatching the properties can have a negative influence on the strength or the corrosion resistance of the welded component. When undermatching, the tensile strength of the base metal violates applicable design concepts. Failure to match certain chemical properties can reduce notch toughness, creep resistance, or the ability to resist certain corrosive environments such as sulfur, oxygen, or high-temperature hydrogen. When weld deposits overmatch the base metal properties, metallurgical notches can be created, and the resistance to corrosive environments, such as wet H_2S, can be lowered, leading to premature failure.

Electrodes or filler metals are many, and with the development of arc welding, a means of classifying electrodes had to be adopted. For example, for carbon-steel electrodes, the American Welding Society (AWS) adopted a series of four- or five-digit numbers prefixed with the letter "E," which stands for "electrode," in a fashion similar to a terminal in an electrical circuit. The first two numbers of a four-digit number, and the first three of a five-digit number, usually show the electrode's minimum tensile strength in kilopounds per square inch of the deposited metal in the as-welded condition. Thus, E6010 = 60,000 lb/in^2 minimum; E7028 = 70,000 lb/in^2 minimum. The next-to-last numeral indicates the position in which the electrode is to be used. For example, in E6010, the "1" means that the electrode can be used for welding in all positions, while the "2" in E7024 indicates that the electrode is suitable for welding in the flat position. The last numeral in the AWS classification system can range from 0 to 8 and covers items such as power supply, type of electrode covering, arc characteristics, and degree of penetration and whether or not the electrode contains iron powder in the coating. Figure 3.6 shows the meaning for the symbols on flux-coated electrodes.

The AWS classifications and specifications provide a standard label for electrodes, indicating known and proven mechanical and chemical properties, and thus eliminate dependence on manufacturers' individual classifications. This system simplifies the ordering of electrodes into the particular types of welding they can do. The AWS specifications also establish standards for length, diameter, packaging, and marking. This permits changing from one brand of electrode to another without altering welding manufacture or repair procedures. There is still the need to consider ease of handling, arc stability, uniform weld bead deposition, and other desirable features of a weld, and these require control by establishing good welding procedures and using skilled, qualified welders.

Electrode and wire storage

Flux, the coating on an electrode, or the filler in flux-cored wire, forms a gaseous shield over the molten metal during welding. It also coats and protects the weld pool from the atmosphere while the metal solidifies.

If this protective shield has been rendered ineffective as the result of wet or chipped flux, hydrogen from moisture in the atmosphere or flux could enter the molten metal. In a rapidly solidifying weld, this hydrogen may become trapped within the weld. In low-alloy and high-alloy steels, the presence of hydrogen gas can lead to underbead cracking and failure in the joint.

Low-hydrogen rods have specific storage requirements to ensure dryness and to maintain their effectiveness in keeping hydrogen input to a minimum. Flux-cored wire should also be protected from moisture contamination. Electrodes should be dried before being used in welding in order to prevent possible hydrogen entrapment in the weld pool. This is especially applicable to any weld erection or repairs in the field, where environmental conditions are not always ideal.

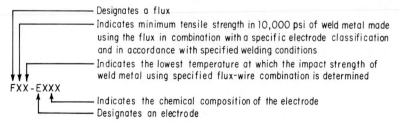

Figure 3.6   Flux electrode classification system.

The welding environment

Welding in adverse weather. Good engineering practice, as well as ASME recommendations, stresses that no welding be done when surfaces are wet or covered with ice or snow, when winds are heavy, or when other conditions exist that could adversely affect the welds being made. It is necessary under such conditions to provide suitable cover or shelter for the welding work and the operator. The ASME also recommends that no welding be done when:

1. Base metal temperature is below 0°F.
2. Temperatures are from 0 to 32°F. However, welding can be done under these circumstances if all surfaces within 3 in of the weld zone are preheated, before welding is started, to a temperature that is at least warm to the hand when touched (about 60°F).

Welding surfaces. Clean surfaces are required of the parts to be welded. Parts to be welded must be clean and free of scale, rust, oil, grease, and other deleterious foreign material for a distance of at least $\frac{1}{2}$ in from the welding joint, for ferrous material, and at least 2 in for nonferrous materials. Detrimental oxides must be removed from the weld-metal contact area. When weld metal is to be deposited over a previously welded surface, all slag must be removed by roughing tools, chisels, air chipping hammers, or other suitable and acceptable means that will prevent impurities from inclusion in the weld metal.

ASME welding code

Section IX of the *ASME Boiler and Pressure Vessel Code* covers the many details that are involved in getting a welding procedure technically certified as well as in obtaining certification for the designation-qualified code welder. All plant people involved with pressure vessel systems must recognize the need for establishing criteria for producing code welds, so that sound welding practices can be followed in erection or repair work. There are many variables to consider in a welding process, and it is the intent of the ASME code to establish a systematic method to control these variables so that a proven, successful weld is made.

The basic program for ASME code welding emphasizes properly prepared and adequate procedures that are to be employed in fabrication and repair work. These procedures must take into account all the variables that may be involved, such as material, thickness, and welding process, to name a few. Fully qualified welding personnel must be used for all the welding-related operations. Documents must be prepared and certified to show that the procedures and welders meet code requirements. A line of communication, including check points, must be established between welding engineers who are responsible for preparing the specifications, the people who will do the welding, and the quality control personnel who are involved in checking and verifying that the welding meets all criteria of acceptability. These procedures must have full management support.

Qualification responsibilities of manufacturers and contractors. Manufacturers or contractors who are to do code welding on boilers and pressure vessels (or nuclear vessels) are responsible for conducting procedure qualification tests and welding operator's qualification tests for work to be done by their organization.

Operators' qualifications remain in effect for as long as they are employed by the same manufacturer or organization and do welding on a continuous basis. But if they change employment, they are no longer considered qualified and thus must take the test again. If they have not done any welding for a period of over six months in the position, material, etc., for which they are qualified, they must be requalified.

The manufacturer or contractor is responsible for keeping all records of procedure and operators' qualification tests. These are needed as evidence of the shop's, or welder's, ability to do acceptable code work. Inspectors have the right, however, to ask for retests if they believe the welding is not acceptable by code requirements. Figure 3.7 is a typical recommended recording form for an ASME welding procedure qualification test. Figure 3.8 is a typical recommended recording form for an ASME welder performance qualification test.

It is standard practice for jurisdictional and commissioned insurance company inspectors to request of the manufacturer or contractor proof that the welding procedure to be employed for new construction, repairs, or alterations has been confirmed as adequate by a procedure applicable to the welding to be performed. Written proof must also be submitted to the inspector that only qualified welders and welding operators are to be used for the welding of code pressure parts. Owners must insist that the manufacturer or contractor meet and comply with these requirements in order to have no delays in the construction or repairs that are to be made.

Quality control. Quality control is a prerequisite in any good manufacturing or contracting business. The ASME has formalized the methods to be used in maintaining good welding, fabrication, and testing procedures. This is now also being applied to repairs on pressure equipment. A quality control program has the following main features:

1. Management support and designation of a manager in charge of all quality control matters, with full authority to implement, maintain, verify, and, if required, change procedures and methods to assure a quality product.

Specification No. .... Date ....
Welding Process .... Manual or Machine ....
Material—Specification .... to .... of P-No. .... to P-No. ....
Thickness (if pipe, diameter and wall thickness) ....
Thickness Range this test qualifies ....
Filler Metal Group No. F. ....   FLUX OR ATMOSPHERE
Weld Metal Analysis No. A- .... Flux Trade Name or Composition ....
Describe Filler Metal .... Inert Gas Composition ....
.... Trade Name .... Flow Rate ....
For oxyacetylene welding—State if Filler Metal is silicon or   Is Backing Strip used? ....
aluminum killed.

                                                               Preheat Temperature Range ....
            WELDING PROCEDURE
Single or Multiple Pass   Postheat Treatment ....
Single or Multiple Arc
Position of Groove
(Flat, horizontal, vertical, or overhead; if vertical, state whether upward or downward)
                               For Information Only
Filler Wire—Diameter   WELDING TECHNIQUES
Trade Name   Joint Dimensions Accord with ....
Type of Backing   .... amps ...   volts ..   inches per min. ....
Forehand or Backhand

                    REDUCED SECTION TENSILE TEST

Specimen No.	Dimensions		Area	Ultimate Total Load, lb.	Ultimate Unit Stress, psi	Character of Failure and Location
	Width	Thickness				

                    GUIDED BEND TESTS

Type and Figure No.	Result	Type and Figure No.	Result

Welder's Name .... Clock No. .... Stamp No. ....
Who by virtue of these tests meets welder performance requirements.
Test Conducted by .... Laboratory—Test No. ....
        per ....
    We certify that the statements in this record are correct and that the test welds were prepared, welded and tested in accordance
with the requirements of Section IX of the ASME Code.
                                            Signed ....
                                                                (Manufacturer)
Date ....   By ....
(Detail of record of tests are illustrative only and may be modified to conform to the type and number of tests required by the Code.
NOTE: Any essential variables in addition to those above shall be recorded.

Figure 3.7 Data for manufacturer's record of welding procedure tests.

2. An organizational chart showing where quality control is to be enforced in design, material selection and inspection, fabrication, testing, final inspection, record documentation, and similar interdepartmental responsibilities.

3. The usual quality control steps in manufacturing, repairs, and alterations.

   a. Design and calculations to code requirements. The code requires

Welder Name...Clock No............................Stamp No............

Welding Process..

Position (If vertical state whether upward or downward).....................
(Flat, horizontal, vertical, or overhead)

In accordance with Procedure Specification No...

Material—Specification................to.................of P-No.................to P-No...........................

Diameter and Wall Thickness (if pipe) otherwise Joint Thickness..

Thickness Range this qualifies............................

FILLER METAL

Specification No..............................Group No. F.................

Describe Filler Metal ....

....

Is Backing Strip used?..................

For Information Only

Filler Metal Diameter and Trade Name.................... Flux for Submerged Arc or Gas for Inert Gas Shielded Arc

.... Welding...

GUIDED BEND TEST RESULTS

Type and Figure No.	Result	Type and Figure No.	Result

Test Conducted by...................................... Laboratory—Test No...................................

per...

We certify that the statements in this record are correct and that the test welds were prepared, welded and tested in accordance with the requirements of Section IX of the ASME Code.

Signed..
(Manufacturer)

Date .... By..

(Detail of record of tests are illustrative only and may be modified to conform to the type and number of tests required by the Code.
NOTE: Any essential variables in addition to those above shall be recorded.

Figure 3.8 Data for welder performance qualification tests on groove welds.

complete documentation that the design meets code requirements; thus, this is a quality control point.

b. Material control. This is to ensure that only code-permitted material is used and that it is free of defects, is of the proper thickness, and is properly prepared for fabrication.

c. Hold points. After each manufacturing step, hold points are desired so that inspections can be made on the material to note if any deleterious changes have taken place.

d. A welding procedure and the qualification of the welders for the job to be done.

e. Monitoring and control of the corrections of discovered defects to ensure compliance with the code and good engineering practice. Quality control personnel are deeply involved in inspection at hold points, including nondestructive tests, and interpretation of these tests.

f. Periodic checking of the accuracy of gages, instruments, and similar measuring or testing devices.

g. Keeping of forms and recorded quality control data. These should be in a safe place for future reference.

Welding procedure qualification. Because of the many variables in welding, such as type of material, thickness, and type of welding to be used, the code requires manufacturers and contractors to have written welding procedures. The code requires testing of the welds to be made using reduced-section test specimens and guided-bend specimens. The variables requiring a new procedure and new test plates are very numerous. Among these are changes in base materials, grouped in ASME Welding Qualifications (Section IX) into P numbers. For example, P-1 includes mostly carbon steels, P-2 consists of wrought iron, P-3 consists of chrome-moly steels with chromium content below 0.75 percent and with a total alloy content not exceeding 2 percent. The P numbers range to P-10, so the base-material variable in procedure qualification is large.

The next variable is the electrode and welding-rod selection, which ranges from F-1 to F-7. Any change in electrode or welding-rod selection requires a new set of test plates or new procedure qualification. Weld metal is classified by weld-metal analysis numbers A-1 to A-8. These are related to equivalent P numbers of the base material. Again, changes of weld metal from equivalent base metal classification require a new set of test plates or procedure qualification.

The thickness of the plate, or pipe, to be welded is another variable. Classification is from $\frac{1}{16}$ to $\frac{3}{8}$ in, from $\frac{3}{8}$ to $\frac{3}{4}$ in, and over $\frac{3}{4}$ in. Each classification requires new test plates as shown in the code. The ASME welding code specifies other variables to consider in requiring a new procedure qualification test, and this should be consulted for specific variables.

Whenever a change in conditions requires requalification, ASME refers to such a change as an "essential variable." Any less drastic change requires a modification of the text or the instructions of the welding procedure specification (WPS); such a change is referred to as a "nonessential variable" and does not require requalification. For example, a WPS qualification of one base metal within a P number qualifies any material within the same P number regardless of special mechanical or chemical properties, except for those procedures that require notch-toughness testing because a weldment will operate below $-20°F$ ($-29°C$). These require separate qualification when switching from one group to another group within the same P number.

Qualification of welders. Before welders are permitted to work on a job covered by a welding code or specification, they must become certified under the code that applies. Many different codes are in use today, so the specific code must be referred to when qualification tests are taken. In general, the following types of work are covered by codes: (1) boiler and pressure vessels and pressure piping, (2) highway and railway bridges, (3) public buildings, (4) tanks and containers for flammable or explosive materials, (5) cross-country pipelines, and (6) aircraft ordnance.

Recertification under the code. Certification differs under the various codes; thus a welder qualified under one code may not be qualified to weld under a different code. Also, in most cases, certification for one employer will not qualify the welder to work for another employer. If the welder uses a different process, or if the procedure is altered drastically, recertification is required for that procedure in most codes. However, if the welder is continuously employed as a welder, recertification is not required, providing the work performed meets the quality requirements. An exception is the military aircraft code, which requires requalification every six months.

Qualification tests may be given by responsible manufacturers or contractors. On pressure vessel work, the *welding procedure* of the fabricator or contractor must also be qualified before the *welder* can be qualified. Under other codes this is not necessary. To become qualified, the welder must make specified welds using the required welding process, type of metal, thickness, electrode type, position, and joint design. Test specimens must be made to standard sizes and under the observation of a qualified person. For most government specifications, a government inspector must witness the making of welding specimens. Specimens must also be properly identified and prepared for testing. The common test is the guided-bend test. However, x-ray examinations, fracture tests, or other tests are also used. Satisfactory completion of test specimens will qualify the welder for specific types of welding. Code certification, in general, is based on the range of thicknesses to be welded, the positions to be used, and the materials to be welded.

Note: Qualification of welders is too extensive to be fully covered here. See Section IX of the *ASME Boiler and Pressure Vessel Code.*

Welder positions and qualifications. The code has classified five welding positions for qualification (see Fig. 3.9). These positions are flat, horizontal, vertical, overhead, and (for groove welds only) horizontal fixed. A welder who qualifies in the horizontal, vertical, or overhead position automatically qualifies for the flat position. For groove welds, qualifying in the fixed horizontal position automatically qualifies the welder for the flat, vertical, and overhead positions. Qualifying in the horizontal, vertical, and overhead positions automatically qualifies the welder for *all* positions. Angular deviations from these positions are plus or minus 15°.

Test plates. Test plates must be prepared when qualifying a welding procedure or a welder. See the ASME code, Section IX, for typical weld test specimens.

Procedure qualification requires:

Two face-bend tests

Two root-bend tests

Two reduced-section tension tests

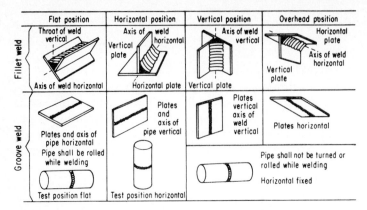

Figure 3.9 Welder qualification is based on the ability to perform good welds in the flat, horizontal, vertical, overhead, and (for groove welds only) horizontal fixed positions.

The test plates for each position that a *welder must pass* include the following:

For groove welds, one face-bend test and one root-bend test are required.

For fillet welds, a test plate is required, but passing the groove weld for each position will also qualify a welder for fillet welding.

The *reduced-section tension test* is used for qualifying the procedure that the shop, or contractor, is to use in welding. When broken in tension, it must have an ultimate tensile strength equal at least to that of the minimum range of the plate that is welded (base material) and the elongation of stretch must be a minimum of 20 percent.

The *side-bend test* is used for qualifying operators. The specimen is subjected to bending against the side of the weld. In the *face-bend test,* the specimen is subjected to bending against the surface, or face, of the weld. In the *root-bend test,* the specimen is subjected to bending against the bottom, or root, of the weld. The *free-bend test* is a shop- or contractor-qualifying procedure test. The test consists of bending the specimen cold; the outside fibers of the weld must elongate at least 30 percent before failure occurs.

In order to pass each test, guided-bend specimens must have no cracks or other open defects exceeding $\frac{1}{8}$ in measured in any direction on the convex surface of the specimen *after bending,* except that cracks occurring on the corners of the specimen during testing are not considered unless these occur from slag inclusions or other welding-technique defects.

Tack welds. Tack welds, which are welds used to hold parts together temporarily, must be made by qualified welders using proven, qualified welding procedures. If tack welds are left on the welded work, they must be feathered in to prevent any stress-concentration points.

The code requires *traceability on welders* and welding operators doing code welding. This may include stamping a joint with welder identification symbols at intervals of no more than 3 ft along a welded joint, unless the manufacturer or contractor has a quality control program that permits tracing the welder by some other procedure.

Welded-joint efficiencies. Section VIII, Division 1, pressure vessels may have different joint efficiencies, depending on the type of joint (single or double butt, double fillet lap, etc.; see code), and also on the degree of radiography examination used to check the soundness of the joint. The code recognizes *full* radiography, *spot* radiography, and *no spot* examination. For a double butt joint, the corresponding efficiencies would then be:

Fully radiographed, 100 percent

Spot radiographed, 85 percent

No spot radiography, 70 percent

If no spot radiography is employed, Section VIII, Division 1, permits only 80 percent of the allowable stress values to be used in design and calculations of strength in code equations, except for calculations concerning unstayed flatheads and covers, thickness of braced and stayed surfaces, and flange designs. The code should be consulted for full particulars. The decrease in joint efficiency from 100 percent to 70 percent when no spot examinations are made on the welded joints means that a fabricator must provide more thickness.

Certain joints *must* be fully radiographed per code requirements: for example, those with butt welds made in vessels that will contain lethal substances, joints with plates over $1\frac{1}{2}$ in thick, and Category B and C butt welds in nozzles and communicating chambers that exceed 10-in nominal pipe size or $1\frac{1}{8}$-in wall thickness. The latest ASME code should be consulted whenever questions of welded joint efficiency arise while making allowable-pressure calculations that require analysis of welded joints and their strength.

Later chapters on calculating allowable pressures will also demonstrate the use of welded-joint efficiencies. A good rule to follow when the welded-joint efficiency is not known is to assume that no spot radiography was used on a vessel, and calculate the allowable pressure using a corresponding 70 percent joint efficiency if a butt joint is under consideration. This will give a conservative allowable pressure.

Welding defects

Types of defects. Welding defects can be many and include porosity, inclusion, undercut, lack of weld penetration, or lack of fusion, to name a few. Striking or breaking the arc can cause craters or cracks if care is not taken. The arc should be struck in the joint, but not at a tack weld. Breaking the arc requires a slow backward motion so that cavity formation is minimized. With thin-section welding, burn-through may occur. Narrowing the welding gap or decreasing the heat input will help reduce this problem. It is always important in shielded-metal-arc welding (SMAW) or submerged-arc welding (SAW) to remove the slag or flux. This requires the welding to be done in such a way that slag is not trapped in the molten metal, or in any other areas in the welding zone; thus, slag inclusions in the finished weld are prevented.

Porosity may occur because of gases evolved from the following sources: a joint surface contaminated with a volatile material or material that becomes volatile upon the application of heat, damp electrodes that give off hydrogen, and excessive flow of shielding gas.

Repair of weld defects. The code permits defects such as cracks, pinholes, and even incomplete fusion to be repaired unless restrictions exist in other sections of the code, as might exist for cryogenic service or similar special considerations. The defects must be removed to sound metal by mechanical means or thermal gouging before rewelding of the joint can be made. Usually dye-penetrant inspection is employed to make sure that a defect has been removed. Postweld heat treatment is required on the repairs.

Grinding is one method used for cleaning or eliminating weld defects such as cracks, slag, spatter, oxides, and general surface defects. It should be done with care so that ground areas are smooth, with rounded edges, and no heat tint should develop. In order to avoid stainless-steel problems, a grinding wheel that has been used on carbon steel should not be used on stainless steels.

The reverse side of double-welded butt joints must, by code rules, have the top weld on the bottom chipped, ground out, or melted out to sound metal in order to secure good penetration or fusion between the upper and lower part of the butt weld.

Welding inspections

The following are some areas that plant personnel should cover when monitoring the welding of pressure vessels in their plant:

1. Conformity of the welding process being used to specifications or written procedure

2. Extent of cleaning of joint prior to welding in order to ensure a good weld

3. Preheat and interpass temperatures maintained and how these compare with the code requirements for the material being welded

4. Joint preparation and conformity to code specifications and print dimensions

5. Filler metal being used and suitability for the welding procedure and the material being welded per code requirements

6. Chipping, grinding, or gouging that is being carried out after each welding pass in order to remove slag and impurities

7. Bracing of the plate during welding in order to control distortion

8. Postweld heat treatment being conducted and compliance with code requirements for temperature and time held

9. Qualification of the welders doing the job and corresponding documentation on file with the manufacturer or contractor

Nondestructive Testing and Examination

Modern inspection techniques to assure good quality of material, to detect hidden flaws in fabrication, to recheck welds and repairs, and to enhance a plant's loss-prevention effort require the use of nondestructive inspection (NDT) methods. The terms "nondestructive testing" and "nondestructive examination" denote that a test is made to find flaws or defects but that no damage is done to the material being tested. A visual examination is a nondestructive test. Pressure vessels in certain types of service require periodic inspections to determine if the harmful effects of overtemperature, overstress, creep, fatigue, and similar long-term actions on materials have caused defects that require repair in order to avoid sudden breakdowns while in service. Nondestructive tests are also used to predict the future life of equipment. For example, periodically checking the thickness of a shell in corrosive service will assist in determining when pressure may have to be reduced as the vessel thins down; this also determines what the wear rate is per year.

Selection of NDT in maintenance inspection

Nondestructive testing is an important tool in maintenance inspection and field repair and is not confined only to thickness testing. The four most used methods at the plant level for checking pressure vessel systems are ultrasonic testing (both for thickness and flaws), magnetic-particle testing, dye-penetrant testing, and radiographic testing. The NDT method to be selected is dictated by several factors: the type of defect to be located, the geometry of the component, access (whether on one or both sides of the component), and the type of metal to be tested. Other considerations may exist, such as whether testing must be performed under on-line conditions or during a shutdown.

Before beginning any component test work, it is important to know what kind of potential defect or flaw is sought. Both "primary" flaws (those produced in the basic metal production or during fabrication such as welding) and "secondary" flaws (those developed in operation) must be considered.

For evaluation of primary flaws, some knowledge of the basic metal production and fabrication processes is helpful, because some types of defects are inherent in the various production processes. For example, the pouring of castings, rolling of steel bars, or extrusion of tubing all carry risks that require NDT monitoring during this production phase. Subsequent fabrication processes, such as forming, heat treatment, and welding of the raw materials, have their own potential defects inherent in the processes.

Secondary flaws develop during operation of the equipment in the plant. These may be due to normal operation, unusual operating conditions, or improper equipment design. Frequently, secondary defects result from excessive primary defects that should have been detected in quality control tests during manufacture of the equipment.

Selection of the best NDT method to use is dictated not only by the type of defect but also by other factors as well. Magnetic-particle testing is used for defects on or close to the surface but not for nonmagnetic materials such as stainless steel. Subsurface defects can be found using the ultrasonic method or radiography.

Ultrasonic testing has the advantage that testing can be performed even if only one side is accessible.

Types of NDT

The most commonly used types of NDT are (1) visual inspections, including the use of borescopes and fiberscopes on sections of a structure that are difficult to see because of blocked lines of vision; (2) radiography; (3) magnetic-particle tests; (4) dye-penetrant tests; (5) ultrasonic tests; (6) eddy-current tests; and (7) acoustic emission tests.

Visual inspections. All inspections start with a visual examination for leaks, wear, cracks, bulges, loosening, and similar signs of trouble (see Fig. 3.10). A borescope or fiberscope can be used in hard-to-see vessel areas, such as in tubing in a heat exchanger. The borescope is a slender, hand-held tube consisting of a series of lenses and internal light sources, which

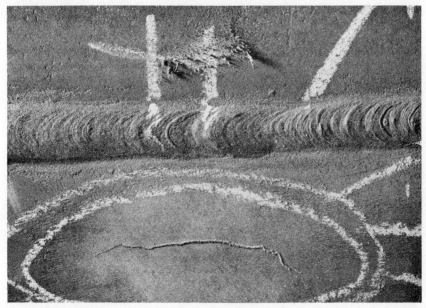

Figure 3.10 Visual examination revealed this longitudinal crack on a urea reactor. (*Courtesy of Hartford Steam Boiler Inspection and Insurance Co.*)

permit viewing an area of a vessel that normally cannot be seen. The advent of fiber optics has led to the development of the fiberscope, which consists of bundles of lights and image-transmitting fibers within a long flexible sheath having an eyepiece lens at one end. Cameras can be attached to the eyepiece to obtain a permanent record of what is seen. Defects found with such instruments include corrosion, blockage of fluid and gas passages, dents, bulges, and similar problems that would be noted if unaided visual inspection were possible.

Radiography. Radiographic inspection uses x-ray and gamma rays obtained from isotopes, such as cobalt 60 and iridium 192, for which the resultant radiant energy can be safely controlled. The radiographic method of testing basically involves passing rays through materials to be tested. The rays impinge on a film or screen, and by noting the contrast of the film, it is possible for an experienced radiographer to detect and detail the internal structure of the object under test. The focal spot is a small area in the x-ray tube from which radiation emanates. In gamma radiography, an isotope like cobalt 60 is the radiation source. When radioactive isotopes are used, strength in curies is important, as is the physical size of the source. The smaller the radiation source, the closer to the material it can be placed; at the same time, the smaller the size, the weaker the source in curies and the longer the exposure time needed.

Code radiography requirements—weld inspections

1. The first requirement is visual inspection of the weld. Joints must have complete joint penetration and be free of undercutting, overlaps, or abrupt ridges and valleys. Weld reinforcement is specified not to exceed that shown in Table 3.3.

2. Welded joints to be radiographed must be sufficiently free of ripples or weld surface irregularities that any radiographic contrast due to such irregularities cannot mask or be confused with the image of any objectionable defect.

3. Gages called "penetrameters" must be used for every exposure of a film. The penetrameter serves as a comparison gage on the film to compare faults in the weld. This is done by making a strip of metal for each exposure and drilling holes in the strip prior to exposure; these strips serve as guides to detect flaws within 2 percent of the plate thickness being welded. These holes are usually drilled with a minimum hole diameter of $\frac{1}{16}$ in specified. The proper method of penetrameter location is shown in Fig. 3.11. When an x-ray picture is taken, the penetrameter will be included, and the holes will serve as a guide right on the film for detecting faults. For comparing the holes on a penetrameter with the holes noted on a weld when the film is developed, an immediate comparison gage is included on each film. The code stipulates tolerances for weld rejects based on penetrameter data.

The code also provides maximum porosity indications and porosity charts that serve as guides for the permissible number of pores (voids) and sizes of voids permitted in each 6-in length of weld. These should be referred to in determining unacceptable voids.

Defects such as cracks, slag inclusions, lack of penetration, voids, and others appear as darkened areas on the film because they have a lower density than solid metal. Typical faults are noted as follows: *Cracks*

TABLE 3.3   Weld Reinforcement Specifications

Plate thickness	Maximum thickness of reinforcement, in
Up to $\frac{1}{2}$ in inclusive	$\frac{1}{16}$
$\frac{1}{2}$ to 1 in inclusive	$\frac{3}{32}$
1 to 2 in inclusive	$\frac{1}{8}$
Over 2 in inclusive	$\frac{5}{16}$

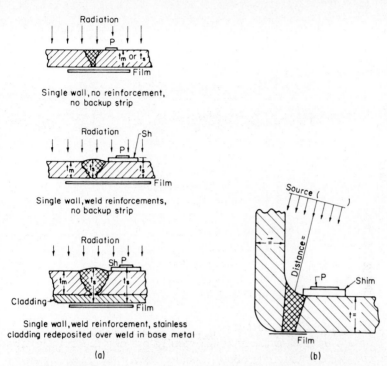

Single wall, no reinforcement, no backup strip

Single wall, weld reinforcements, no backup strip

Single wall, weld reinforcement, stainless cladding redeposited over weld in base metal

(a)

(b)

Figure 3.11 Radiography requires penetrameters to be placed in line with film. (*a*) For butt welds; (*b*) for nozzle welds. P = penetrameter, Sh = shim, t_m = design-material thickness on which the penetrameter is based, t_s = specimen thickness. (*From Spring and Kohan*, Boiler Operator's Guide, *courtesy of McGraw-Hill.*)

appear as dark, irregular lines. *Slag inclusion* shows up as small, dark spots with irregular outlines. *Gas pockets* appear as small, dark spots with smooth outlines and with occasional teardrop fails. *Lack of penetration* is evidenced by a smooth, dark line most often located in the middle of a weld. Figure 3.12 shows a vessel being prepared for x-ray examination.

Limitations of radiography. In conventional radiography, the part to be inspected is illuminated by an x-ray tube or a gamma ray source. Whatever radiation is not absorbed impinges on a film, resulting in a two-dimensional image of the internal structure of the part. Any regions, such as porosity or a crack, that do not absorb radiation will leave a dark area on the film. In general, radiography can only detect discontinuities that have a substantial thickness in the direction parallel to the x-ray beam. Consequently, planar discontinuities such as fatigue cracks, can be detected only if the radiation beam is properly oriented with respect to the

Figure 3.12   Large cylinder being prepared for radiographic examination. (*Courtesy of Struthers Wells Corp.*)

crack. Therefore radiography has only limited applicability in detection of cracking, because of the orientation problem. In other words, if enough is known about the crack to properly orient the x-ray beam, it can probably be detected without the use of radiography. In addition, radiography has several other drawbacks. It is relatively expensive in comparison with other inspection techniques, with up to 60 percent of the total inspection time being spent on setup. Because of the health effects of penetrating radiation, the shielding of workers and operators may also be costly and time-consuming and will probably require substantial downtime for the equipment being inspected. Finally, the member being inspected must be accessible from both sides to accommodate placement of the source and the film.

Magnetic-particle testing.   Magnetic-particle testing is used to detect surface faults by means of setting up a magnetic field or magnetic lines of force between two electrodes. Powdered magnetic material is sprinkled over the work to be tested. The magnetic field will affect the magnetic powder, and these particles will align themselves in a fault, as shown in Fig. 3.13. The correct interpretation of the gatherings of the magnetic powder requires experience and practice. Magnetic-particle inspection is a practical means for spotting close-lipped discontinuities at or near the surface of a part. Both wet and dry methods are currently available.

Discontinuities in a magnetized material give rise to localized leakage fields, and these fields attract finely divided magnetic particles. The particles point a finger at the defect and, on the surface of the part under inspection, mark its extent.

Both direct and alternating currents are used for magnetizing. Direct current (dc) is useful in finding subsurface discontinuities and is commonly used for inspecting welds and castings. Alternating current (ac) is usually employed when highly finished machined parts are checked.

Generally dc magnetization is considered where subsurface and surface cracklike defects must be found. With dc, the magnetic field extends within the part itself, and magnetic leakage fields are produced on the surface by interruption in the magnetic path below the surface.

The principal limitation of the magnetic-particle method is that it applies only to magnetic materials, and is not suited for very small, deep-

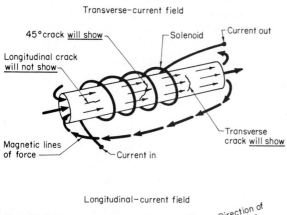

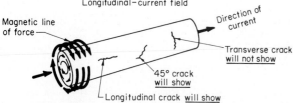

Figure 3.13 Magnetic-particle inspection requires a magnetic field at right angles to the defect in order for it to be detected by the magnetic particles. (*Courtesy of* Power *magazine.*)

seated defects. The deeper the defect is below the surface, the larger it must be to show up. Subsurface defects are easier to find when they have a cracklike shape, such as lack of fusion in the weld. On large, heavy objects, when extremely sensitive inspection is desired, the operation takes more time. This holds true for medium-sized critical parts such as aircraft propellers. With magnetic-particle testing, the surface to be inspected must be available to the operator. This means shafts or other equipment cannot be inspected without removal of pressed wheels, pulleys, or the bearing housing.

Magnetic-particle testing equipment is available in fairly lightweight (35–90 lb, or 16–41 kg) power source units that can be taken to the test site. This equipment has not been routinely used in industry by plant personnel but has been used extensively by NDT contract service personnel. As with ultrasonic inspection, the magnetic-particle method requires that the operator be well trained. This is especially important for reliable interpretation of the results.

Yoke-type magnetizing units (see Fig. 3.14) are extensively used in field inspections. The yoke generates a magnetic field, and if a defect is present, a leakage field is created around the defect. The closer the defect is to the surface, the sharper will be the indications, because the field will become more distorted. For deeper cracks, the indications get wider, a fact that can be used as a guide in judging the depth of the defect. Standards are used for reference purposes. The Betz ring has holes drilled at different depths. As a magnetic field is induced into the Betz ring, magnetic particles are applied to the surface. The number of visible line indications determines the sensitivity of the particles. Code or other requirements may specify the smallest hole that must be visible in the Betz ring as a check on the procedure.

Liquid-penetrant inspection. Liquid (dye)-penetrant inspection is used somewhat like magnetic-particle testing, except that it is used primarily on nonmagnetic material; it can, however, be used on magnetic material. The dye penetrant contains a visible dye, usually red. Indications of de-

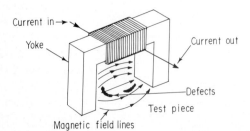

Figure 3.14 Magnetic yokes are extensively used in field inspections because of their portability.

fects appear as red lines or dots against a white developer background. It is primarily a surface-defect indicator.

In the process, a dye penetrant is applied to the part by dip, brush, or spray and is allowed to sit for a specified time. After suitable penetration time, the excess penetrant is removed from the surface, and a developer is applied. The penetrant becomes entrapped in a defect and is brought to the surface by the action of the developer. Cracks are detected by noting the contrast between the white color of the developer and the red penetrant. A perfectly white or blank surface indicates freedom from cracks or other defects that are open to the surface.

Another penetrant method uses a fluorescent penetrant containing a material that fluoresces brilliantly under black light. Indications of defects appear as fluorescent lines or dots against a nonfluorescent background.

The advantages of the dye-penetrant method are as follows: It provides fast, on-the-spot inspection during overhaul or shutdown periods, and the initial cost of the test is relatively low. The disadvantages are that it is not practical on very rough surfaces and color contrast is limited on some surfaces. Also, it detects only defects open to the surface.

Penetrants are easy to use on a wide variety of components, but it is essential that the user follow the directions. Some precautions to observe are:

1. Make sure the surface to be tested is clear of contamination so that the penetrant can wet the interior surface of cracks or defects.

2. Allow sufficient time for complete penetration (usually 5–15 min).

3. Wipe the penetrant clean and wash carefully and sparingly so that all the penetrant is not washed from within the defect area.

4. Perform the tests on surfaces within the working-temperature limits of the system.

Sometimes the metal surface may be too rough to detect defects, such as on welds. It may be necessary to grind the surface smooth before a reliable test can be made.

Ultrasonic testing. Ultrasonic testing makes use of high-frequency sound waves in the range of 0.5 to 10.0 megahertz (MHz) for the inspection of material for flaws and also for thickness testing. The basic principle used in an ultrasonic system is the transformation of an electric impulse into mechanical vibrations, and then the transformation of the mechanical vibrations into electric pulses that can be measured or displayed on a screen called a "cathode-ray-tube (CRT) screen." The transfer of mechanical energy to electric energy is performed by means of a transducer,

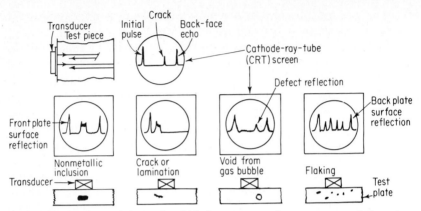

Figure 3.15 In ultrasonic inspection, high-frequency sound waves are passed through a material, with a cathode-ray-tube (CRT) screen used to monitor the reflections, as shown. (*Courtesy of* Power *magazine.*)

which is capable of transforming one form of energy into another (see Fig. 3.15).

Ultrasonic tests are grouped into two basic categories: pulse-echo testing and resonance testing. The pulse-echo method involves transmitting a short burst of high-frequency sound (well above human hearing) through the piece being tested and then detecting the echoes that are received from either a construction detail, such as a shoulder or hole, or a defect in the material with a separation or void in the material sufficiently large that sound cannot be transferred across the interface. In operation, a pulse-echo unit will produce, through an electronic pulser, a short burst of high-frequency electric signal. This is transmitted to the transducer, which is forced to vibrate, usually at its resonant frequency. The probe must be coupled to the test piece with oil, water, or some other liquid or grease. The sound wave train then travels through the test piece until some form of discontinuity or boundary (or the back side of the test piece) is encountered. This discontinuity in the medium causes the sound wave to be reflected to the receiver transducer. The vibrational energy of the sound wave sets the transducer in motion to produce an electric impulse, which is fed into an amplifier. The output of the amplifier is displayed on a cathode-ray tube, which shows the signals on a linear time baseline. If a linear amplifier is used, the amplitude of the returned echoes can be used as a measure of the area producing the reflected signal.

Resonance testing makes use of a tunable, continuous-wave system. This method is usually employed for measuring small or thin walls to 2 or 3 in. The resonance of a crystal is tuned to the piece under test. In practice, a loud pip is heard and can also be seen because the electronic circuit is also in resonance electrically. By proper calibration, direct thickness readings can be made.

Ultrasonic inspection is better suited to detecting crack damage than radiography is. The use of ultrasound requires that a transducer be swept over the surface of the member being tested. The transducer generates pulses of high-frequency acoustic energy (like sound waves, only at frequencies much above the human hearing range). These acoustic pulses are reflected from boundaries within the material. If no boundaries are encountered by the acoustic pulse, it is "echoed" back to the transducer by the back side of the member. Planar discontinuities such as fatigue cracks are easily detected by the presence of the reflected pulses. In a simple "A scan" the reflected pulses appear as blips or peaks on the oscilloscope display screen of the ultrasonic equipment. More sophisticated (and more expensive) equipment can produce displays of two dimensional images of the flaws in a member (a "B scan"), or displays that define the three-dimensional picture (a "C scan").

Ultrasonic inspection requires access to only one side of the member being inspected, and although the equipment can be relatively expensive to purchase, it can be portable and easily used in the field. However, a relatively skilled technician is required to operate the equipment and to interpret the results of an inspection; this requirement, coupled with the equipment cost, may make it difficult for an individual company to do its own in-house ultrasonic inspection. Fortunately, there are field inspection services available at a reasonable cost from many vendors, and the work is generally of high quality since the vendor technicians are usually well trained and experienced in use of the equipment.

Eddy-current testing.   The underlying principle of eddy-current testing is the measurement of impedance of electron flow in the part being tested. Weak electric currents or eddy currents are induced by a probe containing inducing and sensing coils. Any changes in the geometry of a test part, such as a pit, crack, or thinning, will affect the flow of the eddy currents. This change in the flow of the eddy currents will be detected by the sensing coil and displayed on a strip chart, a CRT display, or both. The accuracy of characterization of the detected flaw depends on the quality of the standard used in calibrating the eddy-current instrument. A piece of tubing of material identical to the tubes being inspected should be used for calibration. The standard should have a range of defects similar to those expected to be found in the tubes being tested.

Eddy-current testing is an appropriate NDT for loss-prevention work on heat exchanger tubes. Regular eddy-current testing of tubes in heat exchangers will help detect some of the common problems found in large heat exchangers, such as feedwater heaters, condensers, chillers in air-conditioning units, and numerous process heat exchangers, such as those used in the petrochemical industry. Common causes of tube failures that may be detected include:

1. Tube wear at intermediate supports, usually due to tube vibration. Gas and liquid flows across the tubes can cause this vibration if near the natural frequency of the tubes.

2. Corrosion from acid formation when noncondensable gases are trapped in the exchanger. For example, CO_2 becomes carbonic acid.

3. Fatigue cracks from overrolling, expansion and contraction of the tubes with load, and similar repetitive stresses that may be imposed on the tubes.

4. Stress-corrosion cracking. Nonferrous tubes may be affected by ammonia and mercury.

5. Bulged tubes in air-conditioning chillers, caused by control failures that permit the water in the tubes to freeze.

Eddy-current testing of tubes, if done periodically, can detect early problems that could have required very expensive repairs if they had gone undetected.

Acoustic emission testing.   Acoustic emission is another developing NDT method that could be very effective in the future in the inspection of large pressure vessels, such as those in the petrochemical industry. This method of nondestructive testing is being developed to monitor large pressure vessels such as digesters and nuclear reactors for crack growth. Acoustic emission is based on the principle that a growing defect releases bursts of energy or stress waves that can be detected by sensitive and suitably designed transducers. If a transducer is strategically installed in known highly stressed areas of the pressure vessel, it can convert the minute sound emitted when a material "gives" into electric signals. The signals can be recorded on a computer for immediate analysis or future use. When transducers are installed in a triangular mode, the source of any abnormal sound can be determined by quick trigonometric calculations, and thus the defect can be located for analysis and repair.

Destructive tests

Like NDTs, destructive tests are also used in analyzing the physical condition of material. Besides the well-known tensile test to determine the strength of different metals in tension or compression, destructive tests can also be grouped into chemical, metallographic, and mechanical tests. During these tests the material being tested is damaged; therefore, they are considered destructive.

Chemical tests are used to determine the amount of carbon, phosphorus, and similar constituents of a metal or weld, especially if there is some question of the component integrity. These tests are also used to determine the corrosion resistance of a material or weld.

Metallographic tests of metals and welds require that the specimen be etched and ground for further visual examination, sometimes by magnification. The purpose of these tests is to determine the presence of cracks, porosity, and similar defects. These defects may be in the crystalline structure of the metal, in the heat-affected zone of a weld or base metal. Analysis can assist in determining the reasons for the defect, such as improper fit-up in fabrication, poor welding technique, incorrect weld-rod selection, or service conditions such as corrosion.

Hardness tests are a form of destructive test, since an impression is left in the specimen. Such tests are also used to check on the welding process as respects heat-affected zone hardening, the need for prewelding heat and postweld heat treatment, the possible effect of cold working, and similar metallurgical concerns. Table 3.1 shows the relationship between some hardness numbers (cols. 1–5) and the ultimate tensile strength (col. 6) of the material under test. The Vickers hardness test is used extensively in Great Britain.

Mechanical tests are considered destructive and include tensile tests, yield-stress determinations, endurance tests, and similar stress tests. See the next chapter.

NDT personnel qualification

In order to promote uniformity in conducting and interpreting NDT tests and inspections, the American Society for NDT has drawn up some minimum requirements for NDT personnel, who are graded as follows, according to their qualifications:

A level 1 person must have experience or training in the performance of the inspections and tests that he or she is required to perform. This person should be familiar with the tools and equipment to be employed and should have demonstrated proficiency in their use. She or he must be familiar with calibration and control of inspection and measuring equipment and be capable of verifying that the equipment is in proper condition for use.

A level 2 person must have experience and training in the performance of required inspections and tests and in the organization and evaluation of the results of the inspections and tests. This person must be capable of supervising or maintaining surveillance over the inspections and tests performed by others and of calibrating or establishing the validity of calibration of inspection and measuring equipment. She or he must have demonstrated proficiency in planning and setting up tests and must be capable of determining the validity of test results.

A level 3 person must have broad experience and formal training in the performance of inspections and tests and should be educated through formal courses of study in the principles and techniques of the inspections and tests that are to be performed. This person should be capable of planning and supervising inspections and tests, reviewing and approving procedures, and evaluating the adequacy of activities to accomplish objectives. He or she must be capable of organizing and reporting results and of certifying their validity.

Personnel involved in the performance, evaluation, or supervision of nondestructive examinations, including radiography, ultrasonic, penetrant, magnetic-particle, or eddy-current methods, must meet the level 3 qualification requirements specified in SNT-TC-1A of the American Society for NDT and supplements. Those personnel involved in the performance, evaluation, and supervision of gas-leak test methods must meet the qualification requirements specified for a level 2 person.

Personnel who are assigned the responsibility and authority to perform project functions must have as a minimum the level of capability shown in Table 3.4. When inspections and tests are implemented by teams or groups of individuals, the one responsible must participate and must meet the minimum qualifications indicated.

A file of records of personnel qualifications must be established and maintained by the owner. This file should contain records of past performance, training, initial and periodic evaluations, and certification of the qualifications of each person.

Outside NDT service

As technology advances, so does specialization. NDT inspections started with the instrument manufacturers' supplying skilled and knowledgeable technicians to perform NDT. As instruments became simpler to operate through internal electronic logic circuits, independent NDT inspection firms started to provide NDT service. Process plants and insurance companies specializing in boilers and pressure vessel insurance also started to conduct NDT, with thickness testing being the most prevalent form.

TABLE 3.4   Minimum Levels of Capability for NDT Project Functions

NDT project function	1	2	3
Approve inspection and test procedures			X
Implement inspection and test procedures	X		
Evaluate inspection and test results		X	
Report inspection and test results		X	

Process plants have the option of training their own NDT personnel or contracting the work from a service company. It is best to contract NDT services when complicated or extensive testing is involved. However, for occasional in-plant work, the plant should rely on its own expertise within the operator's capability as obtained by training and experience. All NDT operators, whether in-plant or from an NDT service firm, should receive formal instruction for the type of testing they are to perform. This includes the seemingly simple test procedures such as thickness or penetrant testing.

During a mill shutdown, when the bulk of NDT work is performed, the mill should contract the work with a suitable firm specializing in NDT work. A project engineer should be appointed to work with the NDT service company. The project engineer should be familiar with the testing to be performed, the reason it is needed, and the specific equipment to be tested. It is possible that the NDT team may not be familiar with either the plant or the equipment to be tested, and the project engineer will bridge this information gap.

When extensive NDT testing is planned, such as a survey of pressure vessel thickness, the plant project engineer should meet in advance with the designated NDT manager, who will be on the job to assure that the work will go as planned. This is the time to discuss matters such as supporting equipment or personnel that may be required. The plant project engineer should assure that the NDT service company has adequate and qualified technicians to do the work.

QUESTIONS AND ANSWERS

3-1 Define the body-centered cubic structure of pure iron.

ANSWER: In this structure, the unit cell of atoms is arranged as a cube with an atom at each of the eight corners and one atom in the center of the cube. The structure is called a "space lattice."

3-2 Describe the face-centered cubic structure of pure iron.

ANSWER: In this structure, there is an atom at each of the eight corners, with another atom in the center of each of the six cubic faces.

3-3 What is the effect of grain size on crystal properties?

ANSWER: Grain or crystals can be defined as fine- and coarse-grained. ASTM specifications differentiate one from the other with index numbers based on the number of grains per square inch. Steels with index numbers 1 to 5 are considered coarse-grained, and those 6 or above are considered fine-grained. Fine-grained steels have a greater toughness, lower hardenability, and lower internal stresses. Coarse-grained steels have lower ductility and are prone to embrittlement, because the larger grains permit more penetration of contaminants at grain boundaries. A coarse-grained steel, however, has greater creep strength.

3-4 What is the purpose of adding silicon to steel?

ANSWER: Silicon is added to all steels to remove the oxygen (i.e., it is a deoxidizing agent). If added in amounts up to about 2.25 percent, the strength of the steel is increased without affecting ductility.

3-5 What is the effect on steel of adding chromium?

ANSWER: Chromium added to steel improves the resistance of the steel to oxidation and improves its high-temperature strength and creep properties. Hardenability is improved, and the chromium acts as a carbide stabilizer, so that chromium-molybdenum steels do not form graphite.

3-6 Differentiate between quenching and normalizing.

ANSWER: *Quenching* is rapid cooling of a heated part, while *normalizing* involves heating steel into the all-austenite region and then cooling it in still air to room temperature in order to produce the "normal" structure of ferrite and pearlite steel.

3-7 Define stress relieving.

ANSWER: This is a process to reduce residual stresses in metals by the application of heat to induce redistribution of ductile stress.

3-8 What is the transition temperature as applied to steelmaking?

ANSWER: This is the temperature at which a material changes from ductile to brittle behavior. The usual method of determining transition temperatures is conducting Charpy impact tests over a range of material temperatures.

3-9 Name some typical mechanical properties that describe the ability of a steel material to resist loads.

ANSWER: Mechanical properties that describe resistance to loads are ultimate strength, elastic limit, endurance limit, hardness, and modulus of elasticity.

3-10 What are the three most prominent methods used in measuring the hardness of steel material?

ANSWER: The Brinell, Rockwell, and Shore scleroscope methods are most commonly used in describing the hardness quality of a steel material.

3-11 How is a space lattice defined?

ANSWER: A *space lattice* is the geometric arrangement of atoms in the crystal formation of a given metal, such as face-centered cubic lattice or body-centered cubic lattice, describing the arrangement of atoms in an iron crystal.

3-12   What is the space lattice of alpha iron?

ANSWER: Alpha iron is composed of body-centered cubic lattices.

3-13   What is meant by a eutectic?

ANSWER: A *eutectic* is the alloy of a solution that has the lowest melting point in the solution.

3-14   What is the transformation temperature in carbon steels?

ANSWER: This is the temperature within which austenite (nonmagnetic steel) forms on heating and also the temperature interval within which austenite disappears in cooling in an iron-carbon solution. The temperatures of transformation are dependent on the carbon content, as noted in iron-carbon equilibrium diagrams for the solution.

3-15   Name some typical physical defects that may be found in cast iron.

ANSWER: Blowholes, cracks, segregation of impurities, and coarse grain or lack of uniform grain structure are the most common defects.

3-16   What are the two major chemical steps in the manufacture of steel from ore?

ANSWER: (1) To remove oxygen from the ore in producing pig iron and (2) to remove excess carbon from the pig or cast iron to produce steel.

3-17   How does steel differ from cast iron?

ANSWER: The carbon content is much lower, the tensile strength is much higher, and the ductility and resistance to shock load are higher.

3-18   Briefly, how is excess carbon removed from iron to form steel?

ANSWER: By blowing air through the molten iron. The oxygen in the air unites with the carbon by combustion.

3-19   What is the common measure of ductility?

ANSWER: The percentage elongation as found in making a test of tensile strength.

3-20   What is hardness?

ANSWER: It is a material's resistance to surface deformation on application of an external force.

3-21   What is brittleness?

ANSWER: It is the tendency of a material to fracture on application of shock load.

3-22   What determines the hardness of steel?

ANSWER: The carbon content, the alloy content, and the grain structure (often due to heat treatment).

3-23   How may grain structure and hardness be changed?

ANSWER: By heat treatment at temperatures above the lower critical point.

3-24   What are the major differences between cast iron and malleable iron?

ANSWER: Malleable iron is more ductile. It will withstand safely a considerably greater shock load than cast iron. It possesses a higher tensile strength.

3-25   What are ductility and malleability?

ANSWER: These terms are practically synonymous. They describe the ability of a material to withstand a comparative degree of deformation without failure or impairment of its strength or other physical properties.

3-26   What is mill scale, and what harm does it do?

ANSWER: It is an iron oxide scale formed on the surfaces of a plate when, after rolling at high temperatures, it is exposed to the air. It may, of itself, be a cause of plate deterioration in boiler service by promoting and localizing corrosion.

3-27   Where can the exact tensile strength of a metal product be found?

ANSWER: From the mill test report.

3-28   What else does the mill test report show?

ANSWER: The thickness; the chemical properties; other required physical properties, namely, elongation and yield point; the manufacturer's name; and the heat and slab numbers.

3-29   What is the effect of sulfur and phosphorus when they are melted into steel?

ANSWER: Sulfur makes the steel brittle at high temperature; such steel is often referred to as "hot-short." Phosphorus makes steel brittle at low temperatures, so it is undesirable in parts that are subjected to impact loading when cold. This brittleness when cold is sometimes called "cold-shortness."

3-30   In making a shop inspection of plate material for ASME-constructed boilers, what checking procedures should a code inspector follow?

ANSWER: 1. Examine mill test reports and note if the plate complies with the code specification for that material.

2. Check plate stamping and note if the stamping corresponds with data in the mill test report, including tensile strength, thickness, and grade and quality of finish.

3. Make a visual examination of the plate for defects such as cracks, laminations, inclusions, pits, and similar questionable conditions.

3-31 How is the highest-quality steel produced?

ANSWER: By the electric-furnace method, owing somewhat to more accurate control and to a minimum of oxidizing factors.

3-32 Why is the electric-furnace method not always used?

ANSWER: The expense of manufacture by this method is greater than for other methods.

3-33 When does a steel become a stainless steel?

ANSWER: Stainless steels are classified as alloys of iron having a chromium content exceeding 11 percent. The term "stainless" was derived from their reputation of being highly resistant to corrosion. If the chromium content exceeds 30 percent, these steels are classified as "heat-resisting alloys." Stainless steels are corrosion-resistant because of the formation of a tightly adhering surface film of chromium oxide that does not react with the usual environmental substances, such as air and water. The film also acts as a barrier to protect the metal against further attack from chemicals in contact with the vessel.

3-34 How may welds be broadly classified?

ANSWER: They may be classified into groove, fillet, and plug.

3-35 How may welded joints be classified?

ANSWER: Joints may be classified into butt, tee, lap, corner, and edge.

3-36 How are groove welds classified?

ANSWER: First, into single or double grooves, which indicates whether the welding is to be applied only from one side or from both sides of the joint. Groove welds may be square, bevel groove, V groove, J groove, and U groove.

3-37 Define SMAW.

ANSWER: This is an abbreviation for "shielded-metal-arc welding," also called "stick electrode welding." As the electrode heats, the conducting core melts to provide filler metal for the joint to be welded. The coating on the electrode evolves as a gas with heat, and this forms a gaseous shield for the arc and weld puddle to protect it from contamination as it cools.

3-38 Which welding process uses a nonconsumable electrode?

ANSWER: GTAW, or gas-tungsten-arc welding, develops an arc by a nonconsumable tungsten electrode, which is the electric conductor. The filler metal is added separately from a rod or continuous wire, while inert gas flows around the arc and weld puddle to protect the hot metal.

3-39 Name two gas welding processes.

ANSWER: Oxyhydrogen and oxyacetylene. The first uses hydrogen for combustion, while the second uses acetylene. Both processes depend on burning a gas with oxygen to obtain melting temperatures as high as 6300°F.

3-40 Name some welding defects.

ANSWER: Poor welds will result from the following: misalignment of the parts being welded, cracks in welds or heat-affected zones, pinholes, slag inclusion, porosity from escaping gases, incomplete fusion or lack of penetration, and undercutting of the weld or the leaving of a groove adjacent to the weld because it was not filled by metal.

3-41 Why are heavily coated rods not used when welding by the oxyacetylene process?

ANSWER: Since a gas is being burned, the weld is shielded by the products of combustion that surround the weld and its puddle.

3-42 Differentiate between preheating and stress relieving.

ANSWER: Preheating is the application of heat to the base metal prior to welding, with the purpose of preventing expansion and contraction cracks from developing in the base metal owing to the sudden application of high welding temperatures. Stress relieving is postweld heating of the components to a temperature below the critical metal range in order to remove the residual stresses that may exist in the welded joint due to the welding process.

3-43 Who is responsible for conducting tests of welding procedures and for qualifying welding operators?

ANSWER: The manufacturer or contractor involved with the welding or the person who employs the operator.

3-44 What are "locked up" stresses in welding?

ANSWER: These are stresses caused by the difference in temperature of the base metal and the weld, which results in nonuniform expansion and contraction of weld and base metals. The stresses produced are considered to remain in the joint, unless relieved by stress relieving, during which the joint is reheated at a very controlled rate and temperature.

3-45 Define "reinforcement of weld."

ANSWER: This is the extra weld metal placed above the surface of the base metal in order to assure full penetration to the full thickness of the base metal by the welding operation.

3-46 Explain the purpose of the (*a*) reduced-section tensile test, (*b*) free-bend test, (*c*) root-bend test, (*d*) face-bend test, and (*e*) side-bend test.

ANSWER: (a) The reduced-section tensile test is one used to qualify the welding procedure and consists of testing a properly prepared sample in a tensile-strength machine to develop its ultimate strength so that it can be compared with code requirements. (b) The free-bend test is used to qualify the welding procedure and consists of bending a specimen cold, with the outside fibers required to elongate at least 20 percent before failure may occur in the joint. (c) The root-bend test is used in qualifying welding operators and consists of bending the weld specimen against the bottom of the weld. No cracks or other open defects exceeding $\frac{1}{8}$ in may be present. (d) The face-bend test is also used for qualifying operators and consists of bending the specimen against the surface of the weld. No cracks or other open defects exceeding $\frac{1}{8}$ in may be present. (e) The side-bend test is another test for qualifying operators and consists of bending the welding specimen against the side of the weld. The same restrictions on defects apply, as were described under the root- and face-bend tests.

3-47   What is a mill test report?

ANSWER: This is a report by the steel mill attesting to the chemistry and physical properties of the material. In the case of plate steel, it shows the heat, or slab, number from which the plate was made, stamped on the plate. It also gives the specification and thickness. The mill test report should be compared with code requirements or purchase order specifications, or both, for the pressure vessel system.

3-48   How is nondestructive testing defined?

ANSWER: *Nondestructive testing* is a broad technology of testing that is used to find flaws or defects without damaging the parts under test.

3-49   Name some NDT methods.

ANSWER: Visual, liquid-penetrant, magnetic-particle, eddy-current, radiographic, ultrasonic, and acoustic emission testing are the most prominent NDT methods in use or under intense development.

3-50   What determines the NDT method to be used in a plant?

ANSWER: The determinants can be many and include: the nature of the defect to be found or evaluated and its possible location on the surface versus within the part, the capability of plant personnel, plant operating status during testing (on-line versus shut down), personnel conducting the test (in-plant employees versus an NDT service company). In general, surface defects are analyzed by visual, dye-penetrant, and magnetic-particle tests, while the other tests are used to detect deeper, hidden defects.

3-51   On what principle is radiographic testing based?

ANSWER: Radiographic testing is based on the differential absorption of penetrating radiation by the substance or part being tested. The practice consists of placing an appropriate film behind the area to be tested and exposing it to x-rays or gamma or neutron rays, depending on the source of the radiation, with the rays passing through the area under test. Defects show on the film as changes in density. Radiographic testing is limited to licensed firms or individuals, because of the potential hazard from radiation.

3-52   Name three types of waves used in ultrasonic inspections.

ANSWER: The three waves are straight-through, angle beam, and surface.

3-53   Define the half-life of an isotope.

ANSWER: This is a term used to describe the decrease in strength of radiation. Thus, the half-life of an isotope is the number of years it takes for the isotope to decrease to one-half its present strength.

3-54   Why is it required code practice to remove weld ripples before radiographing a weld?

ANSWER: This is required so that actual defects are not masked or hidden by the radiographic image caused by the weld ripples. If this were not done, it would be difficult on the film to distinguish unacceptable weld defects from images caused by the weld ripples.

3-55   What is the purpose of a penetrameter?

ANSWER: A *penetrameter* is a reference gage that is placed next to the area to be radiographed. It has in it defects, or holes, depending on the thickness of the parts being welded, with certain minimum defects which are considered acceptable, and which must be visible on the film. Defects larger than this minimum are thus visible on the film and may be used as a basis for rejection of the weld, material, etc., unless code repairs are made.

3-56   Name the three types of liquid penetrants acceptable under the ASME code.

ANSWER: The three types of penetrants are (*a*) water-washable, (*b*) postemulsifying, and (*c*) solvent-removable penetrants.

3-57   In prod magnetic-particle testing, what is the maximum prod spacing permitted by the ASME code?

ANSWER: Section V permits a maximum of 8 in.

3-58   What frequency of alternating current is used in eddy-current testing?

ANSWER: The frequency to be used to excite the electromagnetic field will depend on the depth of penetration required. Frequencies vary from 500 to 20,000 Hz. The higher frequencies are used for small-diameter, nonmagnetic tubing.

3-59   To what type of material can magnetic-particle inspection be applied?

ANSWER: Magnetic-particle inspection can be used only on ferromagnetic material.

3-60 What is meant by "passive," as applied to stainless steels?

ANSWER: The stainless steels develop a corrosion-resistant film on the surface, which during usage is classified as passive to corrosion. The films form quickly and are tight in formation, consisting usually of chromium oxide or an adsorbed oxygen film, neither of which is usually visible.

3-61 Describe "passivation."

ANSWER: *Passivation* is the artificial method used to provide the protective film to stainless steel, usually produced by a strong oxidizing agent such as a solution of nitric acid in water.

3-62 What stainless steels are usually employed in pressure vessel systems?

ANSWER: The following major groups are usually employed:

1. Straight chromium with a chromium content up to 30 percent, called the 400 series
2. Chromium-nickel alloys, or the 18-8 stainless steels, called the 300 series
3. Chromium-nickel-manganese stainless steels, called the 200 series

3-63 What is a clad-steel plate?

ANSWER: A *clad-steel plate* is a composite plate made by uniformly and inseparably bonding a layer of a corrosion-resistant metal, usually a stainless steel, to a carbon- or low-alloy steel backing plate. Cladding thickness usually ranges from 15 to 30 percent of the base- or backing-metal thickness. Stainless steels commonly used are the 300 and 400 series.

3-64 Compare ferritic and austenitic stainless steels.

ANSWER: As a class of materials, the ferritic stainless steels are weldable, formable, and highly corrosion-resistant. For the most part, they find applications where improved corrosion resistance is desirable, but where the higher price of austenitic stainless steel is unacceptable. The ASME code recognizes the use of ferritic stainless steel, but its use is limited to temperatures below 700°F.

The ferritic stainless steels have a body-centered cubic crystal structure similar to the room-temperature form of iron. They are magnetic from room temperature up to about 1400°F (760°C). These alloys contain from 11 to 27 percent chrome, and the low-chromium grades are among the least expensive stainless steels.

Austenitic stainless steels contain sufficient chromium to give them good corrosion resistance, and nickel to retain the face-centered cubic crystallographic structure of high-temperature iron down to room temperature. Austenitic stainless steels are nonmagnetic. The most common austenitic alloys are the familiar 18-8 chromium-nickel types, of which types 304, 321, and 347 are the stainless steel alloys frequently used. Both chromium and nickel contents may be increased to improve corrosion resistance, for example in the type 309, 310, and 316 alloys. Additional elements, usually molybdenum, may be added to further enhance corrosion resistance.

4

Imposed Stresses
and Service Effects
in Pressure Vessels

Importance of Defect Analysis

During plant inspections and maintenance, conditions will be found in pressure vessels that require analysis as to cause. This is especially important when determining the reasons for cracks, bulges, thinning, erosion, and even rupture in a localized area, that can cause severe leakage of contents. Failure or defect analysis includes:

1. Compiling a history of operations, maintenance, and past inspection activity

2. Reviewing the stress equations that were used in ordering the vessel, or reviewing the specifications for requested working pressure, material, thickness, and similar strength considerations for the vessel

3. In some cases where the reason for the failure is not clear, performing a metallurgical analysis in order to characterize the nature of the failure

4. Reviewing the controls, their settings, and the setting and functioning of safety devices

5. Proposing from the evidence gathered potential causes of failure

Plant operating, inspection, and maintenance personnel involved with pressure vessel systems should have a working knowledge of the causes of stress on materials, and even have a knowledge of the different types of stress. This will enable them to analyze conditions and defects as they appear in the plant. Some defects develop during fabrication and must be evaluated during initial inspections as to their impact on future service of the pressure vessel. These may include porosity and shrinkage cracks in castings, and inclusions and laminations from the rolling operation in forming plate steel. Similar defects may be caused by welding. Thus, new equipment may have some defects that will magnify stresses in service. New equipment should always receive initial or baseline inspections. Subsequent inspections, and a knowledge of the effects that stress may have on the equipment under service conditions, will assist in analyzing a developing problem and taking corrective action, in order to prevent a future unexpected failure.

Loads on Pressure Vessel Systems

A pressure vessel must resist various loads on it that cause internal stress to the material. Among the loads required to be considered by the ASME code are:

1. The load caused by internal and external pressure.

2. Impact loads such as those caused by rapidly fluctuating pressure or temperature.

3. The load caused by the weight of the vessel and its contents, which would include the weight of water in the vessel during hydrostatic tests.

4. The load caused by structures or machinery mounted on top of the pressure vessels. Such loads would include vibration effects, which can cause repetitive-type stresses on the pressure vessel.

5. Wind and earthquake loads. Tall structures or pressure vessels, such as distillation columns that are installed outdoors, must be able to resist such loads.

6. Superimposed loads of various types used on pressure vessels, such as internal baffles, support reactions against the vessel walls, and similar local loads that cause additional stresses to the vessel above the normal stress of pressure loading.

7. Stress caused by temperature gradients that may exist from one side of the vessel to the other. These stresses result from differential expansion effects, such as may exist in a heat exchanger between the tubes, tube sheet, and shell.

Pressure vessels may also have stresses caused by fabrication such as locked-in stresses in welds from lack of heat treatment, discontinuities that act as stress-concentration points. For example, in welds the term "discontinuity" is used to denote a flaw or an abrupt weld condition that interrupts the continuity of the weld. These flaws may include porosity, slag inclusions, and incomplete fusion or penetration (Fig. 4.1). These magnify normal stresses and can cause fatigue cracks to appear at service loads that are below design or operating loads.

Resistance of Materials to Loads—Stress

Any body or material subject to external forces will resist these external forces. This resistance of the material comes from *within* the material. The internal structure of the material is subject to intercrystalline loading when an external force is applied. Thus, *stress* is defined as the internal force per unit area on the material that resists external forces on the material. It is expressed in pounds per square inch, but the notation "psi" is not used for stress as it is in pressure notation. Stress is always expressed by engineers with the notation lb/in^2 to differentiate it from the psi designation of pressure, which is an *external* force per unit area on the material.

Types of stress

There are several general classifications of stresses that affect materials. A *normal stress* is a stress on an area of a material produced by a force at

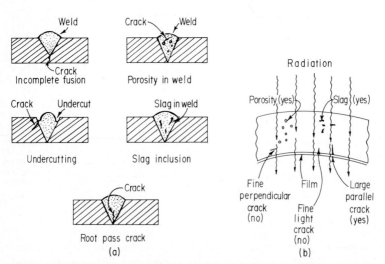

Figure 4.1 Defects may be present in welds and material. (*a*) Welding defects may produce cracks; (*b*) radiographic examination may not reveal some material defects.

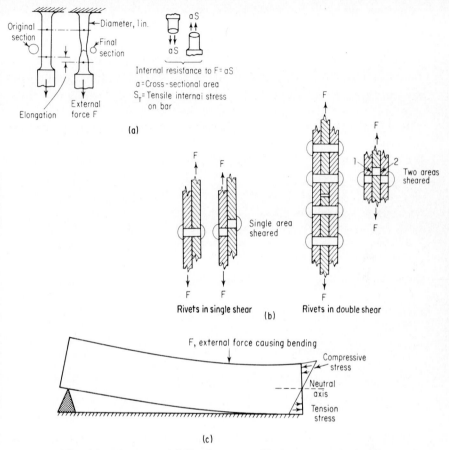

Figure 4.2 Material under stress. (*a*) Tension stress; (*b*) shear stress; (*c*) bending stress.

a right angle to the area acted on. Normal stresses are further classified as *tension stresses* or *compression stresses*. In Fig. 4.2*a* a 1-in-diameter bar is pulled by a force *F*. This force produces a normal stress at a right angle to the cross-sectional area of the bar. Since the force tends to pull (stretch) the bar apart, it is called a "tensile stress." Within the material, the intercrystalline structure (assuming it is steel) is also being stressed. The tensile stress of the bar is found by the following equation, based on the definition of stress as pounds per square inch, or internal force per unit area within the material:

$$F = aS_T$$

where F = external force
$\quad\quad\ \ a$ = cross-sectional area of material resisting F
$\quad\quad\ S_T$ = tensile (internal) stress on the material

From this, the equation for stress S_T is

$$S_T = \frac{F}{a}$$

This equation shows that the tensile stress S_T is found by dividing the external force F, acting normal to the cross-sectional area, by the cross-sectional area of the material resisting the force.

If the force were acting in the opposite direction, a compressive stress (known also as a "bearing stress") would be imposed on the material, but the compressive stress would be found by the same equation,

$$F = aS_C$$

where S_C = compressive stress.

Example Assume that in Fig. 4.2a the tensile force is 15,000 lb and the rod is of 1-in diameter. What would be the tensile stress on the bar?

answer

$$S_T = \frac{F}{a}$$

$$F = 15,000 \text{ lb}$$

$$a = \pi\left(\frac{1}{2}\right)^2 = 0.7854 \text{ in}^2$$

Substituting, we get

$$S_T = \frac{15,000}{0.7854} = 19,099 \text{ lb/in}^2$$

If a force acts on a tangent (sideways) to the area of a material, a shear stress is produced. This is illustrated in Fig. 4.2b, in which a force F acts on the area tangent to its cross-sectional area, thus producing a shear stress.

The shear stress is found by the equation $F = as$, but a is the cross-sectional area resisting shear. In the drawings on the left-hand side of Fig. 4.2b, only one area of the rivet is resisting the external force F; it is in single shear.

In the drawings on the right-hand side of Fig. 4.2b, a butt-riveted joint is shown with butt straps on each side of the butting plates. The rivets in this illustration are in double shear, as *two* areas of the rivet are resisting the load F. An example will illustrate the significance of single shear and double shear.

Example Assume that only one rivet is being considered for analysis, but in one case it is in single shear, in the other in double shear. Assume the rivet to be of 1-in diameter in each case and the load F to be 15,000 lb on each rivet. What is the shear stress for each rivet?

answer

Single-shear rivet

$$S_S = \frac{F}{a}$$

$F = 15{,}000$ lb

$a = \pi r^2$

$$= \pi\left(\frac{1}{2}\right)^2 = 0.7854 \text{ in}^2$$

$$S_S = \frac{15{,}000}{0.7854} = 19{,}099 \text{ lb/in}^2$$

Double-shear rivet

$$S_S = \frac{F}{a}$$

$F = 15{,}000$ lb

$a = 2\pi r^2$

$$= 2\pi\left(\frac{1}{2}\right)^2 = 1.5708 \text{ in}^2$$

$$S_S = \frac{15{,}000}{1.5708} = 9549.5 \text{ lb/in}^2$$

This shows that a rivet in double shear is *twice* as strong as a rivet in single shear.

Another stress is that due to bending. A beam supported on each end and loaded in the middle will develop a bending stress. The beam, when bending, will actually be under a tension stress on one side and a compression stress on the opposite side. This is illustrated in Fig. 4.2c. Flat plates and stayed surfaces in boilers and pressure vessels are some elements that are subjected to bending stresses.

Note: Stress due to torsion, such as in an axle being rotated and transmitting power, is another stress considered in an analysis of resistance of materials. Analysis of this kind of stress is beyond the scope of this book. However, a knowledge of tension, compression, shear, and bending stresses is essential in understanding how pressure is contained in a pressure vessel, pipe, or any other apparatus made of material designed to confine that pressure within safe limits.

Strain and stress-strain diagrams

When a body or material is subjected to external forces, internal stresses resist these forces, but there is always some deformation with load. For example, a steel rod will stretch when pulled on by an external force. The *total stretch* is expressed in a length measurement such as inches or centimeters. *Strain* is defined as the stretch per unit length, or deformation of a body per unit length, and is always expressed as inches per inch, centimeters per centimeter, etc. For example, assume that a steel rod 10 in long stretches 0.010 in with load; then the unit strain will be 0.010/10 = 0.001 in/in.

Some of the fundamental properties of all structural materials that must be determined are found by means of stress-strain diagrams. Structural properties determined from the stress-strain diagram are proportional limit, yield point, ultimate strength, and modulus of elasticity.

Modern engineering practice requires testing of materials so as to specify and identify their physical properties. This is particularly true for

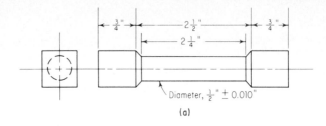

(a)

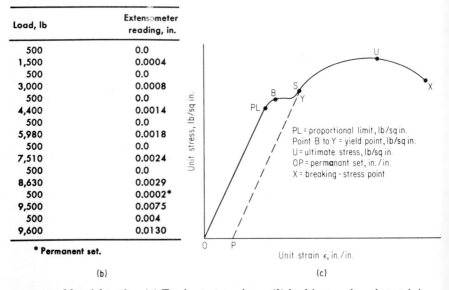

Load, lb	Extensometer reading, in.
500	0.0
1,500	0.0004
500	0.0
3,000	0.0008
500	0.0
4,400	0.0014
500	0.0
5,980	0.0018
500	0.0
7,510	0.0024
500	0.0
8,630	0.0029
500	0.0002*
9,500	0.0075
500	0.004
9,600	0.0130

* Permanent set.

(b)

(c)

Figure 4.3   Material testing. (*a*) Tension-test specimen; (*b*) load in pounds and stretch in inches are taken during tension test of material; (*c*) calculated stress and strain are plotted on a stress-strain diagram.

materials intended to be used in boilers, pressure vessels, and nuclear reactors. Steel manufacturers' laboratories run tests; the ASME code shows sketches and requirements for preparing samples to run tensile tests and bending tests for code materials.

Measuring strain.   A test specimen of a specified grade of steel is cut out from a rolled stock, then machined to fit a test machine. A test specimen for a tensile test is shown in Fig. 4.3*a*. The square ends are clamped in jaws of a tension-test machine. The $\frac{1}{2}$-in round center piece is marked off in a 2-in gage length. An extensometer is attached to the 2-in gage length. This tension-test machine has a dial indicating the force F applied to pull the rod apart. Incremental loading is applied. At each increment, the force F is recorded and the total length l is measured at that incremental loading (see Fig. 4.3*b*). This procedure is followed until rupture occurs.

The data are tabulated, and then by the relationship F/a, where a is the original $\frac{1}{2}$-in-diameter area, stresses are found for each increment of load.

The strain, or stretch, is found by calculating the amount of stretch from the original 2-in length to obtain the unit strain ε. These values are then plotted as in Fig. 4.3c, showing a stress-strain diagram for ductile steel.

Proportional limit. The stress-strain diagram shows a sloping straight line extending from zero upward to the point marked *PL*. The reason is that as the load is increased, the imposed stress S increases, as does the strain ε. The increase for both is in the same ratio, meaning that if the load is doubled, the stress is doubled, and so is the strain. This is why a straight line is drawn, not a curve. The *proportional limit* of a material is thus the maximum unit stress that can be developed in the material without causing a deviation from the law of proportionality of unit stress to unit strain. Of even more significance, it implies that if the load is removed, the material will return to its original length without having a permanent *set* as a result of the loading. The material will not be permanently deformed, but will return to its original shape so long as the proportional limit is not exceeded.

Yield point. As the load on the test specimen is increased further, causing a stress greater than the proportional limit, a unit stress is reached at which point the material continues to stretch *without* an increase in load, assuming it is ductile steel. The unit stress at which this stretch without load occurs is called the "yield point" and is represented by the short horizontal line from B to Y on the stress-strain diagram. The *yield point* of a material is defined as the minimum unit stress in the material at which the material deforms or stretches appreciably without an increase of load.

If a material is stretched or loaded slightly beyond the yield point, a permanent set or deformation occurs in the material. For example, in Fig. 4.3c, if the load is reduced to zero after just passing the yield point, the extensometer will show a permanent stretch or deformation. This is found by drawing a line parallel to the proportional-limit line; the set will be length 0 to P per inch of test specimen.

If the loading on our test specimen is increased, as indicated by the curve S to U, a point of maximum unit stress is reached. Then the unit stress declines with slight additional loading and stretches until it breaks. This is particularly true of ductile material, which "necks down" very rapidly after reaching its maximum unit stress because of reduced area in the neck section, which requires less load to cause rapid stretching to complete breakage.

Ultimate strength.   The *ultimate strength* of a material is defined as the maximum unit stress that can be developed in the material as determined from the *original* cross section of the material. It is point U in the stress-strain diagram. The curve from U to X is a rapid, unstable testing condition, with point X called the "breaking-load point." The maximum unit stress is at point U, and this is the ultimate stress designated for the material.

Example   Data supplied on a tested specimen are: total length is 19 in; sections at each end of the specimen are 1 in × 1 in × 6 in long; center section is $\frac{7}{16}$-in diameter by 7 in long with center section concentric with the square sections at each end; center 7-in section has a 2-in gage length marked.

The specimen being tested here broke when the load reached a maximum of 11,274.75 lb, and the break was through the original $\frac{7}{16}$-in-diameter. But the diameter was then $\frac{1}{4}$ in. The cross-sectional area of a $\frac{7}{16}$-in-diameter bar is 0.15033 in². The length of the 2-in gage section had stretched to 2.55 in.

1. Find the ultimate strength of the material.
2. What is the ultimate strength in pounds per square inch of the 1-in × 1-in section at each end?
3. What is the percentage elongation of the 2-in gage section?

answer

1. In this problem the break was in the $\frac{7}{16}$-in-diameter section; the stress S is

$$S = \frac{F}{a}$$

where a = original cross-sectional area.

$$S = \frac{11{,}274.75}{0.15033} = 75{,}000\text{-lb/in}^2 \text{ ultimate stress}$$

2. The ultimate stress at the 1 in × 1 in section is the same (75,000 lb/in²) because it is still the same material. It did not break at this section because the cross-sectional area is larger than at the $\frac{7}{16}$-in-diameter section.
3. The percentage elongation is found as follows:

$$\frac{\text{(Final length} - \text{original length)} \times 100\%}{\text{Original length}} = \%\text{ elongation}$$

$$\frac{(2.55 - 2) \times 100\%}{2} = \frac{55\%}{2} = 27.5\%$$

Modulus of elasticity.   *Hooke's law* states that the unit stress in a material is proportional to the accompanying unit strain, provided that the unit stress does not exceed the proportional limit. In different words, it states that the ratio of stress to strain for a certain material is always a constant E, called the "modulus of elasticity," or in equation form,

$$E = \frac{\text{stress}}{\text{strain}} = \frac{S}{\varepsilon} = \text{constant}$$

For steel, the modulus of elasticity is usually taken as 30,000,000 and written 30×10^6 lb/in². This is the modulus of elasticity for normal, or axial, loads. There is also a shear modulus of elasticity. For steel it is 12,000,000 lb/in².

Strain gages are used to determine stresses at critical areas of boilers, nuclear reactors, and pressure vessels for which exact calculations cannot be made. With the following relationship between stress, strain, and the modulus of elasticity of the material, the stress can be calculated. It is much easier to measure strain, or the deformation of a material under load, than to measure stress.

As explained above, stress for normal loads is F/a, where F = imposed load and a = original area of material resisting the load; also, unit strain ε is e/l, where ε = strain in inches per inch and e = amount of strain from original length l.

Now,

$$E = \frac{\text{stress}}{\text{strain}}$$

Substituting the above values gives

$$E = \frac{F/a}{e/l}$$

$$E = \frac{S}{e/l}$$

as $$S = \frac{F}{a}$$

Rewriting this in terms of stress S

$$S = \frac{Ee}{l} = E\varepsilon$$

as $$\frac{e}{l} = \varepsilon$$

It can be seen that if strain is measured, stress can be calculated by knowing the modulus of elasticity of the material, which is a constant for the class of material being considered.

The modulus of elasticity is a measure of the *stiffness* of a material. For example, if one material has a modulus of elasticity twice as large as that of a second material, the elastic unit strain in the first material, for a given unit stress, is half as large as that in the other material. Thus, the first material is considered twice as stiff as the other. Some common E values are steel, 30 million; cast iron, 15 million; aluminum, 12 million, concrete, 3 million.

The *elastic limit* is the maximum unit stress that can be developed in the material without causing a permanent set. Test results show that for most structural metals the elastic limit of the material has about the same value as the proportional limit, and in most technical literature the elastic and proportional limits are considered identical. A small difference is apparent in testing work, but for practical purposes they can be treated as identical quantities.

Longitudinal and circumferential stresses in shells

Pressure vessel shells are most often of cylindrical shape. Internal pressure in a cylindrical shell closed at each end tends to burst the vessel along two distinct axes. First, the total pressure acting on the shell tends to cause rupture along a longitudinal axis. The total pressure acting on the heads tends to cause fracture of the shell around its circumference.

Thin-walled cylinders, meaning those in which the thickness of the shell does not exceed half the inside radius, have two stresses, *longitudinal stress* and *circumferential stress*. The latter sometimes is called the "transverse stress." Thick-walled cylinders have these stresses also, but

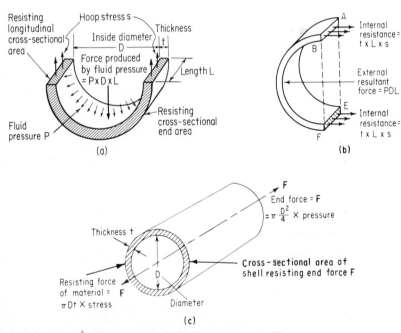

Figure 4.4 Forces and stresses on cylinder due to pressure. Figure parts (*a*) and (*b*) show that forces imposed on shell by pressure are resisted by the two cross-sectional areas shown; (*c*) end force on cylinder due to pressure is resisted by circumferential area of cylinder.

they are determined differently. Both stresses are known by these names because of the loading they resist in a cylinder. Both are fundamentally tensile stresses, and there is a fundamental relation between the longitudinal stress imposed on a shell versus the stress imposed circumferentially.

Longitudinal stress.   Figure 4.4a shows a seamless cylinder with an inside diameter D, shell thickness t, and length L, with a uniform pressure P acting inside the cylinder. Pressure acts on the cylinder walls, so the resultant force created tends to split the cylinder along its long axis. Thus the first stress to be considered is the longitudinal stress resisting this force tending to split the cylinder along this axis. The pressure acts in all directions. But if we cut the cylinder as in Fig. 4.4b, which shows the external force on one side and also the internal material stress resisting this external force, the following force relationships are developed, for a condition of equilibrium to exist.

The force tending to split the cylinder is given by area times pressure.

Force acting on one side $= D \times L \times P$

where $D \times L =$ projected effective area. The internal force of the material resisting this force is

Stress $\times$ material area

or   Resisting force $= S_L \times t \times L \times 2$

where $S_L =$ longitudinal stress
$t \times L =$ one area of the material

But since there are two material areas resisting the force (one on each side), $t \times L$ is multiplied by 2. Equating the two forces gives

$D \times L \times P = S_L \times t \times L \times 2$

Solving for S_L

Longitudinal stress $S_L = \dfrac{DP}{2t}$

Circumferential stress.   The force tending to split the cylinder endwise, or around its circumference, is shown in Fig. 4.4c. Pressure acting on each end creates a force that is equal to the end area (circle) times the pressure, or

End force $= \pi\left(\dfrac{D}{2}\right)^2 \times$ pressure

The material resists this by a force equal to the end area of the material times the stress, or

Resisting force $= \pi DtS_c$

where S_c is circumferential stress. Equating the two forces for equilibrium gives

$$\pi \left(\frac{D}{2}\right)^2 P = \pi \, DtS_c$$

Solving for S_c, circumferential stress,

$$S_c = \frac{DP}{4t}$$

If we compare this with the longitudinal stress, we find the circumferential stress is half the longitudinal stress.

The two equations for longitudinal and circumferential stress are fundamental strength-of-material equations. They are modified somewhat by the *ASME Boiler and Pressure Vessel Code* to take into account manufacturing and experience factors.

The equations developed are for seamless construction, meaning that no welded or riveted joint is present. Later chapters show how the joint efficiency has to be considered to modify these equations. Note that equations for both longitudinal and circumferential stresses (due to pressure) are independent of the length of the vessel. However, if a vessel is very long, the bending stress will have to be added to the stress due to pressure. This is especially true of a vessel filled with a substance of considerable weight.

The significance of the circumferential stress being half the longitudinal stress in a cylinder is seen in many problems in code design and calculation. For example, riveted circumferential joints do not have to be as strong in this direction as they do longitudinally. But in many calculations it is extremely important to make sure that the strength circumferentially is at least half the strength longitudinally. This is brought out in other chapters.

The effect of temperature

Temperature above designed limits has the immediate effect of lowering the permissible stress on a material. For example, SA-30 grade-A-quality carbon-plate firebox steel has an allowable stress of 12,000 lb/in² for temperatures from -20 to $400°$F. At $900°$F the allowable stress is only 5000 lb/in². By assuming the same pressure at both temperatures, it can be seen that a pressure vessel designed for 12,000-lb/in² normal stress will be weakened to 5000/12,000, or 41.7 percent, of its original strength with a temperature increase to $900°$F.

Stress on pressure parts can also be caused by expansion due to temperature rise, and it is pronounced if the parts are restrained because of restrictions or if uneven metal thicknesses are joined abruptly, without transition sections. Temperature increases cause expansion of steel, which can be calculated as follows:

$$e = nl\ (T_2 - T_1)$$

where e = change in length
l = original length
T_1 = original temperature, °F
T_2 = final temperature, °F
n = coefficient of expansion (change in length per unit of length per degree change in temperature)

Example   Steel has a coefficient of thermal expansion of 0.0000065 in per in per °F. To show the possible rate of expansion to be considered, assume that a stay in a horizontal-return tubular (HRT) boiler running from tube sheet to tube sheet is 30 ft long. How much will this rod expand with a temperature change from 70 to 300°F, assuming free expansion?

answer   Substituting into the above equation, we get

$$e = 0.0000065\ (30)(12)(300-70) = 0.538\text{-in stretch}$$

which is over $\frac{1}{2}$ in.

Example   If we assume that the stay rod was fixed at each end and that the tube sheets would *not give*, what compressive stress would be imposed on the rod, neglecting the column effect of a long rod?

answer   This is calculated from the modulus of elasticity equation

$$S = E\varepsilon$$

where $\varepsilon = \dfrac{\text{stretch}}{\text{inch}}$

By substitution,

$$S = 30,000,000\ \frac{0.538}{30 \times 12} = 30,000,000\ (0.001494) = 44,820\ \text{lb/in}^2$$

These examples illustrate the importance of considering temperature effects during operation in order to avoid the rapid stress buildup when a part becomes accidentally heated above design conditions. Remember that the stress developed is not calculated as simply as shown by the illustration. For example, we assumed that the shell and tube sheets would *not* expand because of temperature.

Stress-concentration factors

If a structural material has an abrupt change in a section, for example, a flat plate containing an opening or a sharp corner as shown in the rod in Fig. 4.5a, the stress distribution is not uniform over the cross-sectional area of the material. Near the abrupt change the stress is much higher than calculated. The affected section is said to have a *stress-concentration* section, or area, and the ratio by which the normal stress has to be multiplied (K in Fig. 4.5a) is called the "stress-concentration factor."

Stress concentration plays an important part in structural members subject to routinely repeated loadings, for the stress concentration can lead to cracks and fatigue failures. If the stress concentration is severe enough (even in normal loading), stresses may be induced far above the normal expected stress. Sharp corners in welded joints and other sharply formed shapes must be avoided. Thus openings cut into plates must be reinforced to strengthen the edges around the opening against stress concentration.

The ASME code specifies permissible joint connections to avoid stress concentrations. Fillet radii are specified on formed shapes. Openings must be calculated by code rules. In analytical and design work, stress concentrations are determined by the photoelastic method, stress-coat method, and strain-gage method using the electrically resistant wire gage. In nuclear vessels one must carefully design the elements using the endurance limit and other stress-analysis methods.

Endurance limit (fatigue stress)

The *endurance limit* (also known as "fatigue limit") is the maximum unit stress that can be imposed and repeated on a material through a definite cycle, or range of stress, for an indefinitely large number of times without causing the material to rupture.

The endurance limit is determined by testing a material through a series of stresses, starting near its ultimate strength and repeating the trials using progressively lower stresses. When stressed nearly to its ultimate strength, the specimen will rupture after a few cycles. If a second sample of the same material is again tested but stressed slightly less than before, a larger number of reversals, or cycles, can be imposed. This is continued until a stress value, the endurance limit, is reached, where an indefinite number of stress cycles can be imposed without causing rupture.

Figure 4.5b shows an *S-N* diagram, where stress to rupture is plotted on one side and number of cycles to failure on the other. The horizontal line obtained is the endurance stress for the material. In Fig. 4.5b, this is 22,500 lb/in^2. Endurance limits are widely used in machine design work.

Endurance limits of materials can be modified considerably by environment, temperature swings, corrosion effects, hydrogen embrittlement, and similar conditions generally associated with the study of the fatigue of metals. For example, discontinuities or stress-concentration factors such as sharp corners or cracks can lower or modify endurance limits.

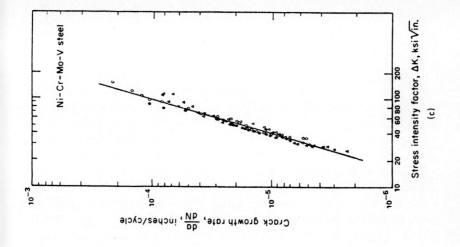

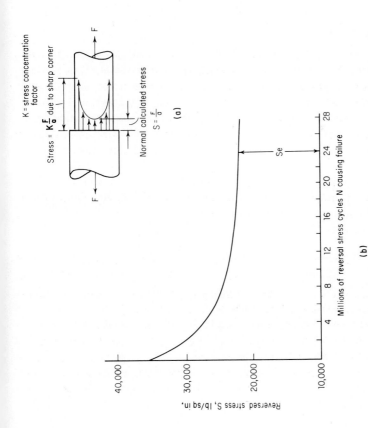

Figure 4.5 (*a*) Sharp corners and abrupt changes in a pressure vessel can increase normal, expected stresses by a factor called "stress concentration." (*b*) Endurance limits are determined by applying repetitive stresses on a material, and then plotting the data on an *S-N* diagram. (*c*) Fracture mechanics uses the stress intensity at the tip of cracks to study fatigue crack growth. Shown is a plot for a nickel-chrome-moly steel.

Cracks

Fatigue cracking is still a major reason for equipment failures. Through the use of visual and other NDT inspection techniques, fatigue cracks or failures can often be detected and repaired before a serious accident can occur. Fatigue damage is due to repetitively applied loads and will usually start at a surface and propagate in a direction perpendicular to the applied tensile load. Thus, inspections of surfaces may detect fatigue damage at its beginning. As was pointed out previously, ductile metals have an elastic limit; if the material is stressed below this limit, it returns to its original shape when the load is removed. When it is stressed above the elastic limit, permanent deformation takes place, known as "plastic yielding." This could occur from a heavy overload. However, unlike an overload failure, fatigue cracks can occur with loads much lower than the elastic limit of the material. This means there is no gross deformation, and this makes fatigue cracks difficult to see, especially in the initial stages of crack growth. The slow, incremental growth of a fatigue crack produces no large noise or large strain to be heard in operation.

Fracture mechanics

The growth rate of fatigue cracks has been quantified as a result of intense developmental work in the space and nuclear stress-analysis field. If a crack is assumed to exist in a material, if it is steel, under stress it will undergo plastic deformation about the crack tip. The crack will grow as the applied stress or load is increased. The distance that the crack front advances with each cycle of loading is a function of the stress intensity applied to the tip of the crack, which is expressed as the stress intensity factor ΔK. The crack growth rate, in inches per cycle of stress application, is expressed as da/dN, where a = crack size in inches and N = fatigue life or number of cycles of stress before failure. An equation for crack growth rate used in failure analysis is

$$\frac{da}{dN} = C(\Delta K)^m$$

where C = material constant determined by test for the class of material (for steels at room temperature, 4×10^{-24} is used)

m = material constant determined from tests (4 is usually used for ferrous material)

See Fig. 4.5c. Crack propagation data can be used to determine the stress intensity ΔK that can be tolerated for a particular design. Also, the number of cycles required to extend an initial crack size a_1 to what is considered a critical crack size a_2 can be established by using the fatigue crack propagation equation, da/dN.

Fracture mechanics and NDT methods are being used to determine whether a flaw needs immediate repair or whether more cycles of stress can be applied before the defect has grown to a size that requires repair or replacement. In the last few decades, there has been a great deal of research done on this failure-prediction method as applied to the basic mechanism of fracture phenomena in solids. The determination of rate of crack growth helps in establishing the number of cycles that can be tolerated before corrective actions will be needed. Nondestructive testing methods are used to periodically check a developing defect in conjunction with fracture mechanics methods.

Prevention of fatigue failures relies on containment of crack growth within what are considered safe limits. Where in-service inspections are possible, a safe service interval is established between inspections. Where in-service inspections are not feasible, a more conservative design is needed so that any anticipated defect will not grow to a dangerous size during the expected life of the component.

The simultaneous action of a part being subjected to repetitive stresses and attacked by the medium in which the part must operate can cause a part to fail as a result of *corrosion fatigue*. The combined effect of these two factors is much greater than the effect of either one alone. The cracking usually begins at surface defects, pits, or irregularities. These points act as stress-concentration points, with the corroding medium intensifying the pitting action. The more repetitive the stress, the faster will be the pitting action, which again will increase the stress concentration. A cumulative effect can materialize that will cause a failure well below a predicted endurance stress.

Pits formed under simultaneous stress and corrosion are always sharper and deeper than pits formed in the same time under stressless conditions. The more repetitive the stress, the faster will be the pitting action. At low-cycle repetitive stress, pitting will proceed entirely by normal corrosion, and the section will fail by the normal tension failure or when the resisting area is thinned down so that the stress rises proportionately, assuming a constant load.

Effect of discontinuities

Notches or other material discontinuities will initiate cracks and act as stress-concentration points, thus magnifying normal stresses above the elastic limit at the notch or discontinuity. Thus, even though the cross-sectional area of the structure is adequate for the loading, the concentrated stress at the root, or bottom, of the notch, or discontinuity, will start localized microyielding of the material at the root of the notch. Any scratch, gouge, nick, or similar discontinuity is a good place for fatigue damage to begin.

Failure of welds from fatigue can occur from internal discontinuities as well as from external geometrical notches. See Fig. 4.1a. The failures will usually start at the toe of the weld reinforcement or at the root of the weld with inadequate joint penetration. In some cases cracks form because of residual stresses that grew with cyclic loading.

Brittle fracture

Brittle fracture is now commonly recognized as an inherent property of ferritic steel. Every ferritic steel has a temperature range below which it becomes notch-sensitive and susceptible to brittle failure with little absorption of energy.

Pressure vessels have failed during hydrostatic tests without warning because of brittle fracture caused by the combination of stress and a water temperature too low for the test. It is important for plant people to be aware of ferritic steel transition temperatures when installing equipment outdoors that may be subjected to low temperatures, or when conducting hydrostatic tests with cold water.

Crack-arrest temperatures.   The manufacturer of a vessel, or the material fabricator, should be asked to supply crack-arrest temperature (CAT) curves. These curves represent the temperatures at which a propagating brittle fracture stops for various levels of applied nominal stress. At temperatures lower than the crack-arrest temperature, fractures start and propagate; at higher temperatures, they do not start because they cannot propagate. The crack-arrest temperature for a stress level equal to the yield strength is called the "fracture transition elastic" (FTE). This is the highest temperature at which fractures can propagate under purely elastic loads. Above FTE, the stress required to propagate fracture is partly plastic, and the fracture is mixed, partly brittle and partly ductile.

Examples of natural crack or dynamic crack tests, which involve the relationship of strain to fracture or energy to failure, are the Robbertson crack-arrest test, the explosion bulge test, and, more recently, the naval research DT (dynamic tear) test. These and other tests simulate in-service initiation of dynamic cracks in brittle materials. There have been more than 50 tests that have been used to evaluate the susceptibility of materials to brittle fracture failures.

Hydrogen embrittlement

Hydrogen embrittlement can occur under high-pressure and high-temperature operations of boilers and pressure vessels. Hydrogen attack on steels can result in severe loss of ductility, with cracks developing that can lead to unexpected brittle fracture. The effect is more severe on high-strength steels. The hydrogen under high pressure and temperature diffuses into the metal as atomic hydrogen. It recombines in the metal as molecular hydrogen in grain boundaries and causes high pressure in these "voids," resulting in bulging and blistering. Hydrogen also combines with carbon in steel under high pressure and temperature and decarbonizes the steel, with embrittlement taking place. Frequently chromium-molybdenum steels are used where resistance to hydrogen embrittlement is required.

For high-pressure plants susceptible to possible hydrogen attack, proper material selection is required, including selection of materials used in repairing vessels. Design and maintenance of process vessels that may be exposed to temperature extremes and partial pressures of hydrogen should be carried out with the assistance of Nelson curves, which are useful in selecting steels that are resistant to such conditions. If these factors are not taken into account, hydrogen embrittlement may result. Other considerations in selecting materials include corrosive service, high-temperature and high-pressure requirements, cyclic service, creep, and similar in-service conditions. Steel selection will normally include considering (1) availability of the material for fabrication and repairs, (2) high-temperature strength needs, (3) notch toughness so that processes can be exposed to frequent starts and stops during stress periods of operation, (4) weldability for ease in fabrication and repair, and (5) ease in applying heat treatment.

Gaseous molecular hydrogen does not readily permeate steel at ambient temperatures, even under several thousand psi, and at low pressures, the gas can reach moderately high temperatures without damage.

The atomic form is what causes hydrogen damage. Atomic hydrogen can be formed by high temperature (above 430°F), moisture breaking down, corrosion, and electrolysis. Hydrogen damage is usually classified as hydrogen blistering, hydrogen embrittlement, decarburization or hydrogen attack.

Weld areas and heat-affected zones (HAZ) have been sites where delayed cracking failures have initiated. Generally, this reflects the higher hardness of the weld metal and HAZ. To minimize high weld-area hardness, one must use base materials and welding consumables that do not have excessively high carbon and combined-alloy contents (i.e., manganese, nickel, and chromium). A proper welding preheat may be required to minimize the rapid cooling in the HAZ, as well as to ensure a dry material, thereby reducing the amounts of hydrogen introduced during welding. Good design and welding will employ low-hydrogen electrodes, full-penetration welds, rounded corners, smooth changes in cross section, and minimum stress raisers.

Blistering results from hydrogen permeation that creates pressure buildup of molecular hydrogen, which cannot diffuse in a metal. Hydrogen commonly collects at laminations, inclusions, voids, and grain boundaries. Such collection can result in local deformation and, in the extreme, total destruction of a vessel wall.

Big blisters found in large storage tanks containing sour crude are probably the most common example of hydrogen blistering. However, blistering of heavy-walled refinery reactors is more serious because of the pressure. Since rimmed steel (low-carbon steel, partially deoxidized) has more numerous voids, the substitution of killed steel (thoroughly deoxidized steel) greatly increases resistance to blistering. The use of alloy linings and multiwalled carbon-steel vessels has minimized this problem, but vent holes are required between the carbon-steel plate and the alloy lining. This will prevent hydrogen accumulation between the plate and the lining and thus prevent the inward collapse of the lining. More important, however, it will prevent the carbon steel's being embrittled by the hydrogen and thereby possibly exposing the vessel to a brittle fracture failure.

Cracking from welding

Cracks from welding are another material strength and stress consideration. Cracks can result from:

1. Insufficient preheating and postweld heat treatment.

2. Improper surface preparation in which all oil, grease, rust, scale, and similar contaminants are not removed.

3. Excessive distortion from locked-up stresses. Peening each weld bead will help to minimize distortion.

4. Excessive moisture, producing hydrogen cracking. Excessive moisture may be due to damp flux or to coated electrodes that are exposed to humid conditions. To avoid this electrodes should be baked in an oven before their use.

5. Variation in weld bead hardness. This is usually due to a lack of uniformity in the composition of the weld deposit.

6. Porosity due to pickup of gases while the deposited weld is molten. The weld deposit must be protected from the air to prevent the absorption of nitrogen into the melt. Holes in the weld can cause cracks to start when loading is applied to the weld.

It is important to establish a periodic inspection program on welds and on the heat-affected zones of welds in order to detect any developing cracks that may lead to an unexpected failure. To avoid problems, qualified technical people should review carefully any proposed repairs on welds to make sure that the proposed procedures and welders are adequate. If not done properly, even minor repairs, such as welding support attachments, can cause stresses and resultant crack initiation. Weld repairs should be made to the equivalent standards of new construction, using qualified or proven welding procedures, trained and qualified welders, and material, thickness, and welding positions required by the code. Heat treatment should be followed per code recommendations to avoid crack initiation from locked-in stresses. NDT methods should be used whenever possible to prove that a weld is free of defects.

Creep Considerations

As high-temperature service vessels are extended in service beyond the expected life of the process, plant people must periodically make inspections in order to determine if creep damage is occurring. Visual and NDT examinations may be required so that unexpected failures do not occur.

Creep is defined as the slow deformation of material with time, with the deformation taking place at elevated temperatures with no increase in stress. Modern material testing includes the determination of how much creep or permanent deformation can be expected in a given length of time so that design can reflect these factors in predictions of life expectancy for pressure vessels. Creep properties are obtained by imposing a constant tensile load on a specimen at a constant temperature and then measuring the strain at intervals of time. Data are plotted to obtain a variety of constant-stress creep-strain curves or constant-time creep-stress curves. In using creep data, the designer tries to establish the expected service life and the corresponding amount of creep deformation that can be tolerated. With these established, a stress can be selected that will satisfy these conditions.

The ASME code Section VIII, Division 1, requires that material that is to be operated in the creep range not be stressed in service beyond the *lowest* of the following values:

1. 100 percent of the average stress permitted for a creep rate of 0.01 percent per 1000 h of service.

2. 67 percent of the average allowable stress that would result in rupture at the end of 100,000 h of service. Note that time of exposure to creep is important.

3. 80 percent of the minimum stress that would result in rupture at the end of 100,000 h.

Radiation

The effect of radiation on steel is especially important in evaluating nuclear power plant pressure vessels. Nuclear pressure vessels may be subjected to neutron irradiation of steel. It is necessary to avoid impurities in the steel as well as large grain sizes in order to limit any vacancy or voids in the material. It is theorized that the radioactivity present in nuclear vessels promotes the formation of helium which, with time, collects at metal grain boundaries to weaken it by a decrease in ductility. Quality control is thus more stringent on nuclear vessels, so cavities may be eliminated.

Corrosion

Corrosion has been defined as the degradation of materials due to their interaction chemically with the environment, resulting in loss of a homogeneous metal structure. Corrosion causes the majority of maintenance and repair problems on pressure vessel systems. If neglected, it can cause severe failures. Figure 4.6 shows an air tank rupture with corrosion evident on the inside surfaces.

Many industries operate with high-pressure and -temperature processes that combine liquids and gases in the process streams and thus

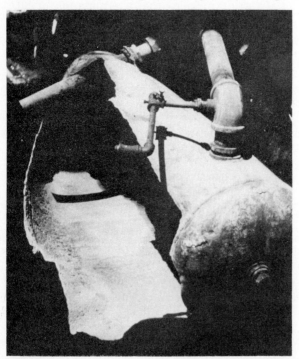

Figure 4.6 Air tank rupture caused by thinning of the shell.

may create corrosion problems in equipment, especially pressure vessels. In the chemical industry, corrosion of metallic materials is a constant concern not only because of possible equipment failure but also because of the possibility of product contamination and, in case of the leaks of flammable or toxic chemicals, the extreme danger to the surroundings. For these reasons, plant people involved with pressure equipment should be familiar with the different types of corrosion so they can interpret the signs pointing to the causes of the corrosion, determine steps that can be taken to control its effect, and then establish periodic inspection programs on the equipment to monitor corrosion.

While alloys and their corrosion resistance have helped, it is important to consider the effect of corrodents on alloy material, because selecting an alloy for corrosive service, if done improperly, may result in accidents and costly equipment shutdowns for repairs. It is important in selecting alloys first to review laboratory and field tests and service records, if possible, from plants that have used the alloy in similar service. Corrosion is a progressive type of deterioration that lends itself to control by periodic inspection programs, such as thickness tests to establish wear rates. In pressure vessels, the areas requiring attention are welds, liquid zones, vapor-liquid interfaces, nozzle connections, and areas with abrupt changes in sections.

A diversified petrochemical plant analyzed data for equipment failure over a 19-year period. Table 4.1 shows the results.

TABLE 4.1 Equipment Failure in Petrochemical Plant
(19-Year Data)

Failure mechanism—rank	Percent of all failure types
1. General corrosion	25.8
2. Fatigue or corrosion fatigue	18.7
3. Stress corrosion	11.7
4. Erosion corrosion	8.7
5. Pitting	6.8
6. Weld and fabrication defects	5.0
7. Overload, mechanical abuse	4.2
8. Brittle fracture	3.9
9. Wrong material	3.4
10. Wear	3.2
11. High-temperature mechanisms (creep, oxidation)	1.8
12. Casting flaws	1.8
13. Crevice corrosion	1.6
14. Intergranular attack	1.6
15. Hydrogen embrittlement	1.1
16. Dissimilar metals	0.6
17. Heat-treatment errors	0.5
18. Dealloying	0.3
19. Other	0.6

SOURCE: Adapted from *Chemical Engineering,* April 1982.

Electrochemical theory of corrosion

In the electrochemical theory of corrosion, metals are considered to be corroded by the local flow of electricity from one metal to another through an electrolyte, or from a negatively charged area to a positively charged area through the electrolyte. The cell shown in Fig. 4.7a represents an electric circuit, with an anode and cathode. An electric current would flow through the metallic plates and the electrolyte, with the anode losing some of its metal by corrosion caused by an oxidation reaction. Hydrogen gas is evolved when a metal corrodes in any acidic solution, and normally the cathode, which is considered polarized, has a layer of adsorbed gas bubbles, which reduces the consumption of metal by corrosion. Therefore, while the anode is losing metal because of oxidation, or rusting, a nondestructive chemical reaction, called reduction, proceeds at the cathode, with hydrogen gas produced. A potential difference is required to make a current flow. The potential difference may be from dissimilar metals, differences in electrolyte concentrations, or a lack of homogeneity of the metal surface, which may cause localized cells to be established. The larger the anodic area in relation to the cathode, the

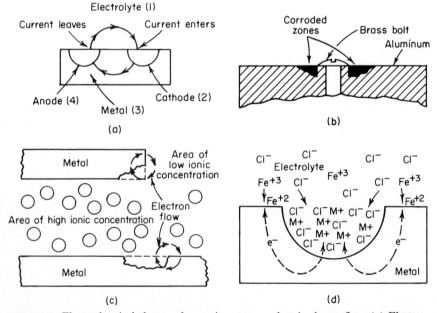

Figure 4.7 Electrochemical theory of corrosion stresses electric charge flow. (a) Electrochemical reaction involves the flow of current during corrosion; (b) in galvanic corrosion, the less noble of two metals is attacked; (c) crevice corrosion occurs at a localized structural fault; (d) chemical pitting is caused by high chloride concentration.

faster the rate of corrosion will be. The well-known chemical equation for rust is

$$4Fe + 3O_2 + H_2O \rightarrow 2Fe_2O_3 \cdot H_2O$$

Atmospheric corrosion proceeds differently from that in an electrolyte, because there is a plentiful supply of oxygen present in the air. Figure 4.7c and d shows some of the paths of current flow for localized corrosion in an electrolyte.

Coatings are used to make metallic surfaces nonconducting, thus stopping corrosion. Stainless steels are passive to corrosion because they form a corrosion-resistant film on their surfaces and thus tend to be cathodic, rather than anodic, to ordinary iron and steel. However, some stainless steels become active, or anodic, in the presence of chloride concentrations, such as seawater.

Types of corrosion

What are some of the types of corrosion (see Fig. 4.8)? Most engineers group them as follows:

Galvanic corrosion

Pitting corrosion

Crevice corrosion

Fretting corrosion

Intergranular corrosion

Oxidation corrosion

Corrosion fatigue

Also classified as material-degradation phenomena are erosion and exfoliation. *Uniform corrosion* is defined as corrosion of a metal at the same rate over its entire surface. Localized corrosion affects only small areas and shows up as pits, crevices, or cracks.

Galvanic corrosion. Galvanic corrosion is considered common to all metals and occurs when two dissimilar metals are in contact in a conducting aqueous solution. Under these conditions, an electric current is set up

Figure 4.8 Some types of corrosion. (*a*) Pitting; (*b*) exfoliation; (*c*) selective leaching; (*d*) stress corrosion (F = Force).

through the solution with one metal being the anode, where corrosion takes place, and the stronger, or more noble, metal being the cathode (see Fig. 4.9). Galvanic corrosion can occur during plant operation when metal from a distant part of the plant is carried by a fluid onto the surfaces of another, dissimilar metallic component, as when copper from a condenser or heat exchanger settles out onto a steel part. When only one element of an alloy is corroded out, selective attack is considered to have occurred, as in the denickelification of nickel from 90-10 or 70-30 Cu-Ni alloys, or the dezincification of brass. The solution to problems of galvanic corrosion is to use metals of the same composition in process streams as much as possible. Insulating lacquers have also been used to form a barrier between two dissimilar metals at the point of contact.

Concentration cell corrosion.   Concentration cell corrosion is a form of galvanic corrosion and is caused by an electric current's being set up between various parts of the same vessel that contain different concentrations of the corrodent. Such corrosion develops in the space of loose joints, such as gasketed assemblies. The corrodent stagnates in this space, thus increasing the ionization of the fluid. The difference in potential between such spots and adjacent areas becomes great enough that a current flows, with the stagnant area being the anode, where, as a result, corrosion starts to appear. Welding or caulking these types of joints will reduce the chance of concentration cell corrosion.

Pitting corrosion.   Pitting corrosion is the formation of holes in an otherwise relatively unattacked surface. Pitting can have various shapes and can act as a stress-concentration point. Pits are also places for corrodents to settle and become more concentrated, thus producing a cumulative damage effect to the material. It is theorized that pitting occurs electrochemically when a small part of the metal surface becomes anodic, with

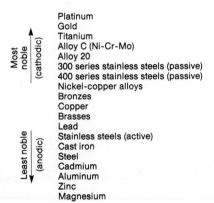

Figure 4.9   Noble metal arrangement in galvanic series.

the cathode being anywhere outside the pit. A small surface defect or a break in the protective layer of the metal is believed to create the anodic spot. It is believed the stagnant solution in a pit may become acidic when metallic ions combine with negative ions such as chlorides in the solution. Small currents are developed in the pit from this electrochemical interaction. A pit, once formed, becomes a concentration cell.

Crevice corrosion. Crevice corrosion is similar to pitting corrosion. Any crevice acts as a concentration cell in which corrodents can settle and become more concentrated and thereby more quickly attack adjoining material. Crevice corrosion has been attributed to one or more of the following: increase in acidity in the crevice, buildup of a damaging ion in the crevice, lack of oxygen in the crevice, and depletion of an inhibitor. Crevice corrosion is fixed by the design of the system but normally occurs on bolts, gasketed areas, and other points of metal contact. Creating designs having a minimum of crevices and keeping the areas clean of corrodents by more frequent maintenance are two of the methods used to combat this corrosion.

Fretting corrosion. Fretting corrosion occurs in areas where metals may slide over each other such as where tubes are in contact with intermediate supports in heat exchangers. The metal-to-metal rubbing causes mechanical damage that removes protective oxide coatings. The freshly exposed surface is then attacked by the corrodents in the fluid. Vibration of long tubes in a heat exchanger may expose these tubes to fretting corrosion. Reduction of vibration is thus one way of minimizing this type of corrosion. Another method is to use harder material. Sliding can also be prevented in heat exchanger tubes by rolling in the tubes at the intermediate supports.

Intergranular corrosion. Intergranular corrosion is another localized form of attack by a corrodent on areas of a material adjacent to grain boundaries. In severe cases of intergranular corrosion, the grain boundaries will appear rough to the naked eye, with particles of loose grain evident. It is believed that intergranular corrosion is due to HAZ problems, such as occur in hot-forming or welding operations, and subsequent poor or inadequate heat treatment. Good heat treatment practices reduce intergranular corrosion. Another approach is to use a modified alloy that has tighter grain boundaries.

Corrosion fatigue. Corrosion fatigue, as previously mentioned, is caused by repetitive stresses on a material in a corrodent fluid. The corrodent's effect may be so severe that it will lower the stress required to cause fatigue failure in still air to a level as low as half the normal level. The best method of avoiding corrosion fatigue is to avoid, in design or fabrication, sharp corners and similar stress raisers that develop into cracks after repeated applied stressing.

Methods to minimize corrosion's effects

Broad strategies in combating corrosion include:

1. Designing to avoid crevices, sharp corners, bends, corrosion cells, and similar areas that have been shown to be prime candidates for corrosion.

2. Designing, operating, and maintaining the system to limit corrodent concentrations. This is possible in some operations, such as water treatment to control water chemistry.

3. Selecting materials that are more corrosion-resistant, if economically justified.

4. Applying a corrosion-resistant lining, such as stainless steel or glass.

5. Establishing a periodic inspection program to track deterioration and establishing minimum requirements for the pressure and temperature of the process so that repairs or replacements are instituted when these minimum requirements are reached, thus avoiding serious accidents and process interruptions. Periodic inspections include NDT inspections where appropriate.

6. Using monitors to keep track of chemistry changes in a process stream and assist in analyzing the concentration of possible corrodents. For example, in the power-generation field, on-line monitoring of contaminants in steam and condensate is being used. Items being measured include pH, O_2, conductivity, cation conductivity, and sodium and chloride concentrations. This is combined with liquid ion chromatography showing ppb of sodium, chloride, ammonium, potassium, fluoride, and sulfate ions. The tests are performed on grab samples from selected loops of the steam-condensate flow. This permits identification of impurities and their sources so that corrective actions can be taken before more damage can result from a corrodent's presence in the process loop.

Stress-Corrosion Cracking (SCC)

Stress-corrosion cracking is a progressive type of failure that causes cracks at stress levels well below a material's yield point. It is caused by a combination of material properties, now recognized as affecting mostly high-strength alloys; it occurs in parts under high tensile strength and operating in a medium or environment that can chemically attack the grain structure of the material, hence the term "corrosion-induced stress cracking." Stress-corrosion cracking is generally intergranular in low-carbon steels, while it is transgranular in austenitic stainless steels. The

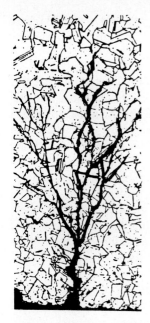

Figure 4.10 Stress-corrosion cracks progress along the material's grain boundaries. The crack shown here started in a surface pit.

stresses involved in stress-corrosion cracking can be imposed by loading, or they may be residual stresses created in the material during forming or welding. Stress-corrosion cracking increases with the temperature of the corroding medium. When one views a stress-corrosion fracture, the break or fracture appears brittle, with none of the localized yielding, plastic deformation, or elongation that are characteristic of a ductile failure. As shown in Fig. 4.10, instead of a single crack, stress-corrosion cracks have a network of fine, featherlike, branched cracks.

While severe general corrosion is not normally seen on the part with SCC, pitting is evident, and these pits serve as stress-concentration points as well as places for corrodents to concentrate and chemically attack the grain boundaries of the material. The total effect can thus become cumulative as the process continues. Research is continuing on determining permissible threshold limits of known corrodents.

Steps that have been recognized as reducing the chance of SCC include:

1. Stress-relieve material after forming or welding operations in order to reduce residual stresses.

2. Maintain corrodents at levels that research indicates are low enough for the material to withstand.

3. Consider protective coatings or inhibitors as prescribed by the material manufacturer or equipment manufacturer.

4. If possible from cost and other considerations, select materials that offer the best resistance to stress-corrosion cracking for the given environment.

Researchers are trying to establish some form of testing standards and interpretation of data in order to determine threshold values for the many variables involved in SCC. All major material manufacturers are conducting continuous research and development work on SCC. Some materials have been developed, but until the exact scientific reason for SCC is understood, periodic inspections must still be maintained on pressure equipment in order to detect SCC at an early stage of development.

Historically, caustic embrittlement cracking on riveted boilers was one of the first indications that chemicals attacking stressed material might alter the material's normal, expected behavior under load. The chemical involved was sodium hydroxide, or caustic, used in boiler water treatment. The caustic would settle around leaking riveted joints and attack the grain boundaries of carbon steel.

Corrosion-fatigue testing specifically uses precracked specimens on the assumption that most materials have some form of invisible cracks. Specimens are immersed in flowing and static aqueous solutions or subjected to dripping liquids in order to duplicate possible service conditions. Cyclic loading is also employed.The thickness of specimens is another variable being considered.

Stainless-Steel Corrosion

Stainless steels resist corrosion over a broad range of operating conditions, but they can be attacked in some environments, such as 50 percent sulfuric and hydrochloric acids at elevated temperatures. Thus, it is still essential to match the corrosion resistance of the particular type of stainless steel to the environment it may have to operate in. There is a profusion of stainless-steel types on the market and more are being developed, along with exotic corrosion-resistant metals, such as tantalum, inconel, graphites, and zirconium. (For additional details on stainless-steel types and properties, see Appendix B.) Pressure vessels in service usually have used the following stainless steels.

Austenitic stainless steels

Sensitization.   Austenitic stainless steels may be susceptible to intergranular corrosion if any heating or cooling is performed between 800 and 1650°F. This causes the chromium along grain boundaries to combine with carbon to form chromium carbides, thereby depleting the grain boundary of chromium, and the steel becomes less resistant to attack from a corrodent along the depleted grain boundaries. This process of depletion of chromium from the grain boundaries makes the stainless steel "sensitized." Sensitization may result from annealing, welding, or stress relieving through the temperatures cited. The stainless steels especially vulnerable to sensitization are the 18-8 stainless steels, grades 304, 316, 309, and 310.

Sensitization can be reduced or avoided in one of two ways: either by maintaining a low carbon level in the alloy or by adding a strong carbide former. If the carbon level is maintained below 0.03 percent, there is limited carbon present to form the undesirable chromium carbides. These grades are referred to as the "L grades," or "extra-low-carbon grades," such as type 304L. In the second way, a strong carbide former may be added to the basic stainless-steel composition to combine with the carbon that is present. The composition of type 321 is similar to type 304, but contains titanium as the carbide former; in type 347, the carbide former is columbium.

In order to take advantage of the carbide formers that are added to types 321 and 347 stainless steel, a stabilization heat treatment is necessary. At temperatures of about 1650°F (900°C), titanium and columbium carbides will form. This temperature is too high for the formation of the undesirable chromium carbides. The titanium and columbium carbides are more stable than chromium carbide; thus, the carbon is essentially removed from the alloy, preventing the formation of chromium carbide at grain boundaries.

Chloride stress-corrosion cracking. Chloride stress-corrosion cracking can occur in stainless steels, especially the austenitic (face-centered cubic) phase. Types 304 and 316, with 9 to 11 percent nickel, are susceptible to stress-corrosion cracking because of their lower nickel content. Generally, for stainless steels to suffer stress-corrosion cracking, the chloride concentrations must be 30 ppm or higher. It is important to realize that high enough concentrations may develop locally, such as in crevices, even though the overall concentration may be below the 30 ppm. The temperature at which stress corrosion occurs is also a consideration, with most failures due to chloride stress-corrosion cracking occurring above 170°F. Some authorities claim that below 120°F SCC from chloride attack does not occur.

High-alloy stainless steels

The highly alloyed stainless steels offer a previously unavailable combination of corrosion resistance, weldability, toughness, and economy. This combination is the result of changes in melting and refining technology. Methods such as argon-oxygen decarburization (AOD), vacuum-oxygen decarburization (VOD), and vacuum-induction melting (VIM) have become standard. The use of such technology improves chromium recovery and provides excellent control of alloying conditions. Also, these methods yield low levels of carbon, nitrogen, and sulfur, resulting in excellent mechanical properties and processibility.

These new steelmaking practices have made possible *chromium-molybdenum-alloy* stainless steels that have the necessary pitting and crevice-corrosion resistance. This sets them apart from conventional stainless steels such as types 304 and 316. Type 316 shows low weight-loss corrosion in seawater; however, it suffers extensive pitting and crevice attack unless kept scrupulously clean.

The higher nickel levels of alloys such as 20Mo-6 and 20Cb-3 resist chloride SCC. At 33 percent nickel and higher there are few, if any, industrial cases of chloride SCC. Alloy 20Cb-3 (33 percent Ni) has survived 864 h without cracking in a glass-wool wick test, which concentrates chlorides drawn from a 1500-ppm solution. The alloy sample was stressed and heated to 212°F (100°C). In this test, type 304 stainless steel cracked after a 72-h exposure. Solutions such as magnesium chloride may require at least a 40 percent nickel content.

Ferritic alloys

The ferritic alloys are also very resistant. Alloy 26-1 has survived 200 h without cracking in boiling 42 percent $MgCl_2$. In comparison, type 304 failed after 8 h, and type 316 lasted only 24 h. Alloy 29-4C has not cracked in U-bend tests after 500 h in boiling 26 percent NaCl.

Duplex alloys

In tests, the duplex alloys also show strong resistance to corrosion. In U-bend tests, alloy 255 survived 30 days in a modified wick test at 212°F (100°C), using 1650 ppm NaCl plus 500 ppm $FeCl_3$.

Other Attacks on Pressure Vessel Systems

Oxidation

Oxidation of low-alloy ferritic steels at temperatures above about 900°F for extended periods of time will generate a protective coating of scale. At some minimum temperature, the scale will lose adhesion, gradually flake off, and be carried away by the working fluid or gas. This flaking off of metal, also called "exfoliation," causes a progressive thinning of the metal, usually tubes in fired service as in boilers. The flaked-off material can be quite hard and brittle and, if carried at high velocity, can cause abrasive wear on valves and nozzles and, in a power plant, to the blades of steam turbines. Most steam generator manufacturers are recommending chrome-moly steels for tubing to resist oxidation. The chromium in these grades reacts with oxygen to form a tight, adherent scale that retards oxidation at elevated temperatures.

Caustic gouging

Caustic gouging, or grooving from loss of metal, is caused by trace amounts of an alkali, such as sodium hydroxide, becoming concentrated in a pit or dent or even between a metal part and scale and attacking the metal in that area.

Erosion corrosion

Erosion corrosion is caused by erosion of material, as when solid particles sweep across a surface, causing mechanical wear and the removal of protective surface films from the metal. This permits a corrodent to attack the metal; thus the action of erosion and corrosion can produce a cumulative loss of metal in the area so affected. It is frequently noted in sharp bends of piping, on pump impellers, on agitators in mixing tanks, and in similar areas of rapid movement of liquids, gases, or solid particles. To avoid erosion corrosion, reduce velocities of flows by using larger pipe sizes, streamline bends and elbows, and consider using baffles to minimize impingement effects. Using harder and more corrosion-resistant materials is another method of avoiding erosion corrosion damage. The ends of tubing can be subjected to erosion corrosion damage; flaring the tubes can help to avoid this.

Heat Exchanger Service Effects and Problems

Heat exchangers develop some unique problems in service. In shell-and-tube heat exchangers the tubes are normally joined to tube sheets or headers by contact-rolling the tubes into holes with roller expanders, or by welding. The tube-to-tube-sheet joint can develop leaks from expansion and contraction effects, and then develop crevices in the tube sheet where crevice corrosion can begin; with high stress, stress-corrosion cracking can develop. Welded tube-to-tube-sheet joints may also develop porosity in the welds because of inadequate cleanliness in the tube weld joint before welding. Other reasons for joint leakage failures include improper rolling, poorly aligned or oval tube holes, and scoring of tube holes in a longitudinal direction.

Galvanic corrosion in heat exchangers

The galvanic corrosion most often observed in heat exchangers does not occur to the tubes themselves but rather to the surrounding tube sheet, since tubing is normally selected to be cathodic (protected) with respect to the tube sheet and water box in waters with high conductivity. Titanium tubes rolled into copper-alloy tube sheets such as Muntz metal, naval brass, and even more infrequently the relatively corrosion-resistant aluminum bronze are likely to cause deep corrosion on the anodic (corroding) copper-alloy tube sheet.

Corrosion to the copper-alloy tube sheets generally seems to be greater at inlet-side locations, where a combination of galvanic differences and water turbulence occurs. A number of coatings have been applied over tube sheets after their initial corrosion, with varying degrees of success. When the tube sheets (anode) are coated, however, small breaks in the coating can cause rapid attack to the tube sheet at this location because of the very small anode (the break in the film) and the very large cathode (the surface of the titanium tubes).

Tube erosion

Tube erosion due to vibration in heat exchangers can occur at intermediate supports of tubes. The vibration causes the tubes to rub against the intermediate support tube sheets. Impingement of high-velocity fluids or gases on tubes is another cause of tube erosion. To prevent this, impingement plates are used for all inlet nozzles on the shell side of heat exchangers. The plates are normally installed in enlarged sections of the inlet line, which provide twice the inlet-pipe area in the annular space around the plates. Impingement plates are also used for two-phase and abrasive fluids (TEMA-B requirements).

Vapor belts with the shell extended to serve as an impingement plate are used to reduce shell-side inlet velocities and to avoid flow-induced vibration.

Tube stress-corrosion cracking

Stress-corrosion cracking of tubing is a particularly troublesome, often spontaneous and unpredictable, failure mechanism that is caused by surface tensile stresses in the tubing alloy that interact with a corrosive environmental species. Stress-corrosion cracking is particularly prevalent in brass tubes exposed to ammonia or ammonium compounds. In steam power plants, for example, the necessary ammonia is produced by the aerobic deterioration of organic matter on the cooling water side of the tube or by the breakdown of cyclohexylamine, hyrazine, or morpholine to ammonia in the condensate or by the direct additions of ammonia to the condensate to control pH levels and, possibly, oxygen levels. Failure initiation is especially prevalent at locations of high tensile stress at the end of the tube-to-tube-sheet roll or at the normal tube bending location near the back of the tube sheet. Stress-corrosion cracking is very similar to fatigue cracking except that there may be more branching, or there may be a preferred intergranular or transgranular cracking direction for any one set of environmental conditions.

Stress-corrosion cracking of 304 stainless-steel feedwater heater tubes has occurred after the failure of saltwater condenser tubing has allowed the intrusion of chlorides into the power plant condensate. Pitting can also occur from chloride excursions such as this. Stress-corrosion cracking of austenitic stainless steel (304, 316, etc.) is typically transgranular, often with traces of chloride still present in the cracks, and normally requires a combination of temperatures above 150°F, chloride ions from sea- or brackish water, and residual or operating tensile stresses in the alloy.

Tube fatigue

Fatigue cracking is normally due to stress cycling (vibrations, etc.) of tubing from nearby rotating machinery, pulsing from pumps, or pulsing of steam turbine exhaust streams. In feedwater heaters, fatigue cracking often occurs at tube support locations when the tubes are exposed to high vibrations such as those caused by a nearby tube failure. In other cases, initiation of fatigue cracks occurs at sites where surface corrosion acts as a stress raiser.

Flow-induced tube vibrations

Flow-induced tube vibrations are more frequent as heat exchangers are designed to handle larger flows than in the past. By necessity, diameters of shells are kept low so that the thicknesses of the shells may be kept low. Thus, to obtain the high volumes, designers make the heat exchangers longer, thus introducing vibration problems to the long, thin tubes within the shell. Excitation forces for the vibration are usually the high velocity of fluid or gas flows across the tubes. Under partial or full loads, resonant vibration of tubes may result.

Flow-induced tube vibration has become a serious consideration in the design of shell-and-tube heat exchangers. Vibration can lead to tubes and tube joints that leak, increased shell-side pressure drop, and excess noise. The result is that the exchangers must be taken out of service for repair and modification.

Tubes are the part of shell-and-tube heat exchangers that is most likely to vibrate as the result of shell-side flow. Failure from vibration occurs from fatigue, erosion, collision between tubes, cutting by the baffles, leaking at the tube joints, etc. Tubes usually vibrate at their natural frequency. Different flow phenomena can cause tubes to vibrate, including vortex shedding, turbulent buffeting, fluid-elastic whirling, eddy formation, and acoustic vibration.

Heat exchanger designers have to solve problems such as: (1) How much vibration is too much? (2) What are the relationships between vibration amplitude, frequency, and tube life for the various tube materials that are available and the media in which the tubes may have to operate?

For critical operations, such as those using a nuclear steam generator or a vital petrochemical exchanger, modeling has been used in solving vibration problems such as determining the effects of stiffer baffling and tube-movement restraints.

Tube pitting

Another problem observed in heat exchanger tubing is pitting. The pitting can occur, for example, in carbon-steel tubes exposed to condensate, where off-line conditions, in the presence of air and moisture, produce obvious rust nodules over pits. The rust nodules characteristically, however, are more massive than the pitting underneath, and the pitting is not normally a cause of tube failure until corrodents continue the attack in the pit.

Most of the serious tube pitting occurs in brackish or saline waters. One common case is the cavernous pitting of stainless steel, usually 316 stainless steel, that occurs beneath tube deposits or under stagnant water conditions. This pitting characteristically results in small pinpoint perforations on the tube surface, while there may be extensive metal wastage inside the tube wall. Another aggravating effect is that only one pit may occur in several hundred feet of tubing, making failure predictions based on laboratory test results difficult.

Dezincification, often referred to as "parting corrosion" or "dealloying," is another form of pitting sometimes seen in aluminum-brass or admiralty tubes. This type of pitting occurs in copper alloys generally containing over 14 percent zinc, with the zinc and copper taken into solution and the copper typically redeposited back into the pit. Detection of the pitting is often simple because of the copper-colored deposits at the pit locations. Dezincification can also occur in layers and is then referred to as "layer-type dezincification."

Effects of mercury and ammonia

Another possible cause of tube failure is mercury stress-corrosion cracking of brass tubing alloys such as admiralty or aluminum brass. Use of mercury manometers or other mercury-containing products near brass tubing should be avoided. Mercury stress-corrosion cracking is seldom reported; this may be because of the difficulty in finding traces of mercury in tubing cracks owing to its high vapor pressure, or because of the inability of mercury to penetrate surface films on the tubing to attack the alloys' grain boundary regions. Stress corrosion caused by mercury occurs on the water side of tubes, such as in surface condensers.

Ammonia is often used for pH adjustment in the treatment of feedwater in power plants; amines, which are used for oxygen scavenging, also produce ammonia-type compounds. These compounds can cause ammonia washing, or general dissolution, of brass, such as admiralty and aluminum brass, and arsenical copper tubes. Ammonia washing, when observed, normally appears as a general dissolution of the tube OD near the tube sheet or tube support plate locations or under other obstructions, such as horizontal baffles, where ammonia can concentrate.

Venting

Venting from heat exchangers of noncondensable gases is necessary to prevent the possible formation of acids that can cause corrosion and stress-corrosion cracking. This is especially important for heat exchangers such as feedwater heaters. Vents should be located at each end of a horizontal feedwater heater to prevent the collection of noncondensable gases in the dead pockets of the shell. The recommendations of the Heat Exchange Institute on venting of these types of pressure vessels should be followed. The vents should be checked at least weekly for proper operation. Corrosion and stress corrosion can occur at the interface between the liquid and the trapped gas, because suspended solids can build up on the tubes at this dead-pocket point. Carbonic acid may form from trapped CO_2 gases.

Tube fouling

If tubes are not kept clean, not only can heat transfer efficiency be lost, but also deposits could cause corrosion attacks. Another problem is that some tubes may become plugged with scale, causing possible physical damage to the tube-to-tube-sheet joint due to differential thermal expansion of the metals involved. The manufacturer's recommendation on temperature drops should be observed and a scheduled program of cleaning established to avoid attacks on tubes from scale or fouling. The water side of heat exchangers used for cooling can also be subject to scaling and corrosion. Water treatment programs are available to prevent these water-side problems. These programs require testing of the water for hardness and pH and for dissolved solids, chlorides, silica, and similar contaminants. Additives are then prescribed by the water treatment specialist; these may include phosphates, chromates, and similar scale preventers or corrosion inhibitors. Acids and alkalis are used to control pH, while biocides help to control organic fouling. Adherence to the recommended treatment, plus periodic inspections to check on its effectiveness, will help in controlling water-side problems in heat exchangers. Changes in the cooling water treatment program can be instituted when inspections indicate that a scaling or corrosion problem may still exist.

Alloy Development

As service conditions in industrial processes have increased in severity, more corrosion-resistant alloys are continuously being developed. Among these are nickel-based alloys with high chromium contents of up to 16 percent, which help in resisting acid attacks. The high nickel content provides resistance to chloride stress-corrosion attack. Molybdenum (up to 16 percent) is used in alloys to resist acid attack and to prevent localized corrosion and pitting. However, the use of these alloys may have resulted in some designs of pressure vessels that are close to the minimum-thickness rules of the ASME code, thus imposing maximum permitted stress on the material. Highly stressed parts are more prone to attack from any contaminants in the working fluid or gas that the system may have to handle. It is essential to maintain a monitoring program for corrosion, cracks, and similar material attacks on these high-alloy vessels, but the frequency of inspections may be reduced.

QUESTIONS AND ANSWERS

4-1 Define "tensile strength."

ANSWER: This is the stress value obtained by dividing the maximum, or breaking, load, in pounds, of a test specimen by the original cross-sectional area of the specimen. This stress, in tension, is also called the "ultimate stress."

4-2 What is meant by the "elastic limit"?

ANSWER: This is the greatest stress that a material can have without permanent deformation after removal of the load.

4-3 Which material is stiffer, one with a higher value of E or one with a lower value?

ANSWER: Stiff materials have a high value for E, or modulus of elasticity. A high stiffness for a material implies high resistance to elastic deformation.

4-4 Define "stress" of a material.

ANSWER: *Stress* is the internal resistance of material to an applied external load and is expressed as pounds per square inch of cross-sectional area of the material.

4-5 What is the ratio stress divided by strain called?

ANSWER: The modulus of elasticity, E.

4-6 Define "endurance limit."

ANSWER: This is the maximum stress that can be applied to a material for a specific number of times (millions of cycles) without failure by fracture. For steel, 10 million cycles without failure is generally the guide used to establish endurance limits.

4-7 What is the relation of vortex shedding to possible tube vibration in a heat exchanger?

ANSWER: Where tube vibration is caused by high-velocity gas flows across the tubes, the force causing the vibration is considered to be caused by the shedding of vortices on the downstream side of the tube from the side where the gas enters the tube bank. As the vortex is shed, the flow pattern of the gases changes, and this also causes a change in the pressure distribution locally around the tube. This results in an oscillation of the gases of a magnitude and in the direction of the fluid-pressure forces acting on the tube. If the frequency of vortex shedding approaches the natural frequency of this tube section, large tube amplitudes of vibration can develop, which can eventually result in tube failures.

4-8 What are some methods of reducing the chance of tube vibration in heat exchangers?

ANSWER: Vibration may be reduced by shortening the distance of unsupported tube spans with additional intermediate supports, by reducing the velocities of flow across the tubes, and by rolling the tubes more firmly in the tube holes of intermediate supports.

4-9 What is Poisson's ratio?

ANSWER: Within the elastic limit of a material, it is the ratio of the *unit* lateral contraction to the *unit* axial elongation of the material. Tensile test specimens, when subjected to axial tension, elongate in the axial direction, but contract at the same time in the lateral direction.

4-10 A $\frac{1}{2}$-in round bar is stretched 0.00195 in in a length of 2 in and decreased 0.000162 in in diameter when a load in tension of 2000 lb is imposed. What is Poisson's ratio?

ANSWER:

Unit strain in direction of stress (elongation) $= \dfrac{0.00195}{2} = 0.000975$

Unit strain in transverse direction (contraction) $= \dfrac{0.00016}{0.5} = 0.00032$

Poisson's ratio $= \dfrac{0.00032}{0.000975} = 0.328$

4-11   Calculate the tensile strength of the specimen in the above question.

ANSWER:

$$s = \frac{P}{a} = \frac{2000}{\pi(0.25)^2} = 10{,}191 \text{ lb/in}^2$$

4-12   A specimen has a modulus of elasticity of 6.5 million. What is the maximum load axially that may be applied on a $\frac{1}{2}$-in rod without stretching the rod more than $\frac{1}{16}$-in in a length of 6 ft?

ANSWER: Use

$$E = \frac{\text{stress}}{\text{strain}}; E = \frac{(F/a)}{(e/l)}; \text{ and solve for } F$$

where F = load
   a = rod area
   e = $\frac{1}{16}$
   l = 6 × 12 = 72 in

$$F = 6{,}500{,}000 \times \frac{0.0625}{72} \times 0.0625 = 325.6 \text{ lb}$$

4-13   How much will the diameter of a steel rod 2-in in diameter change if an axial load of 60,000 is imposed, Poisson's ratio is 0.30, and $E = 30{,}000{,}000$?

ANSWER:

$$\text{Axial stress} = \frac{60{,}000}{\pi(1)^2} = 19{,}091$$

Since $E = \dfrac{\text{stress}}{\text{strain}}$

$$\text{Strain} = \frac{\text{stress}}{E}$$

$$\text{Strain axially} = \frac{19{,}091}{30{,}000{,}000} = 0.000636 \text{ in/in}$$

Transverse strain = 0.3 × 0.000636 in/in = 0.0001808

Change in 2-in diameter = 2 × 0.0001808 = 0.0003616 in

4-14   How is ductility defined?

ANSWER: The *ductility* of a metal is the amount of permanent deformation or strain that it can undergo before fracture occurs. Two methods are used to define or measure ductility: the percentage of elongation and the reduction in area of a specimen tested under tension. Welds are tested for ductility by a bend test. This consists of bending the sample by a specific amount about a plunger of given radius. The increase in distance between gage marks on the tension side of the specimen is noted, and the change is expressed as the percentage of elongation.

4-15 What are brittle materials?

ANSWER: These are materials that can be deformed very little without rupture taking place. This is usually characterized by a sudden, shattering type of failure. Cast iron, concrete, brick, and glass are examples of brittle materials.

4-16 What are resilient materials?

ANSWER: These are materials that can absorb large amounts of energy without experiencing permanent deformation. To express it another way, they return to their original shape after the load is removed. Materials with a low modulus of elasticity and high elastic limit have high resilience.

4-17 Define "tough materials."

ANSWER: Tough materials can absorb large amounts of energy before rupturing. This quality is related to a material's high strength, high ductility, or flexibility. Toughness is a useful measure of the ability of a material to absorb shock loading or sudden blows without rupturing.

4-18 What can the effect of creep be on a structural metal?

ANSWER: Creep causes planes of slow movement in a material's crystalline structure that are also a function of time and temperature. This slow movement or slip can cause sufficient deformation to cause a sudden fracture even when the applied stress is much lower than that which normally could produce a failure under normal loadings. The material has lost its elasticity at that point.

4-19 The term "stress-concentration factor" defines what ratio?

ANSWER: The *stress-concentration factor* is the ratio of the actual stress existing on a given plane of a member under loading to the calculated stress that is needed to resist that loading without taking into account the discontinuity that causes normal stresses to be magnified by the stress-concentration factor.

4-20 What is notch sensitivity?

ANSWER: *Notch sensitivity* or *notch toughness* of a metal is its resistance to the start and propagation of a crack at the base of a standardized notch. Notch sensitivity is measured by the amount of energy absorbed in foot pounds by a specimen as it fractures under impact of a hammer blow that is delivered by a standard weighted pendulum. Brittle materials will absorb little energy before fracturing.

4-21 What type of fracture normally is analyzed by the notched-bar impact test?

ANSWER: Notched-bar impact tests are used to analyze ferritic steels for possibilities of failure by brittle fracture.

4-22 Name the destructive test normally carried out on ferritic steels that provides the nil-ductility transition temperature used in determining the low-temperature ductility properties of the ferritic steel.

ANSWER: The drop-weight test is used to determine the fracture resistance of $\frac{3}{4}$-in and over thick steel at different temperatures, with the steel having a notch that can start a fracture.

4-23   How is "corrosion" defined?

ANSWER: Broadly speaking, *corrosion* is the electrochemical degradation of material caused by the reactions of the material with the surrounding environment. Corrosion may be general, or concentrated on a part of the material. There are various classes of corrosion:

General corrosion

Pitting corrosion

Galvanic corrosion

Intergranular corrosion

Concentrated cell corrosion

Stress-corrosion cracking

4-24   What is a weight-loss corrosion test?

ANSWER: This is a test to establish the suitability of a material for the service intended, or to determine if a material is suitable for the plant's future needs, or to predict the remaining life of a material in service. A clean coupon of the material is measured, weighed, and then exposed to the atmosphere, liquids or gases, or a process stream for a given length of time. The specimen is removed, cleaned of deposits and corrosion products, and reweighed. From this, a corrosion rate can be determined and future life predicted, based on minimum required thickness.

4-25   Explain selective leaching.

ANSWER: Selective leaching is the removal of one element of an alloyed metal, as in dezincification, where zinc is leached out of a copper-zinc alloy to yield a remaining porous metal with poor mechanical properties.

4-26   Describe cathodic and anodic corrosion protection.

ANSWER: When an electrical current flows between anodes and cathodes on a corroding metal surface, the higher the current, the faster the anode areas corrode. By using external circuits, it is possible to impose additional currents on the metal and so change and control its rate of corrosion. That is, one can apply an opposing current to nullify corrosion (cathodic protection) or, in some cases, adjust the metal's potential so that the metal still corrodes but more slowly because it is passive (anodic protection).

There are two types of cathodic protection: impressed current and galvanic or sacrificial anode. The former is accomplished by connecting a large enough dc source to the corroding metal and an electrode that becomes the anode. Galvanic protection uses a more active metal, called a sacrificial anode, to supply the current needed to stop corrosion. However, the cathodic reaction on the surface being protected may produce hydrogen, which can be deleterious in some applications (paint flaking, hydrogen embrittlement). Also, the technique has limited application in complex structures. Neighboring metal equipment may be preferentially corroded as a result of stray currents.

Anodic protection does not stop corrosion completely but reduces it to a very low rate by maintaining a certain potential. Laboratory work is needed to determine this potential and the currents required to achieve and maintain it.

4-27   What alternative method exists to combat corrosion without the use of an expensive alloy?

ANSWER: In the pressure vessel field, cladding or overlaid weld metal of stainless steel is applied to a less expensive metal such as carbon steel. Much work is also being carried out on coatings and different metal sprays in order to reduce the costs of corrosion. Glass and ceramic linings are also used.

4-28   Explain chloride stress-corrosion cracking of stainless steel.

ANSWER: There are two theories about the mechanism of chloride stress corrosion. One theory says that hydrogen, generated by the corrosion process, diffuses into the base metal in the atomic form and embrittles the lattice structure.

A second, more widely accepted, theory proposes an electrochemical mechanism. Stainless steels are covered with a protective oxide film. The chloride ions rupture the film at weak spots, resulting in anodic (bare) and cathodic (film-covered) sites. The galvanic cell produces accelerated attack at the anodic sites, which when combined with tensile stresses produces cracking.

The minimum stress level required for chloride SCC for 300 series stainless steel is extremely low. Most of the commonly used fabrication techniques (welding, cold-drawing, rolling, bending, or crimping) will produce residual stress exceeding the minimum value. Annealing to relieve the residual stresses is seldom effective. Heating to 1000 to 1200°F is required to stress-relieve, within a reasonable time period. Heating in this temperature range, however, severely reduces corrosion resistance. This is because chromium carbides precipitate in the grain boundaries, producing a sensitized structure, which is very undesirable in most corrosive environments. Heating to 1700 to 1900°F prevents sensitization, but may cause other problems. Proper alloying with titanium or columbium will prevent this problem.

4-29   Define "fracture toughness."

ANSWER: *Fracture toughness* is a measure of a material's resistance to brittle fracture. Fracture mechanics is a rapidly developing method used to predict the resistance to fracture of a material by assuming that a discontinuity exists and then calculating crack growth rates to the point of fracture.

4-30   Describe the term "solution anneal" as it may relate to sensitized austenitic steels.

ANSWER: This is a method of controlling intergranular corrosion susceptibility in sensitized austenitic stainless steel through a solution anneal and quench heat treatment. It consists of heating the stainless steel to a temperature of 1950 to 2050°F, maintaining that temperature long enough for all the carbides to go into solid solution, and then quenching or otherwise cooling the steel quickly enough through the sensitization temperature range to avoid reprecipitation of carbides. This treatment will provide a stainless steel with a single solid solution phase not susceptible to intergranular corrosion attack, unless it is again heated to the sensitization temperature range, as by welding.

4-31 Where does fatigue cracking occur in heat exchangers?

ANSWER: Fatigue cracking in heat exchangers is generally due to repetitive stress, usually caused by tube vibration. It occurs adjacent to a support plate or in the span between support plates. Corrosion can accelerate fatigue cracking by establishing a notch or discontinuity. Fatigue cracking in heat exchangers can be reduced by proper spacing of supports, which limits tube vibration.

4-32 How can crack growth be minimized?

ANSWER: Cyclic service for certain designs of pressure vessels should be kept reasonably low for the life of the vessel. The pressure vessel system should be kept clean of possible corrosive agents. Finally, if the service is corrosive, proper material selection must be made, including during welding and repairs.

One of the best means by which designers can supplement analytical design procedures to minimize crack growth in engineering alloys is through proper materials selection. Both stress-corrosion cracking and fatigue crack growth can be strongly influenced, for good or ill, by metallurgical factors. Factors such as alloy chemistry, melting practice, and heat treatment can have pronounced effects on crack growth resistance.

Periodic inspections for crack formation will assist in preventing serious accidents.

Chapter

5

Minimum Code Strength
Calculations

In order to be able to analyze the physical conditions found after a pressure vessel has been in service, it is necessary to return to the ASME minimum code requirements for determining the pressure vessel's remaining strength to resist the pressures that may be imposed on it in future service. This is a must with a pressure vessel that has developed wear and tear from service. Other codes such as those of TEMA may also have to be used in evaluating tube conditions in a heat exchanger, as an example. These references, and others, are excellent sources for reviewing some basic strength considerations and thus making a safe decision on how to proceed. Thus, the codes can be used for making judgments on the condition of equipment in service; they were not developed strictly for new-construction requirements.

Plant-Level Use of Codes

Plant operating, inspection, and maintenance personnel involved with pressure vessel systems should have a working knowledge of how to determine some minimum strength requirements for pressure vessel systems. The previous chapter gave a brief review of the possible stresses that can be imposed on pressure vessel equipment because of internal pressures; bending, "locked-up," stresses from welding; and similar considerations. The basic minimum stress calculations to be reviewed in this chapter conform to minimum code requirements. These will generally involve the following parts of a pressure vessel: tubes, pipes, shells, heads, braced and stayed surfaces, openings and reinforcements, layered construction, extra-thick vessels, jacketed vessels, bolted flange connections, supports, and parts subject to wind loads. The reader is referred to more advanced pressure vessel texts on design calculations from a manufacturing consideration. This chapter is meant to serve as a guide for plant personnel who may be confronted with service problems on pressure vessel equipment and must determine if the vessel is safe to operate, requires repair, or must be replaced. It is the practical application of minimum strength calculations that is being stressed, and not the analysis that is applied by research and development groups, or by the designers involved with new construction, although, in practice, there is a need to consult with designers when unusual plant conditions of abnormal wear or failures materialize.

The *ASME Boiler and Pressure Vessel Code* is continuously being modified as new materials, quality control techniques, and methods of testing material behavior are developed; therefore, the latest edition of the code should be referred to in order to keep current on the latest code requirements. This chapter is devoted to calculations for Section VIII, Division 1, pressure vessels, and not to the more advanced methods for Division 2 vessels.

Allowable Pressure

Any pressure vessel and parts confining pressure must be analyzed per component by carefully considering the strength of the material being used, its physical characteristics as to type and grade, allowable stress, thickness, etc. The forces acting on this material must then be analyzed. These forces are usually created by pressure, but may also include temperature, the weight being supported, and stress concentration, as around an opening. The problem then evolves to comparing the forces acting on the material and determining whether the material is being stressed beyond the allowable stresses governed by the ASME code rules. See Tables 5.1 and 5.2. Elements to be considered depend on the type of vessel but will generally include shells or drums, tubes, tube sheets, heads, flat surfaces, stays, stay bolts, openings, riveted or welded joints, structural supports, and connected piping and valves. Each of these is governed by ASME code rules as to allowable material, allowable stresses, and method of calculating forces to obtain the allowable pressure. Finally, in boiler and pressure vessel application, the weakest element producing the lowest pressure then determines the *allowable pressure* for the vessel.

TABLE 5.1 ASME Allowable Stresses for Some Carbon-Steel Plates and Sheets at Various Temperatures

Spec. no.	Grade	Nominal composition	P no.	Group no.	Specified min. yield, ksi	Specified min. tensile, ksi	Maximum allowable stress, ksi,* for metal temp., °F, not exceeding:											
							−20 to 650	700	750	800	850	900	950	1000	1050	1100	1150	1200
SA-36	...	C–Mn–Si	1	1	36.0	58.0	12.7											
SA-283	A	C	1	1	24.0	45.0	10.4											
	B	C	1	1	27.0	50.0	11.5											
	C	C	1	1	30.0	55.0	12.7											
	D	C	1	1	33.0	60.0	12.7											
SA-285	A	C	1	1	24.0	45.0	11.3	11.0	10.3	9.0	7.8	6.5						
	B	C	1	1	27.0	50.0	12.5	12.1	11.2	9.6	8.1	6.5						
	C	C	1	1	30.0	55.0	13.8	13.3	12.1	10.2	8.4	6.5						
SA-299	...	C–Mn–Si	1	2	40.0/42.0	75.0	18.8	17.7	15.7	12.6	9.6	6.5	4.5	2.5				
SA-414	A	C	1	1	25.0	45.0	11.3	11.0	10.3	9.0	7.8	6.5						
	B	C	1	1	30.0	50.0	12.5	12.1	11.2	9.6	8.1	6.5						
	C	C	1	1	33.0	55.0	13.8	13.3	12.1	10.2	8.4	6.5						
	D	C–Mn	1	1	35.0	60.0	15.0	14.3	12.9	10.8	8.6	6.5						
	E	C–Mn	1	1	38.0	65.0	16.2	15.5	13.8	11.4	8.9	6.5						
	F	C–Mn	1	2	42.0	70.0	17.5	16.6	14.7	12.0	9.2	6.5						
	G	C–Mn	1	2	45.0	75.0	18.8	17.7	15.7	12.6	9.6	6.5						

* Multiply by 1000 to obtain psi.
SOURCE: ASME code, Section VIII, Division 1.

TABLE 5.2   ASME Allowable Stresses for Some Alloyed Steel Material

Spec. no.	Nominal composition	P no.	Group no.	Product form	Grade	Notes*	Specified min. yield, ksi	Specified min. tensile ksi
SA-240	12Cr–1Al	7	1	Plate	405	(7)	25.0	
SA-268	12Cr–1Al	7	1	Smls. Tb.	TP405	(7)	30.0	60.
SA-479	12Cr–1Al	7	1	Bar	405	(7)	25.0	
SA-240	13Cr	7	1	Plate	410S	...	30.0	60.
SA-268	13Cr	6	1	Smls. Tb.	TP410	...		60.
SA-268	12Cr–1Al	7	1	Wld. Tb.	TP405	(4)(11)	30.0	60.
SA-268	13Cr	6	1	Wld. Tb.	TP410	(4)(11)	30.0	60.
SA-268	15Cr	6	2	Wld. Tb.	TP429	...	35.0	60.
SA-268	17Cr	7	2	Wld. Tb.	TP430	(4)(11)	35.0	60.
SA-268	11Cr–Ti	7	1	Wld. Tb.	TP409	(4)(11)	30.0	60.
SA-268	11Cr–Ti	7	1	Smls. Tb.	TP409	...	30.0	60.
SA-268	18Cr–Ti	7	2	Wld. Tb.	TPXM-8	(4)(11)(7)	30.0	60.
SA-268	18Cr–Ti	7	2	Smls. Tb.	TPXM-8	(7)	30.0	60.
SA-240	18Cr–Mo	7	2	Plate	18Cr–2Mo	(12)	45.0	60.
SA-268	18Cr–Mo	7	2	Wld. Tb.	18Cr–2Mo	(4)(11)(12)	45.0	60.
SA-268	18Cr–Mo	7	2	Smls. Tb.	18Cr–2Mo	(12)	45.0	60.
SA-240	13Cr	6	1	Plate	410	...	30.0	65.
SA-240	15Cr	6	1	Plate	429	(7)	30.0	65.
SA-240	17Cr	7	2	Plate	430	(7)		
SA-479	13Cr	6	1	Bar	410	...	40.0	70.
SA-182	13Cr	6	1	Forg.	F6aCl.1	...		
SA-182	13Cr	6	3	Forg.	Cl.F6aCl.2	...	55.0	85.
SA-217	13Cr	6	3	Cast.	CA15	(6)(7)	65.0	90.
SA-193	13Cr	...	...	Bolt.	B6(410)	(5)	85.0	110.
SA-268	15Cr	6	2	Smls. Tb.	TP429	...	35.0	60.
SA-268	17Cr	7	2	Smls. Tb.	TP430	(7)		
SA-479	17Cr	7	2	Bar	TP430	(7)	40.0	70.
SA-479	18Cr–Ti	7	2	Bar	TPXM-8	(7)		
SA-268	26Cr–4Ni–Mo	10E	5	Wld. Tb.	TP329	(4)(13)(14)	70.0	90.
SA-268	26Cr–4Ni–Mo	10E	5	Smls. Tb.	TP329	(13)(14)	70.0	90.
SA-240	26Cr–4Ni–Mo	10E	5	Plate	TP329	(13)	70.0	90.
SA-268	27Cr	10E	5	Smls. Tb.	TP446	...	40.0	70.
SA-412	17Cr–4Ni–6Mn	8	1	Plate	201	...	45.0	95.
SA-182	18Cr–8Ni	8	1	Forg.	F304L	(1)	25.0	65.
SA-240	18Cr–8Ni	8	1	Plate	304L	(1)	25.0	70.
SA-213	18Cr–8Ni	8	1	Smls. Tb.	TP304L	(1)		
SA-312	18Cr–8Ni	8	1	Smls. Pp.	TP304L	(1)		
SA-479	18Cr–8Ni	8	1	Bar	304L	(1)		

* Refer to the ASME code for the notes.

† Multiply by 1000 to obtain psi.

SOURCE: ASME code, Section VIII, Division 1.

								Maximum allowable stress, ksi†, for metal temp., °F, not exceeding:										
	200	300	400	500	600	700	750	800	850	900	950	1000	1050	1100	1150	1200	1250	1300	1350
.0	14.3	13.8	13.3	12.9	12.4	12.1	11.7	11.1	10.4	9.7	8.4	4.0							
.0	14.3	13.8	13.3	12.9	12.4	12.1	11.7	11.1	10.4	9.7	8.4	6.4	4.4	2.9	1.8	1.0			
.8	12.2	11.8	11.3	10.9	10.6	10.3	9.9	9.4	8.8	8.2	7.1	3.4							
.8	12.2	11.8	11.3	10.9	10.6	10.3	9.9	9.4	8.8	8.2	7.1	5.5	3.7	2.4	1.5	0.8			
.7	12.1	11.7	11.3	10.9	10.5	10.2													
.8	12.2	11.8	11.3	10.9	10.6	10.3	9.9	9.4	8.8	8.2	7.2	5.5	3.8	2.7	2.0	1.5			
.8	12.2	11.8	11.3	10.9	10.5	10.2	9.9	9.4											
.0	14.3	13.8	13.3	12.9	12.4	12.1	11.7	11.1											
.8	12.2	11.8	11.3	10.9	10.6	10.3	9.9	9.4											
.0	14.3	13.8	13.3	12.9	12.4														
.0	14.3	13.8	13.3	12.8	12.4														
.8	12.2	11.8	11.3	10.9	10.5														
.0	14.3	13.8	13.3	12.8	12.4														
.3	15.5	15.0	14.4	13.9	13.5	13.1	12.7	12.0	11.3	10.5	8.8	6.4	4.4	2.9	1.8	1.0			
3	15.5	15.0	14.4	13.9	13.5	13.1	12.7	12.0	11.3	10.5	9.2	6.5	4.5	3.2	2.4	1.8			
2	15.4	14.9	14.4	13.9	13.4	13.1	12.6	12.0	11.2	10.4	8.8	6.4							
.3	20.3	19.6	18.9	18.2	17.6	17.1	16.5	15.7	14.4	12.3	8.8	6.4	4.4	2.9	1.8	1.0			
5	21.5	20.7	20.0	19.3	18.7	18.1	17.5	16.7	14.9	11.0	7.6	5.0	3.3	2.2	1.5	1.0			
2	21.2	21.2	21.2	21.2	21.2	21.2	21.2	19.5	15.6	12.0									
.0	14.3	13.8	13.3	12.9	12.4	12.1	11.7	11.1	10.4	9.7	8.5	6.5	4.5	3.2	2.4	1.8			
5	16.6	16.1	15.5	15.0	14.5	14.1	13.6	12.9	12.1	11.0	9.2	6.5							
1	19.1	18.4	18.0	18.0															
5	22.5	21.6	21.2	21.2															
5	21.9	20.5	19.8	19.8															
5	16.6	16.1	15.6	15.0	14.5														
8	20.8	19.1																	
5	15.4	14.2	13.6	13.4	13.3	13.1	13.0	12.9											
7	15.7	15.3	14.7	14.4	14.0	13.5	13.3	13.0											

Allowable Stress

Allowable stresses are used in the design of structures or machine components. The allowable stress is also sometimes called the "allowable working stress." It is the maximum stress considered to be safe when the material is subjected to resisting loads assumed to be comparable to those applied in service. In vessel applications, the term "allowable pressure" is often used. Actually, the allowable pressure is determined by applying the forces acting on a material and then calculating the allowable pressure from the allowable stress on the material.

Safety Factor

In pressure vessel design and usage where life and property may be at stake (see Fig. 5.1), there is a definite need for selecting working stresses and loadings considerably less than the ultimate, or yield, stress for these reasons:

1. There is always some uncertainty in the materials being used: how they were made, how they were assembled, and how they were joined or fabricated with other materials.

Figure 5.1   This acetone storage tank collapsed owing to vacuum. (*Courtesy of the Hartford Steam Boiler Inspection and Insurance Co.*)

2. There is always some uncertainty as to the exact loading that a structure, or part of it, may have to resist and how it is abused in operation.

3. Calculations of all stresses possible in a fabricated structure are never that exact when one considers the variables to be encountered in service through the years.

The allowable stress is determined from the ultimate strength of the material, which is divided by a safety factor. The safety factor used in modern vessels is 4. However, certain elements of a boiler, such as rivets, have to be designed with a safety factor of 5. Other parts have to be designed with a safety factor as high as 12.5, such as the rivets holding lugs on brackets on a horizontal return tubular (HRT) boiler to be suspended from a beam.

In ASME code usage, the factor of safety is the ultimate strength divided by the allowable loadings, or the ultimate stress S_u divided by the allowable stress S_a. In equation form,

$$\text{Safety factor} = \frac{S_u}{S_a}$$

Another method of expressing the safety factor is by dividing the bursting pressure by the allowable pressure. This method is used on state inspection reports and on ASME data reports. In equation form, it is

$$\text{Safety factor} = \frac{\text{bursting pressure}}{\text{allowable pressure}}$$

Example If a vessel is stamped for an allowable pressure of 275 psi and the safety valve is set at 150 psi, what is the safety factor? Assume a code-welded vessel meeting latest code requirements and an *original* design safety factor of 4.

answer The bursting pressure of this vessel would be 4 × 275 = 1100 psi. So,

$$\text{Safety factor} = \frac{\text{bursting pressure}}{\text{allowable pressure}} = \frac{1100}{150} = 7.33$$

Analyzing Pressure Vessel Component Strength

The strength of each component of a pressure vessel needs to be analyzed individually. For plant people evaluating the condition of vessels in service, it is necessary to review the strength of only the component that has a problem, provided all the other components were designed and built to code requirements and show no problems during inspections.

Tubes

Tubes are extensively used for heat transfer in pressure vessel systems and may be bare or finned. The latter are extensively used in air-cooled exchangers, such as condensers for refrigeration and air-conditioning equipment. The variety of available tube materials makes the consideration of pressures, temperatures, corrosion, flow velocities, operating cycles, and similar variables important in tube material selection. Table 5.3 provides some physical characteristics of brass, copper, and steel tubes.

Effect of high-velocity flow. If the design conditions for flow through a tube are consistently exceeded, abnormal tube wear, and therefore tube thinning, can be expected. The constant C can be used in the equation shown at the bottom of Table 5.3 as developed by the Heat Exchange Institute for closed feedwater heaters. This organization recommends a maximum fluid velocity through a tube at normal full-load operating conditions of 8 ft/s at 60°F. The formula provides for a consideration of other temperatures by using the specific gravity of the liquid. This formula is useful in checking what the effect on a feedwater heater would be if it had to handle the water flow of another heater, in addition to its normal flow, because of a plant problem.

Example If a single-pass feedwater heater has 600 $\frac{5}{8}$-in, 12-BWG copper tubes handling feedwater under normal load conditions with a specific gravity of 0.622 and a flow of 500,000 lb/h, what would be the fluid velocity in each tube if the heater had to pass double its normal flow of 500,000 lb/h?

answer Use the equation

$$\text{Liquid velocity in feet per second} = \frac{\text{pounds per hour per tube}}{C \times \text{specific gravity of liquid}}$$

From Table 5.3, $C = 202$ for $\frac{5}{8}$-in, 12-BWG tubes. Each tube is assumed to handle 1/600 of total flow. Substituting,

$$\text{Liquid velocity} = \frac{\dfrac{(2 \times 500,000)}{600}}{202(0.622)} = 13.26 \text{ ft/s}$$

which is above the recommended 8 ft/s. Abnormal tube wear can be expected to result if this high flow is maintained for any length of time.

Tube manufacturing. Three common methods of tube fabrication are used. (1) The seamless tube is pierced hot and drawn to size. (2) The lap-welded (forge-welded) tube consists of a metal strip ("skelp") curved to tubular shape with the longitudinal edges overlapping. Heat is applied and the joint forge-welded. (3) The electric-resistance butt-welded tube is formed as with the second type, but, as its name implies, the joint is butt-welded.

Tubes may be expanded into tube sheets or drums and then beaded or flared, or the tubes may be welded to the tube sheet. It is the usual practice to drill tube holes $\frac{1}{64}$ in greater than the nominal outside diameter of the tube for ease of tube insertion into tube sheets, baffles, and intermediate supports. Triangular-pitch tube layout is recommended with tubes having a minimum center-to-center spacing equal to the tube diameter plus $\frac{3}{16}$ in, or 1.25 times the nominal OD, whichever is greater. The ASME code on pressure vessel tubes and tube-sheet arrangement depends on standards of the heat exchanger industry.

The ASME code requires using the shell equation for determining allowable internal pressures, but the factor SE in the code equation is replaced by the lower allowable stresses given in the code for welded tubes. This is relevant to tube manufacturing, not to joints. Note that the diameter of *tubes* always refers to nominal outside diameter, while *pipe* diameter refers to nominal inside diameter.

Calculating allowable tube pressure. To calculate the allowable internal pressure for a tube in an unfired pressure vessel, use

$$P = \frac{St}{R + 0.6t}$$

where P = maximum allowable pressure
t = required thickness of tube (without corrosion allowance)
R = inside radius of tube
S = Allowable stress, lb/in², for tube material at mean tube temperature

Example An 80-20 copper-nickel tube of seamless construction originally had an OD of $\frac{3}{4}$ in, a BWG of 10, and a nominal thickness of 0.134 in. Eddy-current testing of the tubes revealed that thinning occurred on OD surfaces, so that the remaining thickness in some areas was 0.055 in minimum. The mean tube temperature is under 400°F. The code specifies allowable stress of 9900 lb/in². What is the allowable pressure on the tube?

answer Using the above equation, with $t = 0.055$, $S = 9900$, and $R = 0.482/2 = 0.241$ in (from Table 5.3), and substituting

$$P = \frac{9900(0.055)}{0.241 + 0.6(0.055)} = 1987 \text{ psi}$$

It should be noted that this material has a specified minimum tensile strength of 45,000 lb/in², but at 400°F, the allowable stress is only 9900 lb/in². Nonferrous metals have a steep decline in allowable stresses as temperatures climb. The maximum mean metal temperatures usually recommended for commonly used tube material are:

TABLE 5.3 Properties of Brass, Copper, and Steel Tubes

OD of tubing	BWG gage	Thickness, in	Internal area, in²	External surface per ft length, ft²	Internal surface per ft length, ft²	Weight per ft length—brass, lb	Weight per ft length—copper, lb	Weight per ft length—steel, lb	ID of tubing, in	Constant, C*	OD/ID	Area metal (transverse metal area)
5/8	10	.134	.100	.1636	.0937	.76	.800	.703	.357	155	1.750	.207
5/8	11	.12	.1170	.1636	.101	.70	.735	.647	.385	182	1.62	.190
5/8	12	.109	.130	.1636	.107	.649	.683	.605	.407	202	1.53	.177
5/8	13	.095	.149	.1636	.1142	.581	.61	.54	.435	232	1.44	.158
5/8	14	.083	.165	.1636	.1205	.520	.548	.481	.458	257	1.36	.142
5/8	15	.072	.1817	.1636	.126	.460	.482	.43	.481	283	1.30	.125
5/8	16	.065	.1924	.1636	.1295	.420	.44	.39	.495	300	1.26	.114
5/8	18	.049	.2181	.1636	.1379	.326	.342	.301	.527	340	1.19	.0887
5/8	20	.035	.2419	.1636	.146	.238	.25	.22	.555	377	1.13	.0649
3/4	10	.134	.1822	.1963	.1265	.95	1.04	.882	.482	284	1.56	.260
3/4	11	.120	.2043	.1963	.1335	.87	.918	.81	.510	319	1.47	.238
3/4	12	.109	.223	.1963	.14	.807	.845	.75	.532	348	1.41	.219
3/4	13	.095	.247	.1963	.147	.718	.752	.67	.56	385	1.34	.195
3/4	14	.083	.268	.1963	.153	.640	.67	.591	.584	418	1.28	.174

OD	BWG											
3/4	15	.072		.1963	.159	.565	.392	.522	.606	451	1.24	.153
3/4	16	.065	.302	.1963	.163	.514	.54	.48	.620	471	1.21	.140
3/4	17	.058	.314	.1963	.166	.463	.494	.429	.634	490	1.18	.128
3/4	18	.049	.334	.1963	.1707	.396	.417	.367	.652	521	1.15	.108
3/4	20	.035	.3632	.1963	.179	.289	.306	.267	.680	567	1.10	.0786
7/8	10	.134	.2890	.2297	.158	1.134	1.24	1.06	.607	451	1.44	.312
7/8	11	.12	.317	.2297	.167	1.04	1.096	.968	.635	494	1.38	.284
7/8	14	.083	.394	.2297	.186	.759	.798	.702	.709	615	1.23	.207
7/8	16	.065	.436	.2297	.1952	.608	.636	.562	.745	680	1.17	.165
7/8	18	.049	.474	.2297	.2034	.467	.49	.432	.777	740	1.12	.127
7/8	20	.035	.508	.2297	.210	.340	.357	.314	.805	793	1.09	.0927
1	10	.134	.421	.2618	.192	1.34	1.41	1.24	.732	657	1.37	.364
1	11	.12	.455	.2618	.200	1.22	1.28	1.13	.76	710	1.32	.3317
1	12	.109	.479	.2618	.205	1.12	1.18	1.04	.782	748	1.28	.306
1	13	.095	.515	.2618	.213	.99	1.042	.92	.810	804	1.24	.270
1	14	.083	.5463	.2618	.2183	.88	.923	.813	.834	852	1.20	.239
1	15	.072	.576	.2618	.225	.77	.813	.714	.856	899	1.17	.209
1	16	.065	.594	.2618	.228	.70	.737	.65	.870	927	1.15	.191
1	18	.049	.639	.2618	.236	.54	.568	.50	.902	997	1.11	.146
1	20	.035	.679	.2618	.243	.391	.409	.361	.930	1060	1.07	.106

* Liquid velocity in feet per second = $\dfrac{\text{pounds per tube per hour}}{C \times \text{specific gravity of liquid}}$; specific gravity of water at 60°F = 1.0.

SOURCE: Courtesy of the Heat Exchange Institute.

Material	Maximum metal temperature, °F
Arsenical copper	400
Admiralty metal—types B, C, D	450
90-10 copper-nickel	600
80-20 copper-nickel	700
70-30 copper-nickel, annealed	700
70-30 copper-nickel, stress-relieved	800
70-30 nickel-copper (Monel), annealed	900
70-30 nickel-copper (Monel), stress-relieved	800
Carbon steel	800
Stainless steel	800

Tube-sheet thickness

Bending. The holding power of rolled or welded tube-to-tube-sheet joints must be considered, since they are subject to shearing forces. This problem is the reverse of that found with a press fit. Tube-sheet thickness requirements are not clearly covered by the ASME code. Tube sheets are subjected to bending and can be viewed as flat surfaces requiring staying if not thick enough. Those areas of the tube sheet where no tubes exist, and thus where no staying is provided by the tubes, can be treated as unstayed areas, and the flatplate equations of the code can be used to calculate the minimum thickness that may be required. The Heat Exchange Institute and the Tubular Exchange Manufacturer Association (TEMA) have equations to calculate tube-sheet thicknesses. For illustration purposes, the Heat Exchange Institute equation for bending is:

$$t_b = \frac{FG}{2}\sqrt{\frac{P}{S}}$$

where t_b = thickness (in) of tube sheet required to resist bending
F = design factor; for nonintegral tube sheets, $F = 1.0$ for floating heads and 1.25 for heads with U tubes; for heaters with tube sheets integral with channels, F can be determined from Fig. 5.2
P = allowable working pressure, psi
S = allowable stress of tube-sheet material, lb/in²
G = mean gasket diameter or, if tube sheet is integral with channel, the inside diameter of channel, in

Example A heat exchanger has U tubes operating at a design pressure of 1500 psi. The nonintegral tube sheet is made of material with a code-allowable stress of 11,500 lb/in² and has a mean gasket diameter of 24 in. Inspection disclosed extensive corrosive thinning on the tube sheet, with an average thickness of 3.75 in noted. What is the allowable pressure on this heat exchanger by the Heat Exchanger Institute equation so that the maximum allowable stress in bending is not exceeded as a result of the tube-sheet thinning?

answer Using the above equation, with $t_b = 3.75$ in, $F = 1.25$, $G = 24$ in, $S = 11,500$ lb/in²,

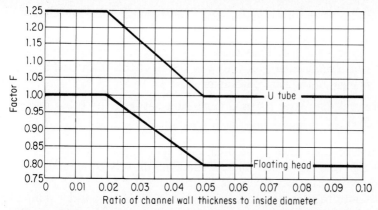

Figure 5.2 Heat Exchange Institute's recommended design factor multiplier to be used in determining the minimum thickness required to resist bending stress in an integral tube sheet.

$$3.75 = \frac{1.25(24)}{2} \sqrt{\frac{P}{11,500}}$$

$$\left(\frac{3.75}{15}\right)^2 = \frac{P}{11,500}$$

$$P = 0.0625(11,500) = 718.8 \text{ psi}$$

The pressure on this heat exchanger would have to be reduced, or, to restore the vessel's allowable pressure to 1500 psi, the tube sheet would have to be replaced. Staying of the tube sheet could also be employed to strengthen the structure.

Shear. The tube-sheet thickness must also be evaluated in terms of shear strength, using the following Heat Exchange Institute equation, and the largest t used:

$$t_s = \frac{AP}{LS_s}$$

where t_s = thickness of tube sheet (in shear) measured at the outer tube limit of area A, in

A = area with largest ratio of area to net metal perimeter, in²

L = net metal perimeter at outer tube center of area; this is the perimeter of area A less number of tube holes in perimeter multiplied by diameter of tube hole, in

S_s = allowable shear stress for tube-sheet material at design temperature. This allowable shear stress is taken as the allowable tensile-stress values tabulated in the ASME code, reduced by the appropriate factors, lbs/in²

Stresses on tube-to-tube-sheet joints. In investigating tube-to-tube-sheet joint leakages, it is necessary to consider the stresses that this joint may be subjected to, such as:

1. Differential pressure stress that exists between the tube- and shell-side fluids

2. Axial stresses on the tubes caused by tension, or compression, as a result of thermal expansion and contraction effects from load or operating swings on the heat exchanger

3. Differential diametric thermal expansion between the tubes and the tube sheet

The tube-to-tube-sheet joint must seal against interleakage between the tube-side and shell-side fluids. Depending on service, in addition to roller expanding of tubes to assure tightness, seal welding and strength-welded tube-to-tube-sheet joints are used, being sure to consider compatibility of the tube metal being welded with the tube-sheet metal. The welding procedure to be used must be qualified to prove that the different metals can be welded together in a joint.

Shells

Field shell calculations.   More field calculations for allowable pressures on shells are made than on any other component. The shell essentially receives the most wear and tear in service and generally is a high-stress component of the pressure vessel.

Most jurisdictions have an equation for those pressure vessels where no code stamping can be found. This equation refers to a shell or cylinder of the vessel. The existing-installation equation for shells is

$$P = \frac{TS \times t \times \%}{FS \times R}$$

where TS = ultimate tensile strength of shell material, lb/in^2
     t = plate thickness, generally minimum, in
     % = efficiency of joint in shell
     FS = factor of safety
     R = inside radius of shell or drum

The factor of safety can range from 4 to 6, depending on the joints in the vessel (welded or riveted, lap- or butt-constructed if riveted), the age of the unit, requirements for periodic hydrostatic testing to prove that the vessel is tight, and similar considerations. A licensed or commissioned boiler inspector of the jurisdiction in which the vessel is located should be consulted.

Code shell calculations.   In general, when calculating allowable pressures for a vessel that has seen service, the best rule to follow is to use the original code method used when the vessel was constructed. The examples to be cited will illustrate current code equations and methods.

Cylindrical shells. In calculating the allowable pressure on shells, it is necessary to make two calculations in accordance with 1972 code changes. One calculation is for the longitudinal joint, using the corresponding welding joint efficiency for that joint. Welding efficiencies are determined from the code and are established by the category of the joint and the amount of radiographic examination employed in checking the quality of the weld. See Fig. 5.3 and Table 5.4. The second calculation involves determining the allowable pressure on the circumferential joints that connect shells to each other, or shells to heads. The category of the joint and the degree of x-ray examination of the weld will determine the weld efficiency to use on the circumferential joint.

Full radiographic examination is required on all butt welds in vessels used to contain lethal substances, in vessels with plate thicknesses at the welded joint that exceed $1\frac{1}{2}$ in, and for other joints specified in the code.

The two shell equations that the code specifies are:

1. For longitudinal welded or seamless joints

$$P = \frac{SEt}{R + 0.6t}$$

2. For circumferential joints on the shell

$$P = \frac{2SEt}{R - 0.4t}$$

where P = maximum allowable internal working pressure, psi
R = inside radius of shell, in
t = minimum required thickness, in, exclusive of corrosion allowance
E = welded or seamless joint efficiency per code; for ligament construction use ligament efficiency
S = maximum allowable stress permitted by the code, lb/in²

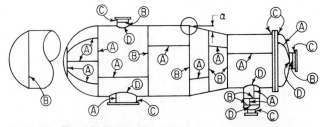

Figure 5.3 The ASME classifies welded-joint *locations* into categories A, B, C, and D, as illustrated. Category A requires welded butt joints of the double-welded type. Category B requires double butt or single-welded butt joints with backing strip. Categories C and D require full-penetration welds on joints. (*Courtesy of the American Society of Mechanical Engineers.*)

TABLE 5.4 ASME-Recommended Arc- and Gas-Welded Joint Efficiencies to Be Used in Code Shell Strength Calculations

No.	Type of joint	Limitations	Degree of examination		
			Fully radiographed* (a)	Spot-examined† (b)	Not spot-examined‡ (c)
1.	Butt joints as attained by double-welding or by other means that will obtain the same quality of deposited weld metal on the inside and outside weld surfaces so as to agree with the requirements of UW-35. Welds using metal backing strips that remain in place are excluded.	None	1.00	0.85	0.70
2.	Single-welded butt joints with backing strips other than those included under (1).	(a) None except as in (b) below. (b) Butt weld with one plate offset—for circumferential joints only, see UW-13(c) and Fig. UW-13.1(k).	0.90	0.80	0.65
3.	Single-welded butt joints without backing strip.	Circumferential joints only, not over $\frac{5}{8}$ in thick and not over 24 in outside diameter.	...	...	0.60
4.	Double full-fillet lap joints.	Longitudinal joints not over $\frac{3}{8}$ in thick; circumferential joints not over $\frac{5}{8}$ in thick.	...	...	0.55

5.	Single full-fillet lap joints with plug welds conforming to UW-17.	(a) Circumferential joints§ for the attachment of heads not over 24 in in outside diameter to shells not over $\frac{1}{2}$ in thick. (b) Circumferential joints for the attachment to shells of jackets not over $\frac{5}{8}$ in in nominal thickness where the distance from the center of the plug weld to the edge of the plate is not less than $1\frac{1}{2}$ times the diameter of the hole for the plug.	...	...	.050
6.	Single full-fillet lap joints without plug welds.	(a) For the attachment of heads, convex to pressure, to shells not over $\frac{5}{8}$-in required thickness, only with use of fillet weld on inside of shell. (b) For attachment of heads, having pressure on either side, to shells of not over 24-in inside diameter and not over $\frac{1}{4}$-in required thickness, with fillet weld on outside of head flange only.	...	...	0.45

* See UW-12(a) and UW-51.
† See UW-12(b) and UW-52.
‡ The maximum allowable joint efficiencies shown in this column are the weld-joint efficiencies multiplied by 0.80 (and rounded off to the nearest 0.05), to effect the basic reduction in allowable stress required by this division for welded vessels that are not spot-examined. See UW-12(c).
§ Joints attaching hemispherical heads to shells are excluded.
SOURCE: Courtesy of the American Society of Mechanical Engineers.

Example Determine the allowable pressure for a vessel with the following original design details, but which now has a minimum thickness of 0.400 in.

Corrosion allowance: $\frac{1}{8}$ in

Nominal thickness: 0.688 in

Outside diameter: 24 in

Carbon steel: allowable stress, 15,000 lb/in²

Design pressure: 500 psi

Longitudinal double-welded butt joint with full radiography

An ellipsoidal head attached to each end of the shell with a single-welded butt joint with a backing strip, and only spot-radiographed

answer Determine the present allowable working pressure without any corrosion allowance. Two calculations must be made, with different weld efficiencies involved. For the longitudinal joint calculation, $E = 100$ percent; for the circumferential joint calculation, $E = 80$ percent. See Table 5.4.

$$R = \frac{24 - (2 \times 0.400)}{2} = 11.6 \text{ in}$$
$$S = 15,000 \text{ lb/in}^2$$
$$t = 0.400 \text{ in}$$

Using equation (1) above and substituting,

$$P = \frac{15,000(1.0)(0.400)}{11.6 + 0.6(0.400)} = 506.8 \text{ psi}$$

Using equation (2) above and substituting,

$$P = \frac{2(15,000)(0.8)(0.400)}{11.6 - 0.4(0.400)} = 842.1 \text{ psi}$$

The maximum allowable pressure per shell requirements is, thus, 506.8 psi, indicating extra thickness was provided for the shell in the original design.

Full radiography is mandatory for certain vessels, such as those containing lethal substances, those of certain thicknesses as mentioned before, and others that are detailed in the code. However, when full radiography is not mandatory, the extent of radiography may be specified by the vessel purchaser or the purchaser's agent according to current code rules. A lack of radiography, however, requires that thicker material be used as a compensating factor.

Example A cylindrical vessel with a 48-in inside diameter has had no radiographic examinations. The shell has a double-welded butt joint, and the circumferential joint is of the single-butt type. The original plate material had an allowable stress of 13,750 lb/in². The data also show an original allowable pressure of 200 psi, with the original thickness being 0.5625 in. Thickness tests are to be made on this vessel by NDT. What is the minimum thickness that should alert the NDT personnel that the allowable pressure may have to be changed?

answer Under some old code methods, the allowable stress would have to be reduced by 80 percent first, but as the third footnote in Table 5.4 indicates, this already has been done by reducing the weld efficiency by 80 percent. It is necessary to check the required t for a 200-psi operation, using the two equations for shells or drums, transposing and solving for t. The result is:

1. For longitudinal welded joints

$$t = \frac{PR}{SE - 0.6P}$$

where P = 200 psi
$\quad\quad R$ = 24 in
$\quad\quad S$ = 13,750 lb/in^2
$\quad\quad E$ = 70 percent (Table 5.4), no spot examination

$$t = \frac{200(24)}{13,750(0.7) - 0.6(200)} = 0.505 \text{ in}$$

2. For circumferential welded joints

$$t = \frac{PR}{2SE + 0.4P}$$

and substituting with E = 65 percent

$$t = \frac{200(24)}{2(13,750)(0.65) + 0.4(200)} = 0.27 \text{ in}$$

Therefore, the minimum thickness required for a 200-psi operating pressure is 0.505 in, excluding any corrosion allowance.

Strength factors for shells or drums under *external pressure* can be calculated by code methods using charts covering various materials. Thin-wall shells under external pressure may fail from bending because of compressive stresses imposed on the shell. The unsupported shell length thus becomes an important factor, as do the outside diameter of the shell and the shell's thickness.

Vessels that may be subject to internal vacuum conditions, and that have thin shells with long lengths, may collapse as shown in Fig. 5.1 owing to buckling. Stiffening rings can be used to strengthen long cylinders against external pressure failures. The ASME code, Section VIII, Division 1, should be consulted for calculating allowable cylinder pressures due to external pressure.

Spherical shells. Large storage vessels in the chemical and petroleum industry use spherical shells to store gases under pressure. Spherical shells are more difficult to fabricate; however, they can be thinner than cylindrical vessels, as the following ASME code equation shows:

$$P = \frac{2SEt}{R + 0.2t}$$

where P = maximum allowable internal working pressure, psi

R = inside radius of sphere, in
t = minimum required thickness, in, exclusive of corrosion allowance
E = lowest joint efficiency in the sphere
S = maximum allowable stress permitted by code, lb/in²

Example A sphere with a 30-ft inside diameter stores a gas at a maximum pressure of 75 psi at 125°F. The sphere is made of a low-alloy steel plate, double-welded butt joints with full radiography are used, and the material has a code-allowable stress of 17,500 lb/in². How thick must the sphere be?

answer Solve the equation from above for t

$$t = \frac{PR}{2SE - 0.2P}$$

where t =thickness of the sphere, in
    $P = 75$ psi
    $E = 100$ percent
    $S = 17,500$ lb/in²
    $R = 15 \times 12 = 180$ in

$$t = \frac{75(180)}{2(17,500)(1) - 0.2(75)} = 0.39 \text{ in required, with no corrosion allowance}$$

Dished heads

The code recognizes the following types of unsupported heads: (1) hemispherical, (2) ellipsoidal, (3) torispherical, (4) conical, without transition knuckles, and (5) toriconical (see Fig. 5.4).

Hemispherical heads. Pressures for hemispherical heads are calculated with the same equation as that shown for a spherical pressure vessel, because a hemispherical head is half a sphere. The E to be used must include the head-to-shell joint. Ellipsoidal heads and torispherical heads must have a thickness at least equal to that of a seamless hemispherical head divided by the head-to-shell joint efficiency.

Example A seamless, hemispherical head is double-butt-welded to a shell, but the shell-to-head weld is not radiographically examined. The head has an inside diameter of 60 in and is constructed of code material with an allowable stress of 18,000 lb/in². The head is $\frac{1}{2}$ in thick and has a 0.125 in corrosion allowance. What is the allowable pressure for this head?

answer Use

$$P = \frac{2SEt}{R + 0.2t}$$

where P = maximum allowable working pressure, psi
    $S = 18,000$ lb/in²
    $t = 0.5 - 0.125 = 0.375$ in
    $R = 30$ in
    $E = 0.70$ (from Table 5.4)

$$P = \frac{2(18,000)(0.7)(0.375)}{30 + 0.2(0.375)} = 314.2 \text{ psi}$$

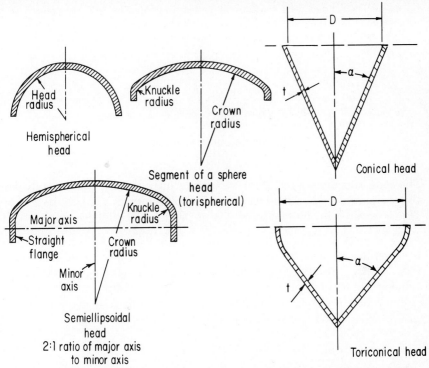

Figure 5.4 ASME formed heads under internal pressure may be hemispherical, semiellipsoidal, torispherical (segment of a sphere), conical, or toriconical.

Ellipsoidal heads. The standard ellipsoidal head is of the semiellipsoidal shape. The minor axis is half the length of the long axis. The code provides the following equation for calculating the allowable pressure for this standard semiellipsoidal head:

$$P = \frac{2SEt}{D + 0.2t}$$

where P = maximum allowable working pressure, psi
$\quad t$ = thickness of the head, in
$\quad S$ = maximum allowable stress, lb/in²
$\quad D$ = inside length of the major axis
$\quad E$ = welding joint efficiency

Example A seamless, ellipsoidal head of standard design has a 48-in inside major axis, operates at 500 psi, and has code material with an allowable stress of 17,500 lb/in². The shell-to-head connection is of the double-butt-weld type, but the welding was not radiographed. Determine the required thickness for this head.

answer   The code requires the allowable stress to be multiplied by 80 percent, because no radiography was employed on the shell-to-head weld. The joint efficiency is 100 percent because of the seamless construction. Then, solving for t in the above equation,

$$t = \frac{PD}{2SE - 0.2P}$$

and substituting, with $S = 17,500(0.8) = 14,000$,

$$t = \frac{500(48)}{2(14,000)(1) - 0.2(500)} = 0.86 \text{ in, without corrosion allowance}$$

The code introduces a K factor for calculating allowable pressures in the above equations when the ratio of major to minor axis may not be 2 to 1. This is covered in the first appendix of Section VIII, Division 1.

Torispherical heads.   See Fig. 5.4. These heads are also called "flanged" and "dished" heads. The standard ASME torispherical head has a knuckle radius of 6 percent of the inside crown radius, and the crown radius equals the outside diameter of the skirt. The code equation to be used in calculating allowable pressures for this type of head is

$$P = \frac{SEt}{0.885L + 0.1t}$$

where P = maximum allowable working pressure, psi
t = thickness of the head, in
S = maximum allowable stress, lb/in^2
E = welding joint efficiency
L = inside crown radius

Example   What would be the thickness required for a torispherical head with the same conditions as given in the previous problem for the ellipsoidal head, and with $L = 48$ in?

answer   Solving for t in the above equation

$$t = \frac{0.885PL}{SE - 0.1P}$$

and substituting, with $L = 48$ in,

$$t = \frac{0.885(500)(48)}{14,000(1) - 0.1(500)} = 1.52 \text{ in, without any corrosion allowance}$$

This would indicate that a semiellipsoidal head of the same thickness and other dimensions as a torispherical head is stronger than the torispherical head.

Conical heads. Conical heads are constructed as shown in Fig. 5.4. The code requires that half the apex angle a be $30°$ or less. Calculation for allowable working pressure involves using the equation

$$P = \frac{2SEt \cos a}{D + 1.2t \cos a}$$

where P = maximum allowable working pressure, psi
  t = thickness of the head, in
  S = maximum allowable stress, lb/in²
  E = welding joint efficiency
  a = half the apex angle
  D = inside diameter of cone head (maximum)

Example What would be the thickness required of a welded cone head with a maximum inside diameter of 48 in, half an apex angle of $30°$, double-butt-welded joints with no radiography employed, and P and S the same as described in the previous problem?

answer Solving for t

$$t = \frac{PD}{2 \cos a \,(SE - 0.6P)}$$

where cos 30 = 0.866
  E = 0.70 (head is assembled by welding)
  S = 17,500 lb/in²
  D = 48 in
  P = 500 psi

Substituting,

$$t = \frac{500(48)}{2(0.866)\,[17,500(0.7) - 0.6(500)]} = 1.16 \text{ in, with no corrosion allowance}$$

Toriconical heads. Toriconical heads are conical heads with transition knuckles, as shown in Fig. 5.4. The knuckle radius, as specified by the code, must be not less than 6 percent of the outside diameter of the head skirt. This type of conical head must be used if half the apex angle exceeds $30°$. The same equation is used as for the conical head, but D is the inside diameter of the head at its point of tangency to the knuckle.

External pressure on dished heads. External allowable pressure for ellipsoidal and torispherical heads can be calculated by using the equations for internal pressure, with E = 100 percent and P being 1.67 times the P shown in the equation.

When pressure acts on the reverse side of a head, the pressure is considered to be on the convex side of the head, commonly referred to as a "reversed dished head." A quick method to determine the allowable pressure has always been to calculate the concave allowable pressure and then multiply this by 0.6. This gives a good approximation of the allowable pressure on heads with pressure on the convex side. See the code for further details.

Shell-to-head juncture stresses

Depending on the geometry of the joint and on the material, an additional stress may be imposed on shells and heads, especially if there is a right-corner head-to-shell joint. A stress concentration may exist at the joint, caused by the differential radial expansion between the head and shell. There are many advanced texts available on pressure vessel design that provide methods to calculate the head-to-shell juncture stresses, and the reader is referred to these.

Determination of these discontinuity stresses is a major task in evaluating noncode cylindrical vessels. Unfortunately, it is not a simple task, since the actual stresses depend on shell geometry, head geometry, and material properties of both the head and the shell. The ASME pressure vessel code does not assist in this task. The code specifies allowable stress levels for a particular material but does not cover the geometry variations that occur in practice. However, the code does permit use of state-of-the-art analytical techniques to calculate design stresses on complex shapes.

Formulas describing the discontinuity stresses for variously shaped pressure vessel closures have been derived by numerous stress analysts. While these formulas serve as useful references in understanding the significance of these stresses, the details of head geometry, changes in thickness, and flanged connections are not modeled in these formulations. It is possible to use a finite-element computer algorithm to take all these effects into consideration. This computer model has been used to investigate how stress is influenced by variations in section thickness, changes in head geometry, and various combinations of material properties (e.g., modulus of elasticity).

Flatheads

The ASME code has several equations for flatheads, depending on whether the head is round, rectangular, or square. For a typical round head, the equation used is

$$t = d \sqrt{\frac{CP}{SE}}$$

where t = minimum required thickness, in
        d = diameter, measured as indicated in code
        C = a factor, depending on method of attachment [Fig. 5.5a(2)]
        S = maximum allowable stress, lb/in^2
        P = maximum allowable pressure, psi
        E = weld-joint efficiency, if welding is done within the head

Example   An unstayed flathead is attached as shown in Fig. 5.5a(2). Welding meets all code requirements. The head is circular with a 16-in diameter and a thickness of $1\frac{1}{2}$ in; the material is SA-285C with an allowable stress of 13,750 lb/in^2. The shell to which the head is attached is $\frac{3}{8}$ (or 0.375) in, and calculations show that the required thickness for a seamless shell is $\frac{5}{16}$ (or 0.3125) in. Determine the allowable pressure for this flathead.

answer   For this method of attachment

$$C = 0.33 \frac{t_r}{t_s}$$

where t_r = required thickness of a seamless shell for the pressure, in
        t_s = actual thickness of shell, without corrosion allowance, in

From the information given, S = 13,750, d = 16, t = 1.5, t_r/t_s = 0.3125/0.375 = 0.833, and C = 0.33t_r/t_s = 0.33(0.833) = 0.275; therefore, with E = 1,

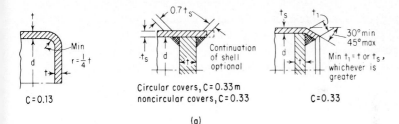

(a)

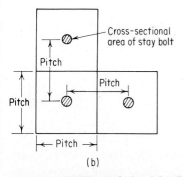

(b)

Figure 5.5   ASME code flathead details. (a) ASME code shell-to-flathead attachments showing C factor to be applied in flathead calculations. In the figure, m = t_r/t_s; t_r = required thickness of a seamless shell for the given pressure; t_s = actual thickness, without corrosion allowance. (b) Stay bolts must resist the force created by the pressure acting on the net area shown. The area the pressure acts on is (pitch)^2—cross-sectional area of stay bolt.

$$1.5 = 16 \sqrt{\frac{0.275P}{13,750(1)}}$$

$$\left(\frac{1.5}{16}\right)^2 = \frac{0.275P}{13,750}$$

$$P = \left(\frac{1.5}{16}\right)^2 \frac{13,750}{0.275} = 439.5 \text{ psi}$$

Bracing and staying

The first point to remember in all problems dealing with bracing and staying is that the stress set up in a stay is due to the unit pressure in pounds per square inch acting on the area of plate supported by that stay. This *total pressure* is resisted by the internal resistance of the brace (unit stress) times the net area of the brace. These facts are the basis for all bracing formulas.

Stay bolts and stays are used in vessels to reinforce flat or other surfaces exposed to pressure loading, because the plate surfaces would have to be made too thick to resist this loading if stays or stay bolts were not used.

To calculate stay bolt problems, it is necessary to establish the code requirement, which states that the required area of a stay bolt at its minimum cross section shall be found by calculating the load on the stay bolt, dividing this by the allowable stress on the stay bolt, and increasing the resultant area by a factor of 1.1. See Fig. 5.5b. In equation form, the following can be developed to use in stay bolt problems:

$$\frac{\text{Load on stay bolt}}{\text{Allowable stress}} = \text{resisting area of stay bolt}$$

Let S = allowable stress, lb/in²
a = area of stay bolt (usually round), in²
p = pitch of stay bolt spacing, in
P = allowable working pressure, psi

Then

$$\frac{(p^2 - a)P}{S} = \frac{a - \text{telltale hole area}}{1.1}$$

In addition to the strength of the stay bolt, the strength of the plate between the stay bolts must be adequate, or the plate might buckle between the stay bolts. The code requires this to be checked by one of the following equations:

$$t = p \sqrt{\frac{P}{CS}} \quad \text{or} \quad P = \frac{t^2 CS}{P^2}$$

where s = maximum allowable stress, lb/in^2
  t = required thickness of plate, in
  p = maximum pitch, in
  P = maximum allowable pressure, psi
  C = factor depending on construction (2.1 for stays screwed through plates of not over $\frac{7}{16}$-in thickness; 2.2 for stays screwed through plates of over $\frac{7}{16}$-in thickness)

Under the code, pitch and C factors depend on construction.

Example (a) What is the maximum allowable square pitch of stay bolts in the flat stayed sheet of a code vessel when the sheets are supported by $\frac{7}{8}$-in screwed stay bolts with $\frac{3}{16}$-in holes (area = 0.0276 in^2) in the outer end? The stay bolts have 12-V threads per inch, and the pressure carried is 115 psi. SA-31 grade A stay bolts and SA-285C plate under 600°F are used. Stay bolt stress = 11,300 and plate stress allowed = 13,700. (b) What is the least thickness of plate permissible in part (a)?

answer (a) Since a = cross-sectional area of stay bolt at bottom of threads = 0.419 in^2

$$(p^2 - a)P = \frac{a - 0.0276}{1.1} S$$

With SA-31 grade A stay bolts, S = 11,300 lb/in^2; P = 115 psi. Therefore, substituting,

$$(p^2 - 0.419)115 = \frac{0.419 - 0.0276}{1.1} \, 11,300$$
$$115p^2 = 4022.8 + 48.185$$
$$p^2 = 35.4$$
$$p = 5.95 \text{ in}$$

(b) Use

$$t = p \sqrt{\frac{P}{CS}}$$

$$t = 5.95 \sqrt{\frac{115}{13,700(2.2)}} = 0.368 \text{ in, or less than } \tfrac{7}{16} \text{ in}$$

For plates under $\frac{7}{16}$ in, we must recalculate with C = 2.1:

$$t = 5.95 \sqrt{\frac{115}{700(2.1)}} = 0.376 \text{ in}$$

Ligaments in drums or shells

In some pressure vessel applications, tube holes are drilled in the shells, and heat-absorbing tubes are inserted in the holes. This pattern of tube holes weakens the shell; the weakening effect is seen as a percent reduction of shell strength and termed "ligament efficiency." See Fig. 5.6. The efficiency of a ligament E_L is found as follows, assuming the pitch between the tube holes is equal

$$E_L = \frac{p - d}{p}$$

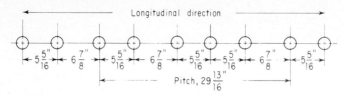

Figure 5.6 The weakening effect of drilling holes in a shell is expressed by the ligament efficiency in strength calculations.

where p = pitch, or longitudinal distances between holes, in
d = tube-hole diameter, in

If the pitch of the tube holes is unequal (Fig. 5.6), a unit longitudinal length in which all the unequal pitches are included should be selected. Then,

$$\text{Efficiency of the ligament} = \frac{P - nd}{P}$$

where P = length selected to include all variations in pitches, in
d = tube-hole diameter, in
n = number of tube holes in longitudinal line in selected length

The ASME code covers methods of finding efficiency of diagonal ligaments by means of charts.

The efficiency of the usual tube ligament is quite low, usually 35 to 50 percent. On riveted vessels, often the ligament is strengthened by riveting a reinforcing strap or doubling plate over the tube-hole section. The tube holes are cut through the entire reinforced section; thus, although the efficiency of the ligament is not increased, because the tube spacing remains unchanged, the thickness is increased.

Another method of increasing the strength of the ligament is to make the drum in two longitudinal halves with two longitudinally welded seams. One half will be somewhat thinner than the other half that contains the ligament. The edges of the thicker half are machined down to the same thickness as that of the other for abutting edges of the longitudinal seams to be welded.

Example The drum of a vessel has a 42-in ID. The tube sheet is $\frac{5}{8}$ in thick and contains $3\frac{9}{32}$-in-diameter tube holes pitched horizontally $5\frac{5}{16}$ in, in banks of three (Fig. 5.6), and two tubes with $6\frac{7}{8}$ in between banks. The shell plate is $\frac{1}{2}$ in thick. The joint efficiency between the tube sheet and shell is 67 percent. What is the allowable working pressure for this drum if the material is SA-285C and the maximum temperature is 650°F?

answer Allowable stress for SA-285C material is 13,750 lb/in². This is a riveted vessel, as can be noted by joint efficiency. The code equation for a shell at the time this vessel was built was

$$P = \frac{0.8SEt}{R + 0.6t}$$

where P = maximum allowable internal working pressure, psi
R = inside radius of shell, in
t = minimum required thickness, in, exclusive of corrosion
E = riveted or ligament joint efficiency
S = maximum allowable stress, lb/in²

It is necessary to calculate the allowable pressures for the two efficiencies (joint and ligament).

1. The allowable pressure based on shell thickness and riveted joint efficiency, with R = 21 in and t = 0.5 in, is

$$P = \frac{0.8(13,750)(0.67)(0.5)}{21 + 0.6(0.5)} = 173.0 \text{ psi}$$

2. In the solution based on ligament efficiency and tube-sheet thickness, ligament efficiency must be calculated first:

$$E = \frac{p - nd}{p}$$

From Fig. 5.6, it may be seen that p = 29.8125 and n = 5; from the data given in the example, d = 3.28125. Therefore,

$$E = \frac{29.8125 - 5(3.28125)}{29.8125} = 0.448$$

$$P = \frac{0.8SEt}{R + 0.6t} = \frac{0.8(13,750)(0.448)(0.625)}{21 + 0.6(0.625)} = 145 \text{ psi}$$

Thus, the allowable pressure is 145 psi.

Thick cylinders and spheres

The code stresses that additional design considerations may be required on vessels that are to be operated over 3000 psi; however, in the mandatory appendix of the code, equations can be applied to thick-walled cylinders, which are generally used for higher pressures. Vessels over 3000 psi pressure may be stamped with the code symbol, provided code material is used and the vessel is designed to good engineering standards. In time, owing to advances in technology, the code will be expanded to cover vessels operating up to 100,000 psi, such as are used in isostatic presses. The thick-cylinder and -sphere equations must be used whenever the pressure exceeds $0.385SE$, or when the thickness of the cylindrical shell exceeds half the radius.

Thick cylinders. Two equations are provided—one for t and the other for P with different Z ratios to be applied. Two examples will illustrate the application of these equations.

Example   A seamless cylinder with 10-in inside diameter is made of SA-182 grade F 321 stainless steel, to be operated at 1000°F and 3000 psi. The allowable stress for this material is 14,000 lb/in² at the designated temperature. What thickness must the cylinder be?

answer   The following code equations are used to find t:

$$Z = \frac{SE + P}{SE - P} \quad \text{and} \quad t = R(Z^{1/2} - 1)$$

where R = inside radius, in
 S = allowable stress, lb/in²
 E = welded-joint efficiency, %
 P = allowable working pressure, psi

From the information given, S = 14,000, P = 3000, E = 1.0, and R = 5. Therefore,

$$Z = \frac{14{,}000(1) + 3000}{14{,}000(1) - 3000} = 1.545$$

$t = 5(1.545^{1/2} - 1) = 1.215$ in, without corrosion allowance

Section VIII, Division 1, of the ASME code also provides the following supplementary design formulas where the thickness t is known and it is required to find the allowable pressure of a thick-walled pressure vessel.

$$P = SE \frac{Z - 1}{Z + 1}$$

with

$$Z = \left(\frac{R + t}{R}\right)^2$$

Example   A seamless forging with an 8½-in ID is 2.25 in thick and is made of code material with an allowable stress of 17,500 lb/in². What is the allowable pressure for this cylinder?

answer   Substituting in the above equations, with $E = 1$, S = 17,500, and $R = \frac{8.5}{2} =$ 4.25,

$$Z = \left(\frac{4.25 + 2.25}{4.25}\right)^2$$

$Z = 2.339$

Then

$$P = 17{,}500(1)\left(\frac{2.339 - 1}{2.339 + 1}\right)$$

$P = 7017.8$ psi

Thick spheres.   Similar equations for spheres exist in the code, but a y factor is used:

1. If P is known, to find t of a thick sphere, use

$$t = R(y^{1/3} - 1) \quad \text{and} \quad y = \frac{2(SE + P)}{2SE - P}$$

where t = minimum required thickness of sphere, in
R = inside radius of sphere, in
S = allowable stress, lb/in^2
E = joint efficiency, %
P = allowable pressure, psi

2. If t is known, to find P, use

$$P = 2SE \frac{y - 1}{y + 2} \quad \text{and} \quad y = \left(\frac{R + t}{R}\right)^3$$

Example What would the thickness of a sphere with a 10-in inside diameter have to be in order for the sphere to be used with 3000 psi allowable pressure per code equations? Assume $S = 14,000$, $P = 3000$, and $E = 1.0$.

answer First, the term y is determined:

$$y = \frac{2(SE + P)}{2SE - P}$$

$$y = \frac{2[14,000(1) + 3000]}{2[14,000(1)] - 3000} = 1.36$$

Then
$$t = R(y^{1/3} - 1)$$
$$t = 5(1.36^{1/3} - 1) = 0.54 \text{ in}$$

Example What would be the pressure allowed on a 15-in ID sphere, 4 in thick, composed of a material with an allowable stress of 17,500 lb/in^2, and with a weld efficiency of 100 percent?

answer Solving for y

$$y = \left(\frac{R + t}{R}\right)^3$$

$$y = \left(\frac{7.5 + 4}{7.5}\right)^3 = 3.61$$

As seen earlier

$$P = 2SE \frac{y - 1}{y + 2}$$

$$P = 2(17,500)(1) \left(\frac{3.61 - 1}{3.61 + 2}\right) = 16,283 \text{ psi}$$

Openings and reinforcements

The code requires considering the weakening effect of holes being cut into shells if the opening is over 3 in for shells of $\frac{3}{8}$-in thickness or less, and 2 in for shells over $\frac{3}{8}$-in thickness. The same rules apply to heads. Excluded are ligament-type openings, threaded openings not over 2 in in pipe size, and large openings, those over half the inside diameter, that require design as a flanged opening. Modern stress analysis can be applied to determine the different loads a welded-in nozzle may cause to the assembly. However, by historical usage, the following code method is still employed to deter-

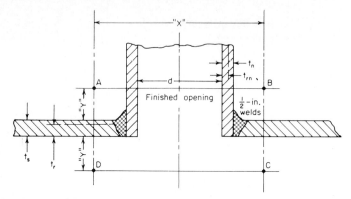

Figure 5.7 The area to be considered within reinforcement credit must be within rectangle $ABCD$. Along the shell wall, this extends out to larger of $2d$ or $d + 2(t_s + t_n)$ shown by X. The limit along the nozzle wall, or Y distances shown, is $2\frac{1}{2}t_s$ or $2\frac{1}{2}t_n$, whichever is smaller. The weld and reinforcement pad in this area can be calculated as reinforcement. t_s = actual shell thickness; t_r = required shell thickness; t_n = actual nozzle thickness; t_{rn} = required nozzle thickness.

mine whether an opening in a shell requires reinforcement to return it to its solid-plate strength. The procedure essentially consists of determining how much area was removed and then noting if the lost area is compensated for by extra thickness from reinforcement of the shell or from the metal from an inserted nozzle, up to the limits of permissible reinforcement, as shown in Fig. 5.7. The code rules illustrated are typical of reinforcement calculations. Failures around openings can also be due to the torque and bending moments that a welded-in nozzle may impose on the connection. This could arise from expansion stresses from connected piping systems. A more advanced text is recommended for a treatment of the analysis of the stresses around nozzle connections.

In shells. The ASME code has detailed requirements on openings cut into shells or headers and on how to calculate whether reinforcement around an opening is necessary. The following procedure generally applies (see Fig. 5.7). The area required to be restored by the finished opening d is

$$A = d \times t_r \times F$$

where d = diameter of finished opening in given plane, in
t_r = required thickness of seamless shell for the pressure
F = factor that considers axis of the nozzle, usually 1.00

To determine if enough metal is available from the shell, nozzle, welds, or reinforcement, the code provides the equations below.

To determine the metal available from the *shell,*

$$A_1 = (E_1 t_s - F t_r) d \quad \text{or} \quad A_1 = 2(E_1 t_s - F t_r)(t_s + t_n)$$

where t_s = actual thickness of shell
t_r = required thickness of a seamless shell
t_n = actual thickness of nozzle attached to shell
E_1 = longitudinal joint efficiency of weld if opening is through the longitudinal weld joint; otherwise, $E_1 = 1$

The larger value is used.

To determine the metal available from the *nozzle,*

$$A_2 = (t_n - t_{rn}) 5 t_s \quad \text{or} \quad A_2 = (t_n - t_{rn})(5 t_n)$$

where t_{rn} = required thickness of seamless nozzle

Use the smaller value.

Example A 5-in, extra-heavy pipe nozzle SA-53B is welded to a shell similar to that shown in Fig. 5.7 ($\frac{1}{2}$-in welds). The shell has an inside diameter of 30 in, a thickness of $\frac{7}{16}$ in, and a working pressure of 200 psi. The material is SA-285C. Assume that all welds are in accordance with the code and that the welds are strong enough for this installation. The outside diameter of the 5-in pipe is 5.563 in, and the thickness is 0.375 in. Does this design meet code requirements for openings in a shell? The allowable stress for SA-285C is 13,750.

answer As seen earlier

$$t = \frac{PR}{SE - 0.6P}$$

Therefore, the shell thickness required is

$$t_r = \frac{200(15.0)}{13,750(1.0) - 0.6(200)} = 0.220 \text{ in}$$

The nozzle thickness required is

$$t_{rn} = \frac{200(2.407)}{13,750(1.0) - 0.6(200)} = 0.035 \text{ in}$$

The area of reinforcement required is

$$A = d \times t_r \times F$$
$$= 4.813(0.22)(1.0) = 1.059 \text{ in}^2$$

The area of reinforcement provided, for $E_1 = 1.00$, is

A_1 (shell) = $[1.0(0.438) - 1.0(0.220)]4.813$ = 1.049 in²
A_2 (nozzle) = $2(2.5)(0.375)(0.375 - 0.035)$ = 0.638 in²
A_s (welds) = $1(0.50)^2$ = 0.250 in²
                                      Total 1.937 in²

Construction meets code requirements, because the area of reinforcement provided exceeds the area removed.

Where additional reinforcement may be required, a pad is fitted around the nozzle opening and the cross-sectional area of the pad, within the defined limits of reinforcements, is calculated and added to the excess area available from the shell or nozzle, as shown in the above example. Section VIII, Division 1, of the ASME code gives some additional examples of reinforcement calculations in the appendix section.

In flat plates. If the opening does not exceed half the diameter or shortest span of a flathead, the area of reinforcement required around the opening is given by

$$A = 0.5dt$$

where d = diameter of the finished opening
t = minimum required thickness of flathead, without corrosion allowance

The same calculations as used above are made to see if excess thickness is available in the flathead and nozzle connections. These calculations can be bypassed if the head is made thicker by increasing the C factor in the equation for minimum t for flat plates on page 230 to $2C$ or 0.75, or doubling the quantity under the square root sign.

Spherical, dished covers of the bolted-head type

Calculations for spherical, dished covers usually involve the strength of the flange against the bending stresses. Extensive calculations are required for this type of closure, and this is usually done by designers and inspection agencies for pressure vessel manufacturers, and not by plant personnel involved with operations and maintenance. Detailed calculations for bolted flange connections are shown in Appendix 2 of Section VIII, Division 1. These can be used to check the strength of these types of closures if flange cracking is noted or bolt problems develop.

The most common flanges are those with ring-type gaskets *inside* the bolt circle, and where no contact exists between mating flanges outside the bolt circle owing to internal or external pressure. The gaskets used are spiral-wound, metal-asbestos-filled gaskets. Flanges may be integral with the connecting shell, such as welding neck flanges (see Fig. 5.8), or they may be loose flanges that are attached lightly to the shell or are not attached, such as slip-on flanges. Slip-on flanges are used for pressures below 150 psi. The basic methods of flange design were developed by the Taylor Forge and Pipe Works Co. in 1949 and have been incorporated into the ASME code as a requirement in Appendix 2. The design must take into account the effect of no-load or no-pressure gasket seating and

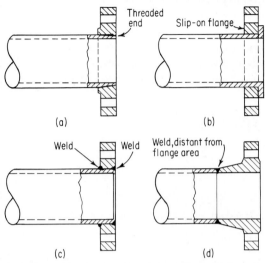

Figure 5.8 Flanges may be of the following types: (a) screwed flange, (b) Van Stone or slip-on flange, (c) welded slip-on flange, (d) weld-neck flange. (*Courtesy of* Power *magazine.*)

the imposed stresses on bolts and flanges from this gasket seating, and the load imposed by operating pressure on the assembly. Initial gasket-seating load is a function of the gasket material and how much it can be compressed to seal the joint, and of the effective gasket or contact area that requires seating. These factors determine the bolting area required. The code has minimum design seating stresses for the different gasket materials and appropriate instructions on calculating bolt loads. Flange stresses must also be calculated for initial gasket-seating conditions, and then for operating design criteria. The stresses on the flanges are mostly due to the moments of the loads acting on the flange. The moment arm is determined by the relative position of the bolt circle with respect to the load producing the moment. Three stresses are produced on the flange: longitudinal hub stresses, radial flange stresses, and tangential flange stresses. The reader should review Appendix 2 of the ASME code. Another source for allowable flange loads and pressures is ANSI B16.5 specifications, titled *Standard Flanges*.

Proof Tests

It is always considered good engineering practice to refer to original code design in checking if a condition on a pressure vessel has violated code requirements, and thus may require repair; lowering of allowable pressure; or, in severe cases of deterioration or operating need, full replacement. The examples cited previously for how to use code requirements to verify that a code vessel is still safe should assist plant people involved with inspecting and maintaining pressure vessel systems, assuming the proper code reference is used in the analysis. Sometimes, however, the calculations we have seen so far cannot be used, and direct testing is necessary.

Proof tests are used to establish the maximum allowable working pressure for pressure vessels, or parts thereof, when the strength cannot be calculated with accuracy. Proof tests can be made, and allowable pressure determined, by: (1) testing to the yield point of the material or (2) testing to the bursting pressure.

Yield testing

Yield testing is limited to material with a ratio of 0.625 or less for minimum specified yield strength to minimum specified ultimate strength. The code requires several tests of the material by the yield-test method to establish the allowable working pressure. The following equation is used in establishing allowable working pressure:

$$P = 0.5H \frac{Y_s}{Y_a}$$

where H = hydrostatic test pressure at which yielding was noted, psi
Y_s = code-specified minimum yield strength, lb/in^2
Y_a = average yield strength as determined from testing the material of the vessel or parts under consideration

The code provides for several other allowable-pressure determinations, which are modifications of the above equation.

Bursting-test method

If the bursting-test procedure is used to establish the allowable working pressure of a vessel and the materials are not of the cast type, the code recommends the following equation:

$$P = \frac{B}{5} \times \frac{SE}{S_a}$$

where B = bursting-test pressure, psi
S = code-specified tensile strength, lb/in^2
S_a = actual tensile strength of material, lb/in^2
E = weld efficiency

Example An SA-240, type 304L, dimpled-constructed vessel failed under a burst test at 425 psi pressure. The code-specified minimum tensile strength for this material is 70,000 lb/in². The actual average strength of the material was found to be 72,500 lb/in². What would be the allowable working pressure? The code requires that a weld efficiency of 80 percent be used for dimpled or embossed construction.

answer Substituting in the bursting-test equation

$$P = \frac{425}{5} \times \frac{70,000(0.8)}{72,500} = 65.7 \text{ psi, without corrosion allowance}$$

For cast parts, the same equation is almost applicable, except that the bursting pressure is multiplied by a casting factor (about 80 percent) and the appropriate tensile strength of the cast material should be used. Note that there is an approximate safety factor of 2 from the allowable pressure based on yield stress and a safety factor of 5 based on ultimate strength.

Stresses Other Than Those Caused by Pressure

Wind loads

Wind loads impose additional stresses on tall outdoor pressure vessels such as are found in large chemical and refinery plants. The velocity of the wind against a structure can be converted into pounds per square foot of wind acting against the projected area of the structure. For example, one equation used to determine wind pressure is based on the wind velocity in miles per hour at a 30-foot-high level:

Wind pressure = 0.00256 × basic wind speed (mph) at 30 ft, as determined by average over time

Average wind speeds vary with terrain. As a result, ANSI A58.1 has been adopted by designers to determine the many variables in calculating the stress imposed on a tall structure, in a certain terrain and with a variable profile on which the wind can act at different heights. Once the wind pressure is determined and the wind load in pressure established, the structure can be treated as a cantilever beam to establish the possible bending stress.

Earthquake loads

Earthquake loads are transient and generally rare. The chief impact of earthquakes is the production of vibratory forces. These are calculated under certain assumed conditions, such as:

1. The earthquake will accelerate only in a horizontal direction and will transmit this acceleration to a structure horizontally.
2. The force on the structure is determined by Newton's law, $F = ma$, where m = the mass of the structure and a = horizontal acceleration.

3. The force acts at the center of gravity of the structure, and thus the resultant overturning moment can be calculated.

4. Again, the structure can be treated as a cantilever beam.

Building codes require a minimum lateral force to be used in calculating possible earthquake loads, and this force depends on the earthquake zones in which vessels are located. The reader is referred to references such as building codes in calculating earthquake loads.

Section VIII, Division 1, design concern is primarily with the principal membrane stress that results from the application of pressure. By the use of a time-proven safety factor, and with rules for detailed construction, including quality control of variables such as welding, reasonable and safe designs have resulted. Failures usually occur because of service factors, such as corrosion, shock starting procedures, and control and safety-device problems, and not because of structural-strength design defects. Design equations should be used, however, at all times to judge whether a piece of pressure equipment is safe to operate or needs repair or replacement.

Section VIII, Division 1, requires that the combined maximum membrane stresses due to pressure and supports, excluding those induced by wind and earthquake, cannot exceed the maximum allowable stresses shown in the ASME stress tables. See Tables 5.1 and 5.2 for a condensation of an ASME stress table. The stresses induced by wind and earthquake, when combined with the maximum primary membrane stresses, cannot exceed 1.2 times the maximum allowable stress shown in the ASME stress tables.

Structural supports

Structural supports hold pressure vessels in place and induce local concentrated stresses on the pressure vessel. On large pressure vessels, the weight of the vessel when filled with water for a hydrostatic test must be considered when analyzing the load on structural supports and the pressure vessel. Supports also require considering the possible temperature changes in operation and how these may produce localized stresses on the pressure vessel.

Vibration

Vibration is another problem that has caused high localized stresses on pressure vessels, such as air tanks with compressors mounted on top of the tanks. Serious cracking can occur from the repeated loading caused by vibration of connected machines, piping movement, and even resonant, or sympathetic, vibration, in which a pipe or support on a pressure vessel is in tune with the operating speed of a distant machine, such as a compressor. Where a manufacturer buys a code pressure vessel, and then mounts machinery on top of it, vibration may cause damage to the vessel even though it was built to code standards.

Quick-Actuating Head Closures

Owing to process needs, many vessels, such as creosoting cylinders, brick-hardening cylinders, and vulcanizers, are opened and closed during processing. The quick-closing door, if not properly tightened, can blow open in operation with possible personnel injury. For this reason, the code requires locking mechanisms "so designed that failure of any one locking element or component in the locking mechanism cannot result in the failure of all other locking elements and the release of the closure." Users and operators should be especially alert in making sure all multibolted-type closures are bolted closed using the full number and types of bolts prescribed in the design. Other closures use a partial rotating flange. These must have some interlock device to ensure that the closure is fully engaged before pressure is applied and that pressure is released before the closure can be opened. The heads and shells for these types of vessels are designed for safety; however, the human element can bypass the design, and this must be guarded against. The code even suggests the use of audible or visible devices to warn of any impending leakage of pressurized contents. Pressure-indicating devices are most essential for the operators and must be positioned so the operators can observe them from their operating stations at all times. The design of the interlocks is left to good engineering practice and has no detailed code requirement, because of the variations possible.

A chief state boiler and pressure vessel inspector of a jurisdiction, after investigating accidents involving quick-actuating closures, has recommended the following inspections and maintenance.*

Periodic inspections of quick-opening door assemblies need to be made by properly trained personnel. General conditions of the moving parts of door assemblies and safety devices should be checked at least monthly. A more thorough annual inspection should be made of the entire vessel. Nondestructive testing of critical elements should be considered on units subject to severe service.

It is necessary that records be kept of all tests, maintenance, and repairs. Some of the features that should be inspected include the following:

- All bearing surfaces should be carefully checked for evidence of excessive wear.

---

* M. L. Snow, Jr., Director, Tennessee Boiler and Elevator Division, "Quick Actuating Closures," *National Board Bulletin,* April 1986.

- Gaskets should be checked for wear, damage, and leakage. It is important that replacement gaskets are always in accordance with the manufacturer's specifications with no deviations.

- Door hinge mechanisms should be checked to ensure that they are properly aligned and that adjustment screws and locking nuts are properly secured.

- Door and locking ring lugs should be examined carefully for evidence of undue stress and for cracks at the junction of the lug and door or locking ring. Locking ring and door wedges should be checked to verify full engagement when closed and should be checked for proper bearing surface contact, wear patterns, and condition. For replacement of missing wedges or securement of loose wedges, the door manufacturer should be consulted.

- Doors using the contracting ring locking device should have the ring checked for loss of springiness, cracks at the points of attachment of the operating lugs, evidence of undue wear on the ring and shear on the pins in the lugs and the operating mechanism.

- With clamp-type doors, the surfaces of the clamps should be checked for wear and the clamps examined for distortion at the portions overlapping the shell ring and door ring. Hinge pins and locking device parts should be checked for wear and evidence of shear.

- With bar-type doors, all bearing surfaces should be inspected for undue wear and the various parts checked for indications of undue stress as well as for distortion. Arm pivot pins should be checked to be sure they are securely held in place and are not bent. Pivot pin mounting brackets should be checked for cracks at the point of attachment to the head and evidence of undue stress in line with the pin holes. Threads of the operating screw should be checked for wear and fit in the nut or handwheel hub.

- It is important that closures of the swing bolt type be checked for missing bolts. If any are missing, they need to be replaced at once. Bolts should be checked for soundness, particularly at the eye, and the threads checked for evidence of stripping or excessive wear. The bolt washers should be flat. Washers that are distorted to a dish shape tend to allow bolt movement out of the slot when nuts are improperly torqued. The closure should also be inspected when closed, to be sure the nuts are fully engaged. Pins should be examined for distortion and secure fit.

- At each inspection of the vessel, the door safety locking appliances should be checked and tested to be sure that they are operating properly and are in good repair.

- Immediate attention should be given any time a locking ring binds or catches at some point during its movement. This is because the point becomes a fulcrum and the entire ring tries to rotate around it. This action may result in shifting of the ring's position, causing unequal overlap on the lugs. Therefore, it is important that any safety device that determines the positioning of the ring, such as microswitches, manually operated pins with two-way valves connected to steam signals, or any other type of device, be located at four equal quadrants of the ring. One safety device at one point is not sufficient to properly indicate the position of the ring. The four devices should be tested.

- The door and the locking mechanism should be checked in both the closed and the open positions. The position of the locking ring, amount of overlap, and any shift in the ring's position should be observed.

- The opening to the vessel should be checked for out-of-roundness at the outer edge. This is the difference between the maximum and minimum inside diameter at any cross section. Under no condition should it exceed 1 percent of the nominal diameter of the cross section under consideration and, preferably, it should be zero.

QUESTIONS AND ANSWERS

5-1 What are the minimum thicknesses specified by the ASME code for shells and heads?

ANSWER: For compressed air, steam, and water service minimum thickness is $\frac{3}{32}$ in. For ferritic steels with tensile properties enhanced by heat treatment it is $\frac{1}{4}$ in. For unfired steam boilers it is $\frac{1}{4}$ in. Except for tubes in heat exchangers, including inner pipes of double-pipe exchangers, the minimum thickness of other shells and heads after forming must be at least $\frac{1}{16}$ in without considering the corrosion allowance.

5-2 If a $\frac{1}{2}$-in plate is ordered, what is the minimum thickness that will still meet code requirements?

ANSWER: Below tolerance of 0.01 in from the $\frac{1}{2}$ in ordered or below 6 percent of the ordered thickness.

5-3 On what basis is the design pressure established for a pressure vessel?

ANSWER: The vessel must be designed for the most severe condition of coincident pressure and temperature that is expected in normal operation, including considering the difference in pressure between the inside and outside of a vessel, or between any connected chambers of a combination unit.

5-4 What is the design temperature?

ANSWER: This is the temperature that is at least equal to the expected mean operating temperature of the metal throughout the thickness under consideration.

5-5 Who must specify the corrosion allowance to be used in a pressure vessel?

ANSWER: The code places the responsibility on the user or the user's designated agent. In practice, manufacturers advertise their products as having a specified corrosion allowance.

5-6 What causes a plate formed into a cylinder to retain its shape?

ANSWER: The plate has been stretched so that the elastic limit has been exceeded and a permanent set of the material has been made by the forming.

5-7 Why are hydrostatic tests limited to $1\frac{1}{2}$ times the allowable pressure?

ANSWER: A greater hydrostatic test may impose stresses beyond the yield or elastic limit, and a permanent deformation may develop in the pressure vessel.

5-8 Why is it necessary to stay or brace flat surfaces?

ANSWER: Flat plates under pressure tend to bulge out to the shape of a sphere; therefore, staying or bracing is required to prevent the flat plate from bulging.

5-9 In a multilayer vessel, what liner must be designed to withstand full negative pressure in the vessel?

ANSWER: The inner shell or head must be designed, and have a thickness, to resist liner collapse from external pressure, as specified by code requirements.

5-10 To what joint is the longitudinal stress applicable on a cylindrical pressure vessel?

ANSWER: The circumferential joint is where stress is created in the longitudinal direction of the vessel.

5-11 What welded-joint efficiency factor is required on pressure vessels containing lethal substances?

ANSWER: Full radiographic examination is required on the mandatory butt weld of the joints. This gives a 100 percent welding efficiency.

5-12 Give the allowable pressure for a vessel, using the following shell data: OD = 36 in; plate thickness = $\frac{1}{2}$ in, with corrosion allowance of $\frac{1}{16}$ in. The material has an allowable stress of 18,000 lb/in². The longitudinal joint and circumferential joints were spot-examined. The longitudinal joint is double-butt-welded; the circumferential joint is a single-welded joint with a backing strip.

ANSWER: The two shell equations must be used with $E = 0.85$ for longitudinal joint calculations and $E = 0.80$ for the circumferential joint. The allowable pressure is 386.9 psi.

5-13 A torispherical head of seamless construction is single-butt-welded to a seamless shell. The allowable working pressure stamped on the vessel is 300 psi. A data sheet shows that spot radiography was employed throughout the welded-joint examination. The material used has an allowable stress of 13,800 lb/in². The inside crown radius is 42 in. Including a corrosion allowance of $\frac{1}{16}$ in, what should the thickness be?

ANSWER: The code requires that the allowable stress be reduced to 85 percent because spot radiography was used on a category B joint. Substituting in the equation for a torispherical head

$$t = \frac{0.885(300)(42)}{0.85(13,800) - 0.1(300)} = 0.953 \text{ in}$$

t required $= 0.953 + 0.0625 = 1.016$ in

5-14 What is the maximum pitch permitted for stay bolts?

ANSWER: The maximum pitch permitted is $8\frac{1}{2}$ in, except for welded-in stay bolts, for which the pitch may be up to 15 times the diameter of the stay bolt.

5-15 What is the permissible out-of-roundness of a welded shell?

ANSWER: The difference between the maximum and minimum inside diameters measured at any cross section cannot exceed 1 percent of the nominal diameter at the cross section being considered.

5-16 When may pneumatic tests be used instead of hydrostatic tests?

ANSWER: For those vessels that were so designed that they cannot safely be filled with water, for those that cannot be readily dried, and for those that will be used in a service for which traces of water for process reasons are not permitted.

5-17 What is the pressure in the standard pneumatic test?

ANSWER: The pneumatic test pressure must be at least equal to 1.25 times the allowable working pressure of the vessel.

5-18 What do the letters "L," "UB," and "DF" on the nameplates of pressure vessels signify?

ANSWER: The letter "L" is for lethal service; "UB" is for unfired steam boilers; and "DF" is for a pressure vessel for direct-firing service, such as a reformer in an ammonia plant.

5-19 Who must complete data reports on pressure vessels?

ANSWER: The manufacturer must fill out Form U-1 or Form U-1a of the ASME data reports, and these must be signed by the manufacturer and the code inspector.

5-20 (a) A vessel's shell has a 40-in diameter; its tensile strength is 55,000 lb/in², its plate thickness is $\frac{1}{2}$ in, and its factor of safety is 5. What is the maximum safe working pressure according to the existing-installation equation? (*b*) If the pressure is increased 10 percent, what is the factor of safety FS? (*c*) If corrosion causes a general reduction in thickness of $\frac{1}{8}$ in, what is the maximum safe pressure if joint efficiency is 94 percent?

ANSWER: (*a*)

$$P = \frac{55,000 \times 0.5 \times 0.94}{20 \times 5} = 258 \text{ lb/in}^2$$

Note that the existing-installation equation is used with a safety factor of 5 because of the riveted construction.

(*b*) $258 \times 1.10 = 284 \text{ lb/in}^2$

Transposing the existing-installation formula and solving for the factor of safety, we have

$$\text{FS} = \frac{55,000 \times 0.5 \times 0.94}{20 \times 284} = 4.56$$

(*c*) A reduction in thickness of $\frac{1}{8}$ in will be 25 percent, which will leave 75 percent of the solid plate. Since this is below the efficiency of the longitudinal seam (94.0 percent), the maximum safe pressure should be reduced as follows:

$$P = \frac{55,000 \times 0.375 \times 1.0}{20 \times 5} = 206 \text{ psi}$$

5-21   How is the tightness of a reinforcing pad around a nozzle checked?

ANSWER: A telltale of at least $\frac{1}{4}$ in must be provided in order to check the welds off the inside of the vessel. Usually a preliminary compressed air and soapsuds test is made for this purpose.

5-22   Why is it necessary to vent heat exchangers using steam?

ANSWER: To avoid the accumulation of noncondensable gases that can become acidic in combination with water.

5-23   What is the possible effect of forming U bends in tubes?

ANSWER: Cold work in forming the tubes may leave stresses that with certain materials may cause cracking in hostile environments. Heat treatment may be required to relieve the induced stresses from the cold working.

5-24   What can cause an impact load on a pressure vessel system?

ANSWER: Fluid or water hammer shock can impose an impact load.

5-25   May welded joints be examined by sectioning the weld as a substitute for spot-radiographic examination?

ANSWER: The code still requires spot radiography even if welds are examined by sectioning.

5-26   What is the difference between a clad vessel and a lined vessel?

ANSWER: Clad plate has a corrosion-resistant material integrally bonded during plate manufacture to a less resistant base metal, while a lined vessel may have a corrosion-resistant liner attached by welding to the vessel's wall at selected or intermittent areas of the vessel.

5-27   Describe the term "nominal thickness."

ANSWER: This is the thickness supplied by the manufacturer for the vessel part under consideration. Another term equivalent to it is "actual thickness."

5-28   What would be the *total* pressure on a column 43 ft high containing a liquid having a density of 70 lb/ft³ if the liquid pressure on top showed 200 psi?

ANSWER: The maximum pressure would be on the bottom of the column, with

$$P = 200 + \frac{43(70)}{144} = 220.9 \text{ psi}$$

This is the pressure that should be used in calculating the required thickness at the bottom of the column. The ASME code requires consideration of all coincident pressure in design.

5-29   If spot radiography is used, what is the maximum distance allowed between individual spot-radiographic examinations and what is the minimum length required to be radiographed under this weld examination procedure?

ANSWER: Spot-radiographic examinations are required at 50-ft intervals of the joint being examined. The spot radiography must be at least 6 in in length.

5-30   What are the principal reasons for using nonferrous materials in pressure vessels?

ANSWER: The nonferrous materials are primarily chosen because of corrosion resistance, ease of cleaning the vessels, hygienic cleaning advantages with regard to food, scale resistance at high temperatures, and notch toughness at low-temperature service.

5-31   A seamless, cast Yankee roll is made of SA-278, Class 40 material. Its original thickness was $2\frac{1}{4}$ in. Inspection showed that because of crown machining in the past, the thickness was now 1.5 in. If the ID of the roll was 16 ft, what would be the allowable pressure for this vessel, based solely on circumferential stress?

ANSWER:

$$P = \frac{6000}{96.9} = 61.9 \text{ psi, without corrosion allowance.}$$

5-32   What is the maximum size of a plug that can be installed in a cast-iron pressure vessel?

ANSWER: The depth or length of the plug cannot be greater than 20 percent of the vessel thickness where the plug is to be installed, and the diameter of the plug cannot exceed the vessel thickness at the plug location.

5-33   What is the minimum thickness permitted for a layer in a multiwall pressure vessel?

ANSWER: The minimum thickness of a layer is $\frac{1}{8}$ in.

5-34   A prototype vessel is made of steel, but its strength cannot be calculated. It is therefore subjected to a bursting proof test and fails at 515 psi. The design allowed a welding efficiency of 70 percent, since no spot radiography was employed. The material had a specified tensile strength of 55,000, and actual tensile test showed an average tensile strength of 58,000 lb/in². What should the allowable pressure be?

ANSWER:

$$P = \frac{515}{5} \times \frac{55,000(0.7)}{58,000} = 97.7 \text{ psi}$$

Chapter

6

Operation, Controls,
and Safety Devices

Integrating Maintenance and Inspection with Operating Results

Most inspection and plant maintenance are geared to preventing production breakdowns. Product impairment is a distinct hazard in the operation of pressure vessel systems; these systems can be affected not only by operation but also by the inspection and maintenance that has been applied to the process stream. There is thus a need to integrate operation, inspection, and maintenance into a team effort in order to prevent forced outages or product impairment.

The main purpose of a pressure vessel system is to produce a marketable product. In most cases, this is the main goal of management, and all technical people involved with the operation, testing, and maintenance of the system realize that this is a necessary prerequisite of having a job. Therefore, most plant operations concentrate on delivering an acceptable product, and the entire plant monitoring system tracks the process to note if changes are occurring in the finished product, so that corrections can be made before the product becomes nonsalable or of inferior quality. However, equipment wears, controllers become sluggish, instruments become defective, and usually these malfunctions also affect the product stream. In this way, equipment reliability is important not only from a safety point of view but also for the production of a satisfactory product.

There are many common concerns in the operation of a pressure vessel system. Some will be cited, but each plant may have its own unique features.

Chemical Plant Considerations

Chemical plants include an extremely broad category, ranging from those plants that produce basic chemicals, such as chlorine, ammonia, and sodium, to those that produce the complex organic and inorganic compound products that are made from various feedstocks in huge continuous-flow plants. It is impossible to cite the many risks that can exist for literally thousands of chemical compounds that are manufactured in industry.

The following general considerations are required in evaluating chemical plant exposures that will affect operation, maintenance, and inspection practices.

1. Is the flow of the multiple-batch-line type or of the continuous-flow type?

2. Is the process corrosive, and has suitable material been used for the objects in the process? Is there a thickness testing program?

3. Are any reactions of the exothermic type that could produce a runaway chemical reaction with a subsequent rapid rise in pressures and temperatures?

4. Are any of the gases or fluids being processed of a volatile, combustible nature such that any process disturbance or leaks could result in an explosive mixture? What instrumentation is there to warn operators?

5. Is there the possibility of catalyst damage from a process disturbance? What emergency procedure is there to minimize the damage?

6. How long is the time from the beginning of a process to a finished product? Is there emergency equipment to expedite the return to normal flow in the event of power, water, or other service interruption?

7. Is there a possibility of goods in process being spoiled or damaged from a process disturbance? What steps are needed to reduce this exposure?

8. How long would it take to clean out lines and equipment after hardening, solidification, or catalyst damage has occurred owing to a process disturbance, and resume operations?

9. Are any pressure vessels of unusual design, making repairs expensive and possibly of long duration? This could include such factors as high-alloy steel construction; multiwall, high-pressure service; and catalyst-removal difficulties.

These are some of the items that must be considered when reviewing chemical plant exposures, and the maintenance and inspection effort that may be required to avoid product impairment or process interruption.

Operating Surveillance

Operating concerns that require continuous monitoring include the following:

1. Are temperature, pressure, and flow gages and charts working and accurate?
2. Is the purge system ready for an emergency with an adequate supply of purge gas? Are all valves in working order?
3. Are cooling-water pumps, including emergency pumps, in operable condition?
4. Is the electric supply and distribution system at proper voltage, not overloaded, and clean and tight on breakers and bus bars? Is the emergency generator suitable for service?
5. Is the steam supply maintaining desired pressures and temperatures? Is the backup boiler operable?
6. Are isolation valves for maintenance and emergency in serviceable condition and not leaking or stuck?
7. Are all pressure vessels operating within their stamped allowable working pressures; free of leaks and vibration; and with vents, rupture disks, or safety relief valves in operable condition, set at the correct pressure, and having the proper capacity?
8. Does the pressure vessel system have adequate drains to remove sludge, and are prescribed blowdown procedures for the process being followed?
9. Is the sequence of starting and shutdown posted so that operators can carefully follow the proper procedure?
10. Are equipment manufacturers' maintenance and testing practices being followed, as well as the company engineering department's setpoint guidelines on pressure, temperature, flows, density, and similar process variables?

Operator Training

The continuous technological changes and increased automation of pressure vessel systems require periodic updating of the training that is being applied to plant operating personnel. This process should include (1) development of new training programs to meet specific changes in the plant,

(2) institution of instruction procedures, (3) development of training programs for dealing with emergencies.

Changes requiring a review of training include such factors as new operating strategies, new processes, new raw or finished process material being used, new equipment, and new supervision techniques.

Operating manuals

It should be standard procedure to have operating manuals brought up to date as needed. In general, the operating manual should treat the following subjects in a distinctive, sectionalized arrangement:

1. General description and background of process

2. Material properties

3. Description of equipment

4. Start-up and shutdown procedures

5. Normal operating procedures

6. Safety and health considerations

7. Precautions and emergency action

Training and preparation for emergencies

All plant personnel, including management, supervisory, and operating personnel, should be trained in controlling plant emergencies, which include large windstorms, earthquakes, floods, power failure, fires, explosions, and accidental release of toxic or dangerous materials. Plant fire squads or brigades should be formed by management. They should be versed in handling all types of plant emergencies.

"Simulated emergency" drills should be performed by plant process operators where the operation is considered to be hazardous. Malfunctions of the process can be simulated and emergency actions carefully undertaken to note the cause and effect.

"Disaster" drills that simulate a major catastrophe should be undertaken periodically. Public police, fire, and other emergency units, as well as those from nearby cooperating plants, should be called in to assist.

In each plant, a nucleus of management personnel should be trained to handle potential disaster situations. They should be familiar with all the rescue and emergency control facilities available in the area.

Training should emphasize the early recognition of process malfunctions that can develop into serious fire, explosion, or health problems. Employees should immediately report to their supervisors and take prescribed actions on any abnormal circumstances, such as:

1. Leak of materials from equipment

2. Release of hazardous gases, vapors, or smoke

3. Defective or damaged equipment

4. Detection of unusual odors

5. Appearance of unusual sounds in the process

6. Abnormal conditions such as excessively high or low temperature or pressure

7. Infraction of plant operating procedures or safety regulations

8. Unauthorized hazardous work

9. Unauthorized vehicles or personnel in hazardous areas of plant

It is also necessary to train operators to handle various disposal situations, such as accidental spillovers, discarding of unsatisfactory products, and the disposal of hazardous waste materials. Serious problems can develop if such materials are highly flammable, toxic, or unstable. In some cases, plans will have to be made for the evacuation of all personnel from the area. Special dump tanks, spot ventilation, and disposal systems may have to be provided. Waste disposal procedures must be in conformance with local, state, and federal regulations.

Identifying possible equipment breakdowns is a serious loss-prevention responsibility. Operators throughout the plant are frequently in a position to anticipate potential losses through their close contact and experience with the situation. They should be encouraged to point out equipment maintenance and replacement needs to their supervisors and maintenance department.

Determining the number of people that may be needed in an emergency is a management responsibility. Operators may be too busy trying to reestablish or maintain operating conditions at their assigned stations to help in fighting fires, cleaning up spills, or covering equipment exposed to the elements in a storm. These types of situations must be evaluated by management to make sure that the help will be available when needed.

Start-Up and Shutdown Procedures

Start-up and shutdown must be done with attention to the possible effects of shocking a pressure vessel system by rapid pressure or temperature changes. Most vessel manufacturers provide detailed instructions for starting and shutting down their equipment, and process engineers should supplement these with specific instructions for the process involved. Thermal shocking can be avoided by slow, uniform heating and cooling of a vessel system. Large, thick rotating vessels, for example, should be rotated slowly, and if they are steam-heated, steam should be admitted to the vessel in a controlled manner to avoid thermal stresses being set up between thin and thick sections, which may not expand at the same rate. The same precautions must be used in shutting down. The rotating pressure vessel should be kept rotating after the heat source has been removed in order to maintain uniform temperatures throughout the vessel while it is cooling.

Because of the variety of pressure vessels, careful planning and training may be necessary in order to avoid shocking a system when starting or shutting down a process line. Typical of such instructions is the following for heat exchangers, as developed by American Standard for their equipment:

1. Be sure the entire system is clean before starting the operation in order to prevent plugging of tubes or shell-side passages with refuse. The use of strainers or settling tanks in pipelines leading to the heat exchanger is recommended.

2. Open vent connections before starting up.

3. Start operating gradually. See Table 6.1 for suggested start-up and shutdown procedures for most applications. If in doubt, consult the manufacturer.

4. After the system is completely filled with the operating fluids and all air has been vented, close all manual vent connections.

5. Retighten bolting on all gasketed or packed joints after the heat exchanger has reached operating temperatures to prevent leaks and gasket failures.

6. Do not operate the heat exchanger under pressure and temperature conditions in excess of those specified on the nameplate.

7. To guard against water hammer, drain condensate from steam-heated exchangers and similar apparatuses, both when starting up and when shutting down.

8. Drain all fluids when shutting down to eliminate possible freezing and corrosion.

9. In all installations, avoid pulsation of fluids, since this causes vibration and will result in reduced operating life.

10. Do not under any circumstances operate the heat exchanger at a flow rate greater than that shown on the design specifications. Excessive flows can cause vibration and severely damage the heat exchanger tube bundle.

11. Protect heat exchangers that are out of service for extended periods of time against corrosion. Heat exchangers that are out of service for short periods and use water as the flowing medium should be thoroughly drained and blown dry with warm air if possible. If this is not practical,

the water should be circulated through the heat exchanger on a daily basis to prevent stagnant water conditions that can ultimately precipitate corrosion.

Clearance for maintenance

Large pressure vessels require special considerations for maintenance needs. All pressure vessels should be installed with sufficient clearance to allow inspection and maintenance to be done without having to disturb adjacent equipment. Ample space should be provided for the removal of covers and shells or bundles of tubes, and for the retightening and, if needed, repair welding of joints. For large vessels, clearance should be provided so that cranes or hoists can be used to service the vessel.

Freeze-up protection

Failure to provide adequate freeze protection, or failure to take the necessary measures of draining lines and vessels during freezing weather conditions, can result in damaged piping, blocked or impaired instrument lines, frozen jackets, and similar mishaps that may idle a plant for days while repairs and restart problems are handled. Usually large interruptions of business are experienced if freeze-up during inclement weather is not adequately considered in design and operation.

Methods used to prevent freeze-up failures include:

1. Providing self-draining lines, warm-up bypasses, and good valve placement for shutoff of flows threatened by low temperatures, and instituting good and tested operating procedures against freeze-up occurrences.
2. Heat-tracing lines and vessels that may be exposed to freezing by the use of electrical heat, steam heat, or circulation of warm liquids to prevent stagnant pockets from forming.
3. Installing internal heating coils within pressure vessels that go on automatically if cold weather threatens the vessel or its contents.
4. Locating process equipment vulnerable to freeze damage within heated buildings.
5. Adding additional insulation to those lines and vessels with low flow rates.

Interaction of Operators, Controls, and Computers

As process plants become larger, and increasingly of the single-train type, the need for automation and better instrumentation and controls is be-

TABLE 6.1 Recommended Start-Up and Shutdown Procedures for Heat Exchangers

| Type of construction | Fluid location and relative temp. | | | | Start-up procedure | Shutdown procedure |
| | Shell side | | Tube side | | | |
	Type of fluid	Rel. temp.	Type of fluid	Rel. temp.		
Fixed tube sheet (nonremovable bundle)	Liquid	Hot	Liquid	Cold	Start both fluids gradually at the same time.	Shut down both fluids gradually at the same time.
	Condensing gas (i.e., steam)	Hot	Liquid or gas	Cold	Start hot fluid first, then cold fluid.	Shut down cold fluid first, then hot fluid.
	Gas	Hot	Liquid	Cold	Start cold fluid first, then hot fluid.	Shut down cold fluid gradually, then hot fluid.
	Liquid	Cold	Liquid	Hot	Start both flows gradually at the same time.	Shut down both fluids gradually at the same time.
	Liquid	Cold	Gas	Hot	Start cold fluid first, then hot fluid.	Shut down hot fluid first, then cold fluid.

U tube; packed floating head; packed floating tube sheet; internal floating head (all these types have removable bundles)	Liquid	Liquid	Hot	Cold	Start cold fluid first, then start hot fluid gradually.	Shut down hot fluid first, then cold fluid.
	Condensing gas (i.e. steam)	Liquid or gas	Hot	Cold	Start cold fluid first, then start hot fluid gradually.	Shut down cold fluid first, then shut down hot fluid gradually.
	Gas	Liquid	Hot	Cold	Start cold fluid first, then start hot fluid gradually.	Shut down hot fluid first, then cold fluid.
	Liquid	Liquid	Cold	Hot	Start cold fluid first, then start hot fluid gradually.	Shut down hot fluid first, then cold fluid.
	Liquid	Gas	Cold	Hot	Start cold fluid first, then start hot fluid gradually.	Shut down hot fluid first, then cold fluid.

CAUTION: Every effort should be made to avoid subjecting the unit to thermal shock, overpressure, or hydraulic hammer, since these conditions may impose stresses that exceed the mechanical strength of the unit or the system in which it is installed, which may result in leaks or other damage to the unit or system.

GENERAL COMMENTS: (1) In all start-up and shutdown operations, fluid flows should be regulated so as to avoid thermal shocking the unit regardless of whether it is of removable or nonremovable construction. (2) For fixed tube-sheet (nonremovable bundle) units in which the tube-side fluid cannot be shut down, it is recommended that a bypass arrangement be incorporated in the system and that the tube-side fluid be bypassed before the shell-side fluid is shut down.

SOURCE: Courtesy of American Standard.

coming clear; increased use of computers for data logging and analysis and, as a minimum, for warning operators that set points are being exceeded is becoming a necessity. The time for decision and action during an emergency has now been reduced by fast-moving process streams to the point that it is essential for the corrective and protective measures in the emergency to be built into a computer program, and for the operators to be trained properly, in order to have effective interaction between computers, operators, and the emergency system for the process. The term "fail-safe" is used to denote a preprogrammed emergency shutdown system. This, of course, requires continuous maintenance and testing of all equipment in order to ensure that all systems will function when needed.

Control and safety systems are evolving and changing quite rapidly, largely because of the development of all types of sensors to measure variables and the rapid development and cost reduction of electronic devices and data processing systems.

A company that wants to use a computer system for process control can purchase a general-purpose computer and convert it into a process computer by writing or acquiring the necessary software and then interfacing the computer with the operators and the process equipment.

Often a company will buy a computer from one manufacturer but the display screens, known as "cathode-ray tubes (CRTs)," from a different source. The reason for this is that CRTs differ in their display capabilities and resolution. In today's process control systems, the color CRT has become the standard for operator interface. Process control CRTs have graphic capabilities that allow them to display controller faceplates, or to represent a schematic of the process on the face of the tube, or to provide similar graphic functions required for an effective operator interface.

Interaction with the process operator involves both presenting information to and accepting commands or directives from the operator. Traditional equipment provided analog meters and status lights to present information to the operator; knobs and switches enabled the operator to take actions.

Process controls are often mounted in panels in densely packed arrangements. Although this saves space, the direct association of a particular instrument with a piece of process equipment is lost, especially in large plants. One alternative is to create a large panel with a graphic representation of the process flowsheet. The instruments are then physically installed at the logical positions on the graphic flowsheet.

CRT-based operator interfaces emulate both of the above. Almost all systems provide emulations of the front panel (or faceplate) of conventional panel instruments. These faceplate displays are generally presented in groups of eight. The configuration of the display basically consists of identification of the eight points that are to appear on a given display. The faceplate presentation will depend on the type of point (process variable, control loop, discrete output, etc.), but the format of each faceplate is fixed by the supplier. Thus, the user can easily configure these displays, and the configuration specifications require very little storage.

CRTs have also proved to be very effective in presenting process graphics. As in the conventional panels, key process points can be presented at locations appropriate for the graphics being displayed. But compared with faceplates, the process graphic displays require more effort to define and more memory for storage.

As for operator inputs, the CRTs have a limitation compared with the conventional panels. On the latter, all knobs and switches are accessible all the time. On the CRTs, only a subset of these is available at any one time.

Computer data acquisition generally encompasses the following three functions:

1. Conversion of process signals to a form that can be used by a digital system. The process interface hardware provides this function.

2. Transfer of the above into the computer's memory. The computer's I/O subsystem provides this function.

3. Conversion of the data from the front end to more meaningful representations. For example, the binary representation of the voltage from a thermocouple is converted to the temperature in degrees Celsius or Fahrenheit. Software modules within the computer perform this function.

Pressure Vessel Instrumentation

To maintain production at reasonable costs requires controls, and controls require measuring instruments and sensors in order to determine operating conditions within a process or piece of equipment. While the field of sensor development is accelerating in order to join robotic devices and computers with process controllers and with similar forms of automation, sensors cannot replace human intelligence in diagnostic capability at this time. New and more accurate equipment is being developed to achieve tighter manufacturing tolerances and better quality, in addition to reducing labor costs. Computer control is becoming more localized instead of being applied at the level of total-plant process flow. The latter can become extremely complex and expensive.

Code requirements

Specific requirements for pressure vessel instrumentation is not clearly defined in ASME Section VIII, Division 1. Establishing requirements is left to the jurisdictions in which specific pressure vessels are to be operated. Reference in the code is made to having a pressure gage on the vessel during a proof test. The gage must have dials graduated to at least $1\frac{1}{2}$ times the intended proof test. Most state laws make reference to "indicating and control devices as will ensure the vessel's safe operation." Good engineering practice appears to be the governing rule.

Instrumentation, as it applies to the process industries, encompasses all the equipment required to monitor and control the operation of a process. Instrumentation can be subdivided into four major categories:

1. Measurement devices

2. Actuators

3. Regulatory controls

4. Supervisory controls

Pressure and pressure gages

Simply stated, *pressure* is the force acting on a surface divided by the area of the surface. *Zero absolute pressure* is usually defined as a perfect vacuum; *barometric pressure* is the force exerted by the atmosphere at a given location and time. *Gage pressure* is a scale based on local barometric pressure, which of course varies with altitude and atmospheric conditions. As an approximation, gage pressure can be converted to "absolute pressure" by adding the value of one standard atmosphere, or 14.696 psi.

Static pressure in a fluid stream is the pressure acting on a sensor moving with the fluid so that it is at rest or static with respect to the fluid. In practice, static pressure is usually measured by tapping into a vessel or pipe perpendicular to the flow direction. This method assumes that fluid disturbances caused by the tap are small enough to ignore. Experience shows that a burr-free tapping made with reasonable care will produce adequate results. The maximum size of static pressure taps recommended by ASME are:

Pipe size, in	Tap diameter, in
2	0.25
3	0.375
4 to 8	0.50
10 and larger	0.75

Types of pressure gages

Bourdon tubes. The bourdon tube is the most common pressure-sensing element. This is a flattened metal tube, sealed at one end and bent in the shape of a C or spiral. Since the inner and outer surfaces of the tube have different areas, a force imbalance caused by pressure will make the tube unwind. This motion can be read directly on a gage or transduced to an electrical or pneumatic signal proportional to the pressure (see Fig. 6.1*a*). Care must be taken to avoid corrosion or deposits in the tube that could affect its characteristics.

Bellows.   Bellows are thin-walled, cylindrical shells with transverse corrugations. Deflection is related linearly to the applied pressure, even over relatively large distances. They are usually specified according to their effective area and spring rate. Effective area is determined by applying a known pressure to the bellows and measuring the resulting force. *Spring rate* is defined as the quotient of deflection and force. As with bourdon tubes, motion may activate a gage or pneumatic or electronic transmitter.

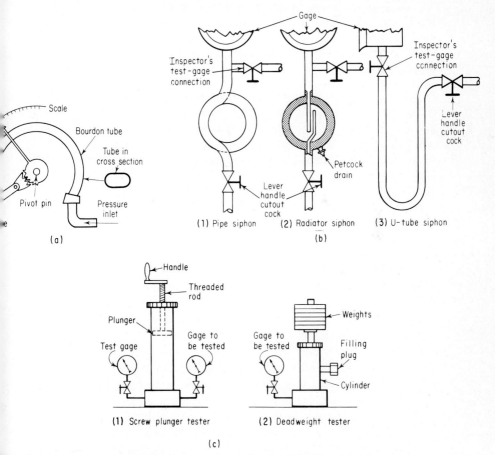

6.1   Pressure gages are a necessity in pressure vessel systems. (*a*) Bourdon tube pressing element; (*b*) siphons used to protect pressure gages from hot steam or liquids; (*c*) methods used to test pressure gages.

Diaphragms. Diaphragms are widely used as sensing elements in high-accuracy instruments. They are either flat or corrugated depending on the pressure range to be handled, and are suitable for measuring pressures from a few millimeters of water to thousands of psi. Significantly, diaphragms can be used as elastic seals or as partitions between two media.

Most frequently, diaphragms are designed to transmit forces or very limited motion. They are superior to bellows or bourdon tubes because they have comparatively few areas where process fluids can deposit and may be made from corrosion-resistant metals or coated with elastomers such as Teflon. On the other hand, diaphragm-actuated instruments are more expensive.

U tubes. The U-tube manometer is the simplest of the devices that rely on a change in liquid height to indicate pressure. Process instruments based on this principle use a float or various electrical techniques to detect changes in liquid level rather than visual observation. For precise readings, corrections must be made for temperature and the local value of gravity acceleration.

Positioning of pressure gages. When measuring liquid pressures the main considerations are the effect of the process fluid on the sensing element and the influence of the sensor's location above or below the measuring point. The sensing element can be isolated with a sealing liquid. The sealer should be immiscible in the process fluid, chemically nonreactive with it, and higher in density. A mechanical seal is a better approach since it does not require selecting the proper sealing fluid.

Steam installations must be designed to keep the sensor's temperature within operating limits. Usually the sensor is mounted below the pressure tap and the connecting line filled with water. Heat losses from the line will condense enough steam to maintain the liquid level.

A siphon is simply a pigtail or drop leg in the tubing to the gage for condensing steam, thus protecting the spring and other delicate parts from high temperatures. Three forms are shown in Fig. 6.1b. If there is danger of freezing during long periods of shutdown, the siphon should be removed or drained.

For temperatures over 406°F, no brass or copper tubing should be used. The pressure gage dial should be graduated to twice the safety-valve setting, but in no case less than $1\frac{1}{2}$ times this setting. A valve connection of at least $\frac{1}{4}$-in pipe size must be installed on a boiler for the exclusive purpose of attaching a test gage and, when the boiler is operating, for checking the accuracy of the boiler pressure gage. This connection is known as the "inspector's connection." The pressure gage must be illuminated and free from objectionable glare or reflection that can in any way obstruct an operator's view while noting the setting on the gage. The pointer on the gage must be in a near vertical position when indicating the normal operating pressure. This is also true of other pressure gages in the boiler room that are used on auxiliaries. Pressure gages must not be tilted forward more than 30° from vertical and then only when it is necessary for proper viewing of the dial graduations.

For liquid service, the pressure sensor can be located at any convenient elevation relative to the measuring point, provided the effect of the location on the "zero" of the instrument is acceptable or can be adjusted out. If the sensor is mounted below the measuring point the static head in the connecting line will require zero suppression. Conversely, a sensor placed above the pressure point will require elevation.

In the measurement of the pressure of gases, the primary consideration is to prevent accumulation of entrained and condensed liquids and solids in the sensor or connecting lines. For this reason the sensor is usually located above the process connection, with lines arranged to facilitate drainage. Because of the relatively low density of gases, the degree of elevation does not have a significant influence on the zero reading.

Temperature gages

Process plants must conduct all types of temperature measurements because of the diversity of products made. A review of the gages available will show that their operating principles are based on:

1. Change in volume

2. Change in electrical resistance

3. Voltage created at the junction of two dissimilar metals

4. Intensity of emitted radiation

5. Resonant frequency of a crystal

Another division of temperature-measuring devices is into mechanical devices (bimetallic and filled-system thermometers) and electronic devices (resistance temperature detectors, or RTDs; thermocouples; thermistors; semiconductors; and noncontact infrared instruments). Mechanical temperature-measuring devices are good for displaying temperatures, but they cannot transmit output signals and thus are not used with electronic or digital control systems. See Fig. 6.2a for the operating temperature ranges of temperature-measuring devices. This figure also shows a bimetallic thermometer (Fig. 6.2b) and a filled-system thermometer (Fig. 6.2c).

Mechanical temperature devices

Bimetallic thermometers. If a composite metal strip, made of two dissimilar metals welded or riveted together, is heated, it bends in the direction

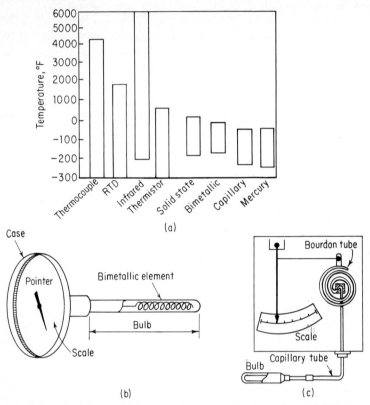

Figure 6.2   Temperature measurement. (*a*) Temperature limits of different temperature sensors; (*b*) bimetallic thermometer; (*c*) filled-system thermometer. (*Courtesy of* Power *magazine.*)

of the metal with the lower coefficient of expansion. The amount of bend for a given temperature change is repeatable and can be used to indicate temperature, provided the purity of the two metals is controlled. Usually the strip is formed into a coil, and a pointer is attached to the inner end. As the coil expands, the pointer indicates temperature on a scale (Fig. 6.2*b*).

Bimetallic thermometers are inexpensive and moderately accurate— typically 1 percent of reading for industrial types. The upper temperature limit is around 1000°F, but 500°F is the usual limit. Because of the scale layout, close resolution is difficult to obtain except for narrow-range thermometers. In industrial models the elements are usually sealed in a tube, since corrosion can degrade the metal strip.

Filled-system thermometers.   Devices of this type (Fig. 6.2*c*) are made with an expandable gas or liquid in a hermetically sealed enclosure. When heat is applied, the pressure of the expanding fluid causes a helical tube to stretch, moving a pointer over a scale. Other variations use the fluid to operate a piston connected to an indicator, which displays temperature on a linear scale. This type is not often used for temperature measurement, but may be used to operate a switch for thermostatic control.

Performance of filled-system thermometers has some limitations. Compensation is needed for ambient temperature effects on the long lengths of capillary tubing. Also, since they must remain sealed, careful installation is essential.

Mercury-in-glass thermometers, although unsuitable for most industrial uses, are extremely accurate—to 0.01°F—and are often used as secondary temperature standards and in laboratory work. Time response is slow and range is limited.

Electronic temperature devices

Resistance temperature detectors. RTDs operate on the principle that an electrical conductor's resistance changes with temperature. Platinum is usually used because of its stability at high temperatures.

Thermocouples. Thermocouples depend on the principle that when two dissimilar metals are joined, an electrical voltage is generated and this varies with temperature. This voltage can be measured with electronic circuitry. Thermocouples have a "hot" junction, which is the measuring point, and a "cold" junction, which is the reference point. Thermocouple sensors can be made in a variety of configurations and sizes and are extensively used for the continuous measurement of the temperatures of hot streams.

Thermistors. Thermistors are resistors that vary in their electrical resistance with temperature changes and thus are similar to RTDs. This quality can be used to measure temperature with the appropriate electrical circuitry. Shortcomings of thermistors in comparison with RTDs and thermocouples is a narrower operating range and the fact that the change of resistance with temperature is not linear.

Infrared sensors. Infrared sensors are used to check temperatures by sighting with an optical lens the area to be measured. The infrared device uses the principle that all objects emit infrared radiation at varying intensity and this depends on their temperature. This radiation can be converted to a usable form by electronic circuitry.

Temperature is based on inferential measure of heat. Scales such as Fahrenheit, centigrade, and Kelvin are arbitrary and defined under standard conditions of pressure. See Chap. 1 for interconversions of scales.

The temperature-sensing element should be located as close as possible to the spot where the temperature reading is desired. A remote location, even if there is no error in the steady-state reading, gives large errors through dynamic dead time, and automatic control may be impossible. A small bulb gives faster response than a large bulb, regardless of the type. To reduce film effects, always install the sensing element where the gas or liquid is moving rapidly.

Thermocouple lead wires should be enclosed in a grounded, separate conduit away from ac power lines. The same is true with resistance thermometers. In addition, when there is a significant distance between the element's resistance and the resistance to the current converter, a third wire should be used for compensation.

Thermocouple extension wire must be specified by the manufacturer. Use of other types will cause unpredictable errors.

Certain chemical-process applications may dictate special requirements, such as electrical or pneumatic transmission, local or remote indication or recording, remote or local measurement, wide or narrow temperature range, high accuracy, and so on. The costs of incorporating these features into each type of temperature-measurement device must be evaluated for the specific application requirements to determine which is the most cost-effective.

Flowmeters

There are many flowmeters from which to select. These include turbine meters, vortex shedding meters, swirlmeters, ultrasonic meters, positive displacement meters, rotameters, and nuclear magnetic resonance meters.

Almost all fluid metering systems have two parts: a primary element that produces measurable phenomena such as pressure difference, turbine rotation, or voltage and a secondary device to detect the phenomena and change them into force, motion, or signal for remote transmission. Secondary devices vary almost without limit. Primary elements, however, depend on a few simple physical principles. Their proper installation and operation are usually quite distinct and independent from those of the secondary element.

Differential-pressure devices. Differential-pressure producers are the most common flow-measuring devices. The orifice plate, a flat disk with a machined hole, is used along with the venturi and flow nozzle. For a given flow rate the differential pressure produced is inversely proportional to the size of the restriction. For example, if the primary element has an area half that of the pipe in which it is installed, a noncompressible fluid doubles its velocity while passing through the orifice. See Fig. 6.3.

Since the change in velocity through an obstruction gives an increase in kinetic energy, there will be a decrease in the corresponding static pressure. Flow rate is related to the pressure differential, fluid properties, size of the orifice, and a discharge coefficient. Values for the discharge coeffi-

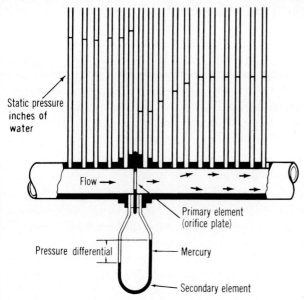

Figure 6.3   An orifice plate is used to create a difference in pressure on the high- and low-flow side of the plate, and this pressure difference can be used to determine the amount of flow in the pipe.

cient have been determined experimentally as a function of meter type, pressure tap location, pipe size, Reynolds number, and ratio of orifice diameter to pipe diameter. The most complete tabulations are published in *Fluid Meters* by the American Society of Mechanical Engineers and *Orifice Metering of Natural Gas* by the American Gas Association.

Magnetic flowmeters.   Many flow measurements are best made without an obstruction in the stream. The magnetic flowmeter, operating on the same principle as an electric generator, is one such device. It consists of a permanent magnet or electromagnet placed so that fluid flow induces an electromotive force across the fluid. A secondary element converts the low-level signal into one suitable for process control. Signal output is linearly proportional to flow rate and is independent of density and viscosity variations.

Magnetic flowmeters cover an extensive size range. One application is on a chemical additive feed line with a diameter of 0.1 in. At the other extreme, a 9-ft meter operates at a sewage plant. Maximum line pressure depends on the rating of the metering tube and flanges. Safe maximum operating temperature ranges from about 170 to 300°F, depending on the magnetic coil insulation and temperature rating of electronic components in the primary element.

Of course, the fluid to be metered must be conductive. Gases and most petroleum derivatives are below the practical threshold conductivity, which is about 10 $\mu\Omega/cm^2$ for most magnetic flowmeters.

Point-velocity devices.   Point-velocity instruments such as the pitot tube, venturi tube, and hot-wire anemometer measure velocity or mass flow at a single point in a moving stream. They may be applied in both open channels and closed circuits and are most often found in large ducts where other devices would be prohibitively expensive.

Level-measuring devices and controls

Level-measuring devices are important in all plants where tanks have fluid pumped into or out of them and depend on level-indicating devices to maintain a level within set points. They are also extensively used to control the water level in boilers and similar fired equipment. The level control for these applications also shuts burners down in the event of low water, thereby preventing overheating damage. Thus, it becomes an important safety device.

Displacement sensors.   Displacement-level sensors detect the change in buoyant force on a body immersed in a liquid. If the body has a uniform cross section along the axis of level change, its suspended weight will decrease linearly as level increases.

A secondary element transduces the force change to a gage reading, recording, or signal transmission.

Displacement sensors are generally installed in a side arm or vessel well to avoid the effect of fluid motion on the float. Compared with other level-measurement systems, displacement units are generally less expensive. Their primary disadvantages are that their calibration span cannot be changed and major zero-point adjustments require relocation in the vessel.

Level sensors in the chemical industry.   For level measurement in the chemical process industries, a particularly important consideration is to have a chemically inert seal between the process fluid and the primary sensor. This has led to wide use of the diaphragm-sealed transmitter, employing a corrosion-resistant process seal. These devices transpose a liquid level to a pressure measurement, usually ranging from 20 to 800 in H_2O.

Density-measuring devices

Density measurements are used to control some process flow parameters in the petrochemical industry. Density varies with temperature; therefore, a base temperature must be used. This is usually 32°F, but the petroleum industry uses 60°F. Different process industries have their own units of density, as can be noted by the following:

1. Specific gravity (SG): the ratio of a material's weight to an equal volume of water at a specified temperature

2. Degrees API: the density unit used by the American Petroleum Institute

3. Degrees Baumé: a dual scale for acid density, with one scale for liquids lighter than water and one for those heavier than water

4. Twaddle: a scale divided into 200 equal parts over the SG range of 1.0 to 2.0; an SG increment of 0.005 equals 1 degree Twaddle

5. Brix: percent sucrose by weight in a pure sugar solution at 17.5°C

6. Richter, Sikes, and Tralles: scales reading directly in percent ethyl alcohol by weight in water

Determination of density can be accomplished by using the equation pressure = height of fluid times density of fluid, wherever it is possible to measure pressure and height of fluid. This cannot be used in pressurized vessels or where the fluid is in motion.

Fixed-volume weighers. With fixed-volume weighers, process fluid passes through a container of known volume. The container's weight gives a direct density measurement.

Radiation. Radiation devices pass gamma rays through the fluid to be measured. The amount of radiation picked up by the detector varies inversely with the material's density. The variable actually measured is the degree of radiation absorption.

Buoyancy. The principle of buoyancy also can be used to determine density, since the force on a submerged body is related to fluid density.

Refractometry. Another method is based on principles of refractometry: the velocity of light traveling through a substance is reduced in proportion to the optical density of the material. Differential density refractometers are limited to clear fluids. Critical-angle types have a tungsten light source centered on the angle at which the beam converts from reflection to refraction. The collumnated light beam is focused on a rotating prism, which causes the beam to sweep the process stream. Refraction, detected by a photocell and amplified into a usable signal, occurs when the light ray sweeps down to the critical angle.

Vibration.   Vibration of a U tube filled with liquid is also a function of the mass of the material within the tube. While the tube is vibrated by a pulsating current through a drive coil, the vibration is sensed and converted into a milliamp output signal. Since the vibration varies with fluid mass, the output signal can be calibrated in density units.

Torque.   The gravimetric method measures torque on an impeller being driven in a fixed-volume sample chamber compared with torque developed by driving the impeller in air. The torques are converted through mechanical linkages and transduced into standard instrument signals. The measuring system divides the torque developed in the gas by that developed in air, giving the same reading as dividing gas density by air density, i.e., specific gravity.

Combustible-gas meters

One instrument which may not be involved with direct operation, but which is critical in many plants where combustible gases are used or processed, is the portable combustible-gas meter. This is a portable instrument for quick determination and measurement of combustible gases or vapors that may be present in air. These instruments come in different models for picking up or sensing different gases. They should be used whenever a leak is suspected and before entering any tank for inspection purposes.

Strip charts versus CRT displays

Strip charts are being replaced by computer printouts; however, they can be extremely useful as a display of changing conditions for the operator so that adjustments can be made manually, or the equipment safely secured before more damage is sustained. They should be part of the operator's tools in intelligently diagnosing if an operation is proceeding smoothly.

CRT displays provide another process-to-operator interface and are displacing strip charts in some applications. These units can be programmed to display many items, such as:

Emergency indicators and alarms

Component diagnosis and maintenance

Process operation information

Process evaluation and diagnosis

Process supervision information

Piping and instrument diagrams

There are many more instruments used in process plants that involve other measurable items on pressure vessel systems. A plant's file on pressure vessels should include piping and instrument diagrams for important pressure vessels. These diagrams are useful when starting an operation and when checking malfunctions in the process control.

An instrument list is useful when checking out a system. This is an overall listing of all instruments by instrument tag numbers for easy identification. Supplementary information on the instrument list could include whether the unit is protected by hot- or cold-type insulation, whether the instrument requires heat tracing to prevent malfunction in subzero weather, and possibly what set points for operation and alarms are necessary, in addition to the usual information and instructions issued by the instrument manufacturer.

Instrument freeze-up

As with pressure vessels in general, outdoor instruments may be exposed to freeze damage. Many outdoor plants in the southern part of the United States are severely affected when unexpected cold weather envelops the plant. Quite often the plant suffers outages from serious instrumentation and control failures due to freezing of entrapped moisture in control lines, such as can occur with outdoor pneumatic systems. Extra air dryers are needed to eliminate as much moisture as possible, or the system may require emergency heat tracing on the outdoor lines.

Instrument maintenance

Instrument maintenance and calibration in modern process plants is a specialized field and is of growing importance as more automation is developed. In most cases, the instrument and control technician must be familiar not only with the instrument and the controller but also with required process phenomena that the instrument may be sensing and controlling. Periodic checking and calibration of these devices will help prevent unexpected shutdowns due to control failure. Calibration involves comparing the reading on the instrument with the value indicated by a reliable, known standard or with lab test results, where grab-sampling testing is done for quality control purposes.

Sensors and controls

Sensors and controls are major features in the operation of automatic pressure vessel systems and have made possible large-scale production of uniform-quality products.

Sensors are devices that have the inherent ability to monitor changes in the medium being measured, such as temperature, pressure, flow, level, draft, percentage of CO_2, and similar quantities. Transducers are used to convert the changes noted by a sensor to an electric or a pneumatic signal. These signals are sent to controls, which are set to regulate a quantity within set points. The signals are thus compared with these set points, and if needed, a signal is then relayed to an actuator so that control of the medium is maintained within the established set points. Some controls incorporate the sensor, transducer, and actuator into a self-contained control device.

Manual, Regulating, and Control Valves

Valves are extensively used for the following basic functions:

1. To provide on-off capability in a pressure vessel system
2. To provide a means of controlling or modulating flow rates
3. To provide a direction in the flow, or to prevent reversal of flow, for example, by check valves
4. To keep pressure, temperature, and similar physical variables within set operating points
5. To provide overpressure protection, such as is performed by safety and relief valves

ASME code requirements

The code requires that valves to be used for pressure vessel operation be marked with the name or trademark of the manufacturer, and with other markings of industry standards, so that the pressure and temperature rating of the valve can be determined from the manufacturer's catalog or similar industry listings. The code makes reference to the ANSI B16.5 standard for flanged valves. Any valve used in a jurisdictionally inspected pressure vessel will have to use code-permitted material for the service or construction. Some code requirements may be worth mentioning.

1. Cast-iron vessels and their valves cannot be used for lethal or flammable service.
2. Cast-iron construction cannot exceed the following values:
    a. 160 psi and 450°F for steam, gas, or other vapor service
    b. 160 psi and 375°F for liquid service
    c. 250 psi for liquids at temperatures less than their boiling points, but not to exceed 120°F
    d. 300 psi and 450°F for bolted heads, covers, and closures

3. The allowable stress permitted by the code for the material and its operating temperature must be multiplied by a quality factor if the material is cast, which many valves are. For those castings examined only to the minimum requirements, the allowable stress would have to be multiplied by 80 percent. If all the requirements for examination of the casting are performed, the casting quality can be as high as 100 percent.

Valve selection

Valve selection according to materials, type, and rating depends on the expected service. Fluid properties, pressure ratings, and expected operating conditions are only a few of the items to be considered. For example, variance in pressure and temperature in the process also may influence valve selection. The following factors must be reviewed in valve selection:

Design temperature limits

Design pressure limits

Control requirements

Allowable pressure drop

Corrosive nature of fluid

Erosion possibilities

Fouling possibilities

Leakage hazards

Heat conservation

Regardless of the type of valve being considered, all valves have characteristics in common, such as:

1. Mating surfaces that act as the seals to stop flow through the valve. This generally requires a stationary seat as well as a movable one.
2. A device that protrudes through the valve body, which moves the movable seat. This is often called the "valve stem."
3. A stem packing or similar seal arrangement to prevent loss of fluid where the stem emerges from the valve.
4. A handwheel or similar device to assist in the movement of the valve stem.
5. A passage for the fluid to flow through the valve. The configuration of this passage defines the type of control that can be expected.

Valve materials.   Material selection criteria are related to the valve selection criteria (see Table 6.2a and b). Table 6.2b lists valve and body trim

TABLE 6.2a Valve Material Selection

ASTM Material Specification and Service Description

ASTM spec no.	Service and description
A216 grade WCB (*A105 grade 11*)	Carbon steel. Regularly used in all 150-, 300-, and 600-lb valves. If long service at high temperature is anticipated, there is the possibility of graphite formation in flanged or screwed-end valves above 800°F and in welding-end valves above 775°F.
A217 grade WC1	Carbon-molybdenum steel. Suitable for slightly higher pressures than carbon steel at temperatures above 650°F. If long service at temperatures above 875°F is anticipated, there is the possibility of graphite formation.
A217 grade WC6 (*A182 grade F11*)	$1\frac{1}{4}$-$\frac{1}{2}$ chrome-moly steel. For temperatures up to 1000°F.
A217 grade WC9 (*A182 grade F22*)	$2\frac{1}{4}$-1 chrome-moly steel. For temperatures between 1000 and 1050°F.
A217 grade C5	4-6-$\frac{1}{2}$ chrome-moly steel for high-temperature corrosive oil service.
A217 grade C12	8-10-1 chrome-moly steel for high-temperature corrosive oil service.
A351 grade CF8M (*A182 grade FBM*)	18-8 chrome-nickel moly steel for temperatures up to 1250°F.
A352 grade LCB	A carbon steel for low-temperature applications to -50°F.
A352 grade LC1	A carbon-moly steel for low-temperature applications to -75°F.
A352 grade LC3	A low-carbon $3\frac{1}{2}$% nickel steel for low-temperature applications to -100°F, or somewhat lower.

The main ASTM spec numbers given are for castings. The alternative numbers (*in italics*) are ASTM specs for forgings, where applicable.

materials that are most often used for the many fluids and gases encountered in service. Trim material may have to be different from the material selected for the body, depending on valve function. For example, in a control valve that modulates flows, and where liquid or gas velocities increase at the point of control, erosion may require a trim material to resist this abnormal wear in comparison with the valve body.

Valves are pressure-containing vessels. They must be designed to confine the fluid being piped under pressure without leakage and without undue distortion of the pressure-containing parts.

As the temperature of the fluid being piped becomes elevated, the problem becomes more complex. The tensile strength of all metals and metal alloys is reduced as operating temperatures increase. The operating temperature is an exceedingly important factor in the selection of the material from which valves are manufactured.

The ability of a material to resist the corrosive effects of the fluid being

TABLE 6.2b Valve Material Selection

Guide for Material Selection

Fluid	Body and major parts	Trim
Acetic acid	Aluminum, Ni-resist, bronze	Bronze, Monel, stainless
Acetone	Brass, steel, iron	Bronze
Air	Iron, steel, brass, plastic	Bronze, plastic
Ammonia	Iron, steel, Ni-resist, stainless	Steel, stainless
Brine	Iron, steel, Ni-resist, brass, Monel, plastic	Bronze, Monel, plastic
Calcium chloride	Iron, steel, Ni-resist, brass, aluminum, stainless	Bronze, Monel, stainless
Calcium hydroxide	Iron, steel, stainless	Stainless, Monel
Carbonic acid	Brass, stainless, aluminum, Ni-resist	Bronze, stainless
Carbon tetrachloride	Iron, steel, Ni-resist, brass, stainless	Bronze, Monel, stainless
Chlorine—dry gas	Brass, iron, steel, plastic	Bronze, plastic
Chromic acid	Iron, stainless	Stainless
Citric acid	Bronze, Ni-resist, copper, stainless, aluminum	Bronze, Monel, stainless
Coal-tar solvents	Iron, steel, Ni-resist	Stainless, Monel
Copper sulfate	Stainless	Stainless
Creosote	Iron, steel	Stainless, Monel
Gelatine	Aluminum, stainless	Stainless, Monel
Glucose	Iron, steel, brass	Bronze, Monel, stainless
Hydrochloric acid	Bronze, Hastelloy C, plastic	Bronze, Monel, Hastelloy C, plastic

(*continued*)

piped is also a major consideration. A material entirely satisfactory for a use from the standpoint of pressure and temperature may be unsuitable because of rapid corrosion when in contact with the fluid being piped. Although the corrosion rate may be low enough for a particular use, objectionable discoloration and change in taste or odor of a product may rule out the use of the material. If a process is only temporary, or if the valves can be considered expendable and some corrosion can be tolerated, a material with less corrosion resistance than another may be selected, providing an economic saving.

TABLE 6.2b   Valve Material Selection (*continued*)
Guide for Material Selection

Fluid	Body and major parts	Trim
Hydrogen peroxide	Aluminum, stainless, Ni-resist	Stainless
Hydrogen sulfide	Stainless, Ni-resist, aluminum	Stainless
Nitric acid	Stainless, plastic	Stainless, plastic
Oxygen	Steel, brass, stainless	Bronze, stainless
Phenol	Iron, steel, brass, bronze, stainless, aluminum	Bronze, stainless
Phosphoric acid	Stainless, Ni-resist, plastic	Bronze, stainless, plastic
Potassium hydroxide	Iron, steel, Ni-resist, stainless	Stainless, Monel
Soap	Iron, steel, brass, bronze, stainless	Bronze, stainless, Monel
Sodium carbonate	Iron, steel, Ni-resist, stainless	Bronze, stainless
Sodium hydroxide	Iron, steel, Ni-resist, stainless	Stainless, Monel
Sodium phosphate	Stainless, Ni-resist	Stainless
Sodium sulfate	Iron, steel, Ni-resist, stainless	Bronze, stainless, Monel
Sulfate liquors	Iron, steel, Ni-resist, stainless	Stainless, Monel
Sulfuric acid	Iron, Ni-resist, bronze, stainless, plastic	Bronze, Monel, stainless, plastic
Sulfurous acid	Bronze, stainless	Bronze, stainless
Tannic acid	Brass, bronze, Ni-resist, stainless	Bronze, Monel, stainless
Water—distilled (demineralized)	Stainless, aluminum, plastic	Stainless, plastic
Water—mine	Iron, steel, Ni-resist, bronze, stainless	Bronze, stainless, Monel
Water—sea	Iron, steel, Ni-resist, plastic, bronze, copper, stainless	Bronze, stainless, Monel, plastic

Choice of material for any use is determined by its ability to withstand the pressure and temperature conditions of the services, its resistance to the corrosiveness of the fluid, and the cost as compared with any other suitable material.

Selection specifications.   Valve standards organizations publish much literature on valve specifications such as material, pressure, and temperature ratings and size dimensions. The following discussion provides some sources of information for selecting materials for various pressure vessel systems.

The American National Standards Institute (ANSI) publishes the rules for design of piping systems. The American Society of Heating, Refrigerating and Air-Conditioning Engineers (ASHRAE) suggests the use of ASME/ANSI B3.1, *Power Piping,* as a standard for high-temperature water systems. ANSI/ASME B31.9, *Building Services Piping,* was issued in 1982 and can be used for some piping systems. ASME/ANSI B31.3, *Chemical Plant and Petroleum Refinery Piping,* should be used for fuel piping; ANSI Z223.1, *Fuel Gas Piping,* or ANSI/NFPA 31, *Fuel Oil Piping,* should also be used for fuel piping. These design rules specify that valves are to be manufactured by ANSI B16 standards, which are both dimensional and design standards.

The American Society for Testing and Materials (ASTM) is concerned with specifying materials and test procedures for the materials that are used in valves. ANSI, API, and MSS specify ASTM materials and tests.

The Manufacturers Standardization Society of the Valve and Fitting Industry (MSS) is an organization of manufacturers that publishes specifications where no other specification exists. MSS fills voids in the specification and standards system. MSS withdraws their specifications when an ANSI or ASME standard is issued to replace them.

The pressure ratings of flanges and valves at various operating temperatures are set by the standards. For example, the pressure rating for a standard Class 300 flanged steel valve operating at 400°F is 635 psi.

Special valves are standard valves subjected to a rigorous quality control program and are rated higher than standard valves. Only end valves can be classified as special.

Valve pressure and temperature limits are different from those for connected piping. High-temperature piping systems depend on yielding to redistribute stresses. Valves are geometrically sensitive, and the excess deformation, even though elastic, would destroy the valve's ability to be an effective closure device. Valves are designed to operate at much lower stress levels than piping and fittings.

Cast-iron valves are not always made to a standard and may be rated by the manufacturer. Cast iron is rated as WOG (water, oil, and gas pressure), CWP (cold working pressure), and SWP (steam working pressure). WOG and CWP are the rated pressures of a cast-iron valve at 100°F; SWP is the rated pressure at saturated steam temperature.

Types of valves

Only the most frequently found valves will be described, and the reader may wish to pursue this subject further by contacting the many valve manufacturers in industry.

Hand-operated valves.   Most valves in the process industries are for flow and no-flow service:

Block valves.   Valves which are normally open to allow full flow but which can be closed to permit rerouting of flow for alternative operation or isolation of a piece of equipment for maintenance (also called "stop valves").

Drain valves.   Valves located at the low point in a pipeline, vessel, or other piece of equipment for the purpose of removing liquids from the system. Under normal operating conditions, drain valves are in the closed position and are operated only occasionally. Most systems require at least one drain valve for removing hydrostatic testing fluid and for emptying process fluids on shutdown.

Bleed valves.   Small valves placed within systems that are expected to be occasionally closed off while under pressure. The bleed valve allows the isolated system to be safely depressurized by discharge of vapor or liquid, before opening the system for inspection or maintenance, or for purging one fluid from the system before introducing another.

Also, bleed valves are commonly installed between double block valves. The bleed valve is turned to the open position to positively prevent mixing of different fluids in two systems separated by the two block valves in series.

Dump valves.   Valves provided for the purpose of rapidly discharging a fluid from a vessel or other piece of equipment. A dump valve is located at the low point of the equipment and differs from a drain valve as to the urgency of removal of the fluid. For this reason, a dump valve is generally a quick-opening device; it is sized for greater flow rate than a drain valve.

Many manual valves are available, not only for flow and no-flow service but also for manual regulation, especially if a controller or regulator may not be working properly.

Gate valves.   A gate valve is a straight-through flow design (Fig. 6.4a). The barrier to flow is a disk or wedge-shaped dam sliding at right angles to the direction of flow and seating tightly in the valve body. When partially open, this type of valve exhibits a crescent-shaped opening for flow, which changes in area quite rapidly upon slight adjustment of the valve handle, thus making such a valve undesirable for partial flow control.

This valve designation is further subdivided to differentiate between rising stem (or spindle) and nonrising stem, solid wedge and double disk, inside screw and outside-stem thread, bolted and screwed bonnet, handwheel- and lever-operated stem, etc.

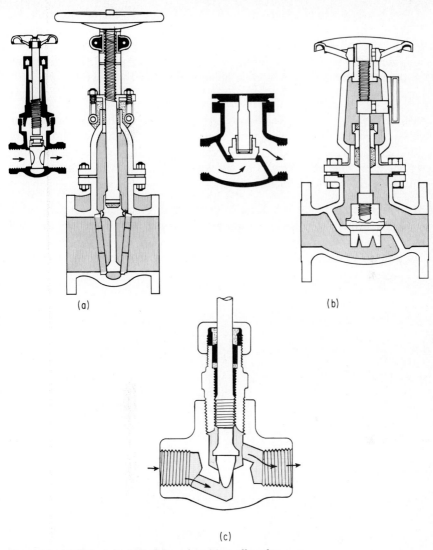

(a)   (b)

(c)

Figure 6.4 (*a*) Gate valve; (*b*) globe valve; (*c*) needle valve.

Globe valves.   The globe valve is so named because of the bulbous shape of the valve body (Fig. 6.4*b*). Flow through this type of valve is directed up or down through a circular opening in the labyrinth, which may be sealed either by forcing a replaceable disk down on a flat seat or by inserting a tapered metallic plug into a conical seat.

Where a plug is used, the seat is made with a different taper from the plug to furnish a line contact for the seal. The stem is sealed with a packing gland.

Globe valves are available with many different details of construction. The stem or spindle may move outward or simply rotate as the valve is opened, the screw threads on the stem may be inside or outside the pressure zone, the seat may be permanent or replaceable, and the valve may be operated by one stroke of a lever handle or by several turns of a handwheel.

Needle valves.   Figure 6.4c shows a special design of the plug-type globe valve, in which the plug is a slender, tapered needle in a small orifice of different taper. The needle valve is especially well adapted for fine control of flow when the flow is at the low range. This valve assures good shutoff if the service is clean.

Diaphragm valves.   A diaphragm valve (see Fig. 6.5a) consists of a straight-through-flow body design, which may or may not be interrupted with a transverse weir. Valve closure is effected by pressing a flexible diaphragm against the interior body wall or against the transverse weir.

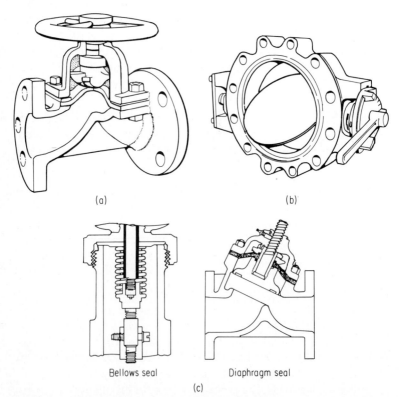

(a)   (b)

Bellows seal   Diaphragm seal
(c)

Figure 6.5   (a) Diaphragm valve has diaphragm acting as plug-and-stem seal. Some type of plastic is used for the diaphragm. (b) Butterfly valve saves installation space. (c) Valve stem seals of the diaphragm type exclude fluid handled from the operating mechanism and prevent outside leakage.

The diaphragm seals the valve body from the stem and the biscuit-shaped compressor so that the valve is leakproof.

The original valve of this type was called the "Saunders valve," named after the patent holder. The Saunders valve has a contoured transverse weir and requires only a small movement of the diaphragm to open or close.

Advantage:
The diaphragm valve is leakproof and requires no packing.

Disadvantages:
The diaphragm valve is not suitable for high-pressure operation.

Flow control is not good at very low flow rates.

Butterfly valves.   The butterfly valve (see Fig. 6.5b) is a straight-through-flow design. The barrier to flow is a tilting disk that pivots on a central transverse shaft and is operated by rotation of the shaft. The disk seats on the walls of the valve body. Rotation of the shaft may be accomplished by means of a lever, bevel gears, or worm and spur gears, and the shaft seal is made by a packing gland. Butterfly valves are generally used to control gas flows. Two valves operating in parallel are sometimes used with liquid or gas flow to provide temperature control with bypass around a heat exchanger.

Check valves.   Check valves are used to permit flow in one direction only. They are extensively used in the service area of a plant (water, steam, air, gas), as well as for process flow applications on a pressure vessel system. See Fig. 6.6.

Swing check valves.   The swing check valve is a straight-through-flow device equipped with a seat on which a disk rests under no-flow conditions. An arm connected to the disk supports it from a pin set in the upper portion of the valve body (or in the side closure), which permits opening the check valve for inspection and maintenance. The supporting pin is a hinge on which the disk may swing freely away from the line of flow upon the exertion of fluid pressure on its upstream side (Fig. 6.6a).

A reversal of flow exerts fluid pressure on the downstream side of the disk, forcing it against the seat, and stopping flow. The greater the downstream pressure, the tighter the disk is seated. Swing check valves are designed for horizontal flow but may also be used for vertical flow in the upward direction.

Advantages:
The swing check valve is simple and easily maintained.

Such a valve may be used in either liquid or vapor service.

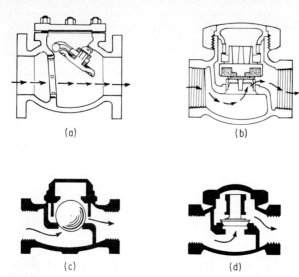

(a)   (b)

(c)   (d)

Figure 6.6   Types of check valves. (*a*) Swing check valve; (*b*) lift check valve; (*c*) ball check valve; (*d*) piston check valve.

Prevention of gross reverse flow is assured in clean service.

The valve is easily cleaned through the side closure when necessary.

Disadvantages:
A swing check valve does not assure tight shutoff, and backflow leakage may cause contamination of the upstream fluid.

Operation of the valve is dependent upon a clean seat and disk, which makes it unfit for slurry service.

The swing check valve is generally intended for horizontal flow; it adds more pressure drop to the system when used vertically.

A sudden reversal of flow causes the disk to slam down on the seat and cause severe water hammer.

Lift check valves.   Lift check valves are designed for horizontal lines or vertical lines with upward flow. The body of a lift check valve for horizontal flow (Fig. 6.6*b*) is similar to that of a globe valve. The body of the check valve for vertical flow is similar to that of the angle valve, except that the fluid exits through the top of the valve instead of the side. The barrier to flow is a free-moving plug with a stem that moves within a guide fastened to the valve body or side closure. The plug drops into a seat by gravity and opens by fluid pressure on the upstream side. Reversal of fluid flow forces the plug back into the seat.

Advantages:
The free-moving plug has no hinge that could cause mechanical failure.

The side closure allows easy access for inspection and cleaning.

The lift plug can be quickly replaced.

The guide pocket behind the stem acts to cushion the motion of the plug as the valve opens.

Disadvantages:
A lift check valve does not assure tight shutoff.

Ball check valves. A ball-check valve (Fig. 6.6c) is similar to the lift check valve except that the plug is replaced with a sphere, usually of solid metal. A guide is used to limit the movement of the ball and direct it back to the seat. Use of this valve is limited to clean service, and the valve does not assure tight shutoff.

Spring-loaded check valves. Spring-loaded check valves are used on swing check valves, with the spring being used to reduce the chatter and water hammer that accompany the rapid opening and closing of check valves. Corrosion or fatigue of the spring has to be watched, because without the spring, the valve will not function properly.

Maintenance of valve parts

Wearable valve parts require maintenance to prevent leakage. Some seats permit removal of the valve cap to provide access to the valve disk so that it can be rotated and reground to its body seat without removing the valve from the line. Some valves have metal disks with renewable seats for easy replacement of worn seats. Most larger valves need to have their seats periodically reground to assure a tight fit.

Packing is another valve part that needs periodic maintenance. All hand-operated valves have stems that protrude through the bodies of the valve. Packing is used to prevent the flowing medium from leaking to the atmosphere. The packing specified for a valve must be compatible with the fluid's corrosive characteristics and temperature. There is a movement to use materials other than asbestos. Several nonasbestos materials are now being offered. Typical packing material temperature limits are as follows:

- Teflon, 500°F
- Straight asbestos, 750°F
- Asbestos-core packing reinforced with Inconel metal wires, 1200°F
- Carbon packing, 650°F
- Polymer fiber packing, 625°F

Valve connections

There are an almost endless number of design variations and combinations available for each functional type of valve.

Butt weld ends. Butt weld end connections are available on steel and alloy valves. ANSI B16.25 specifies machining of butt weld ends that will permit the attachment of valves into a piping system by welding.

Valves ordered with butt weld ends should include specifications of the schedule of pipe to which the valves will be welded so that the ends can be machined with an inside diameter to match the bore of the pipe. Butt-welded valves eliminate the flange, which is the weakest part of a valve. A butt-welded valve can be rated for a higher pressure than a flanged valve can.

ANSI B16.5 specifies the contour of the weld end, and ANSI B16.10 specifies the end-to-end dimensions so that any manufacturer's valve can fit into the piping system without any change in the general piping dimensions.

Flanged ends. Flanged ends are available on bronze; iron; ductile iron; and cast, forged, and alloy steel valves. Flanges are made to mate with all ANSI and API flange sizes.

Flanged ends allow valves to be bolted into a piping system. They are used wherever it is necessary to remove or replace valves without disturbing the piping system and where the use of slip blinds is required for safety.

Bronze flanges are flat-faced with two concentric V grooves between the port bore and the bolt circle.

The 125-lb iron flange is flat-faced; the 250-lb iron flange is machined with a $\frac{1}{16}$-in raised face.

ANSI and API flanges are furnished with a raised face. Dimensions of flanges are specified by ANSI B16.10.

Screwed ends. Screwed end connections are available on bronze; ductile iron; iron; and cast, forged, and alloy steel valves. Valves are furnished with female threaded ends matching the male thread of the pipe to which it is joined. The contour of the tapered threads is specified by ANSI B2.1.

Screwed ends are limited to valves under 4 in. Pipe makeup becomes increasingly difficult if the size is over 4 in.

Socket-weld ends. Socket-weld end connections are available on small forged steel valves through the 2-in size. Socket-weld ends have certain advantages over butt weld ends, particularly in small sizes. The pipe to be welded to the valve slips into the socket and is supported and aligned by the socket. There must be at least $\frac{1}{16}$-in clearance between the pipe and the bottom of the socket. Pipe that touches the bottom of the socket and is welded out will be overstressed and can fail. Piping fabricators have several techniques that ensure that the pipe is not bottomed out.

Regulators

Regulators are still being used in certain applications, as when pressure regulators are used in place of control valves. Control valves depend on sensors to measure pressure, temperatures, liquid levels, and similar variables; the sensors then transfer this measurement by pneumatic or electric signals to an actuator, which maintains the measured variable within desired set points. A regulator is more self-contained and does not depend on transmitters.

Most regulator designs fall into three classes: spring-loaded, direct-operated; pilot-operated; and pneumatic-operated or air-loaded. The direct-operated regulator is the simplest and is intended for rugged service where high accuracy is not especially needed. Droop is typically 25 to 30 percent but can be lessened by choosing a larger regulator.

Pilot-operated regulators use a pilot mechanism to sense a change in set point. They then load or unload a diaphragm or piston to move the inner valve. This type commonly has an accuracy of 95 to 99 percent but is susceptible to dirt and should not be used for intermittent service.

A pneumatic-operated regulator uses an air or steam loader. It combines the durability of a direct-operated model with the accuracy of a pilot-operated one, and should be first choice, provided there is an air supply available.

Control valves

Control valves are used when fine regulation within narrow set points is required. A control valve is also best if any remote feedback or control, derivative action, or automatic reset is required. For example, if steam is used as a heating medium, both temperature and flow should be monitored. A sensor can be installed on a regulator, but not easily.

Because regulators are a bit more susceptible to dirt, they should generally be ruled out for intermittent service. If they are used for such service, care should be taken to clean them after each downtime. Also, a regulator uses the force of the flowing steam or fluid to operate its valve, and needs a pressure drop of at least 5 to 15 psi to function. If the pressure drop in the system is below this range, a control valve should be used.

Accuracy of regulation for a regulator or control valve is the offset pressure expressed as a percentage of the set point. For example, if a valve has a set point of 100 psig at minimum flow and the outlet pressure falls to 75 psig at maximum flow, the accuracy of the regulator is 75 percent. The converse of the regulator's accuracy is its droop. In this case the droop is 25 percent.

The fail-safe concept is an important consideration in installing control valves. Installation of control valves that depend on air or electric power for actuation require a fail-safe approach so that the valve will either close or open by spring action or some other means that does not depend on air or electric power. For example, a valve operated by an electric solenoid and controlling the fuel fed to a fired object should automatically close on loss of power in order to prevent fuel from being continuously supplied to the fired object, which could result in an uncontrolled firing condition. Many accidents have occurred because the fail-safe concept was not followed.

Electrical Equipment

Electrical equipment in hazardous areas requires careful review so that the requirements of the *National Electrical Code* for hazardous areas is complied with at all times. The *National Electrical Code* defines hazardous areas in the following manner:

Class I: Areas with flammable gases or vapors

Class II: Areas containing dust

Class III: Areas containing flying particles, such as fibers

Class I gases are subdivided into the following groups:

Group A: Acetylene—most hazardous

Group B: Hydrogen or similar gases or vapors

Group C: Ethyl ether vapors, cyclopropane, ethylene

Group D: Hexane, gasoline, naphtha, propane, etc.

Division 1 areas may contain ignitable concentrations of gas, dust, or fibers under normal conditions. Division 2 areas may contain ignitable concentrations of gas under abnormal conditions, or where dust or fibers may be present, stored, or handled.

During the design and installation of a process unit, each section of the plant should be classified by a combination of the above definitions. For example, Class I, Group D, Division 1 would be a hazardous area containing a gas such as pentane, under normal conditions.

Any electrical equipment installed must be suitable for use in the specific area. This includes replacements when repairs are made. Explosion-proof installations allow safe operations of electrical equipment in potentially hazardous areas. They accomplish this by being designed so as to confine any explosion (ignited by the equipment) within a strong, heavy, relatively tight enclosure and thereby prevent spread of the explosion outside the enclosure. If the *National Electrical Code* classification system is carefully followed, not all equipment may have to be explosion-proof.

Maintenance and repair personnel must guard against installing the wrong electrical equipment in hazardous areas in their rush to restore production or operations when a piece of electrical equipment requires repair.

Regulation and Control

Operators in fast-moving process streams using high-pressure vessels must be especially alert and constantly aware of the potential risks that may be involved in the process. Classroom training on the plant's process flow and the equipment involved would be extremely useful in laying the groundwork that operations can play in assuring a safe and reliable process. This should be supplemented with field training in similar plants and should include training on the actions that may be needed in the event that regulators and controls malfunction.

Instrumentation personnel also should be trained to understand the interaction of instrumentation with the process and how a malfunction in instrumentation could affect the operation and operating personnel involved. Backup to instrumentation can include (1) a redundant or parallel-flow control system, (2) a return to some regulatory-type operation, or (3) manual control valves, with dependence placed on upper-limit safety switches or hard restraints such as safety and relief valves and rupture disks.

In general, regulatory controls can be divided into two categories: discrete controls, and proportioning controls. Discrete controls cover on-off and other two-state controls. A motor may be either running or not. A two-position valve may be either open or closed. In a batching system, for example, each ingredient is metered through a pump and then introduced to the batching environment by a two-position valve. There is no need for intermediate positions on the valve. Discrete regulatory controls make only these on-off types of decisions.

Continuous or proportioning controls, on the other hand, include most flow, level, pressure, and temperature controls, in which the output from the control logic is used to position a control valve.

Transmitters

Pressure, temperature, and similar measured quantities must be transmitted to actuators, regulatory controls, supervisory controls, and displays in remote control rooms. Transmitters are pneumatic, electric, or electronic.

The pneumatic type usually employs an analog control system and has the advantage of being less expensive than other types. Another advantage is the fact that since there are no electric signals required, there is less risk of electric sparks in sensitive process areas. Pneumatic systems require dry air, with the air supply having a dew point above $-40°F$ to prevent condensation in the loops of the pneumatic system.

Electronic control systems depend on electric-signal transmitters. Essentially, there are three types of electronic control systems.

1. Conventional electronic instruments

2. Electronic systems with all field devices explosion-proof

3. Intrinsically safe electronic systems

The need to have a clear understanding of the differences is important.

Safety considerations for electric controls

Some plant duties involve at least one flammable liquid that is being processed in both the vapor and the liquid phases. Since there always is the possibility of a leak of liquid or vapor, particularly from pump seals, it is essential for complete safety that there be no source of ignition in the vicinity of the equipment. While many instruments such as controllers and alarms can be located in a control room removed from the process, all local electronic instruments must be either explosion-proof or intrinsically safe.

With explosion-proof equipment, electrical devices and wiring are protected by boxes or conduits that will contain any explosion that may occur.

With intrinsically safe equipment, barriers limit the transmission of electrical energy to such a low level that it is not possible to generate a spark. Since explosion-proof boxes and conduits are not required, wiring costs are reduced.

For any intrinsically safe system to be accepted for insurance purposes, Factory Mutual (FM), Industrial Risk Insurers (IRI), Canadian Standards Association (CSA), or local fire department approval usually must be obtained. This approval applies to a combination of barriers and field devices. Therefore, when a loop incorporates such instruments from different manufacturers, it is essential to ensure that approval has been obtained for the combination of instruments.

Regulatory controls

A regulatory controller accepts a signal or signals from one or more sensors and determines what adjustments, if any, must be made in order to maintain a process operation at or near the desired value or target. A "regulatory controller" can be an operator who reads the output of measurement devices and then makes process adjustments using actuator devices. Pneumatic and electronic systems have been and still are being used for this purpose, but more and more manufacturers are using digital control to perform many of these tasks. These devices read sensor outputs and use these values for automatic process control, thereby freeing up operators for areas that need close attention.

Actuators

If an operator or an automatic control system specifies any adjustments to the process, these adjustments are implemented by an actuator. Pneumatic actuators that position the stem of a valve account for the overwhelming majority of the actuators in process plants. Other actuators include variable-speed drives, found, for instance, on many screw feeders and on large pumps and compressors.

Distributed computer control

A distributed computer control system utilizes microprocessor-based controllers that are distributed along a communications network between the process being controlled and the process operator's console or a process computer. The term "distributed" can have two different meanings. One is the use of the term in a functional sense. In the early direct digital control (DDC) computers, as many as 200 loops might be implemented in a single processor, the result being a highly centralized system. Upon failure of this processor, all 200 loops would go out of service. In today's distributed systems, these same 200 loops are functionally distributed to *several* "smart" controllers. Depending on the manufacturer, an individual controller might be performing only 1 loop, or it might be performing 8 loops, or it might be performing 30 loops. In any case, the net result is that the loss of a single microprocessor would have less impact than the loss of a centralized DDC computer.

Alternatively, the term "distributed" can be used in a geographical sense. A typical maximum length of a data communications network is 2 miles. Therefore it would be possible to distribute the controllers over a 2-mile geographical area. Applying this capability to a process plant, it means that the controller for a chemical reactor could be mounted on the side of the reactor, the controller for a distillation tower at the base of the tower, and so forth.

Large-case pneumatic controls were located within the process area in much the same way. Using this technology, the controls were distributed, and so was the operator interface. In large plants the desirability of a centralized operator interface quickly became obvious. Today's distributed systems permit controls to be geographically distributed while retaining a centralized operator interface located in the central control room, where overall process supervision can be done.

This has certain advantages, the one most commonly cited being a reduction in the amount of signal wiring needed. Unfortunately, there are certain disadvantages. Having controls located in a process area does not appeal to many service people, because instrumentation that is located in these areas has always presented more maintenance problems. Maintenance personnel prefer that controls be located in a temperature- and humidity-controlled environment. Service records reveal that maintenance personnel spend a great deal of time working on instrumentation located in process areas, and very little on instrumentation located in a control room.

Environmental considerations within the process area can also be an obstacle to geographically distributed control, as can a number of other factors. In most installations, all the controls are mounted in the control room or in the adjacent instrument room. From a geographic view, the control system is still centralized.

Safety and Safety Devices

Alarms

Alerting process operations personnel to abnormal conditions within the process has long been a function of a control system. The annunciator panels incorporated into conventional control rooms provide this function.

Alarms can be activated by both continuous and discrete variables. For process variables such as flow, temperature, and so forth, the following types of alarms are available:

Absolute-limit alarms. Usually called high-low alarms, these generate an alarm if the process variable exceeds the high limit or goes below the low limit. Example: alarm if the reactor temperature exceeds 350°F.

Deviation alarms. An alarm occurs if the process variable differs from a specified target by more than the alarm limit. Example: alarm if the reactor temperature differs from 250°F by more than 10°F.

Trend alarm. Also called a "rate-of-change alarm," this alarm occurs if the rate of change of the process variable exceeds a specified limit. This alarm is normally applied to the absolute value of the rate of change, although there are a few situations where this is not appropriate. Example: alarm if the reactor temperature is changing by more than 5°F per minute.

For discrete variables, three types of alarms are commonly supplied:

Status alarms.   The alarm occurs whenever the input is not in its specified state.

Change-of-state alarms.   The alert occurs whenever the state of the input changes.

Limit-switch alarms on discrete outputs.   For two-state valves, limit switches can be installed to detect when the valve is fully open or fully closed. When the state of the output to the valve changes, the states of the limit switches should agree within a period of time consistent with the travel time of the valve.

Many variations and extensions to the above are possible.

The presence of many alarms can be a detriment to an operator faced with many alarms ringing at once, because this can cause confusion as to which alarm should be considered first. Therefore, extreme care is needed in selecting alarms and in establishing the sequence in which they alert operators. One of the conclusions in the report on the Three Mile Island nuclear accident was that the operators received too many alarms and thus were faced with a severe diagnostic decision. The term "priority alarm system" is gaining increased attention to counteract this possibility of too many alarms confusing the operators.

Interlocks

Interlocks are used to start or shut down a process or system in a manner that will prevent damage to equipment resulting from the following of improper *sequences* in operations. For example, a water-cooled air compressor should not be started unless water flow is first passing through the jacket in order to prevent overheating in the compressor. An interlock device can prevent this. The interlock could consist of wiring the controls in series so that the water pump supplying the jacket must start before the motor driving the compressor can be energized. Another possibility is to have a flow switch in the waterline that must close a contact before the compressor motor can be started. Thus, a variety of interlock schemes and devices are available to start and shut down equipment properly in a logical sequence. It is essential to review operations and determine where interlocks will prove beneficial, even though they may take away the operator's manual sequence of starting and shutdown options at times.

All interlock circuits consist of the following components:

1. An input device such as a pressure switch, flow switch, selector switch, or relay that measures a variable.

2. A built-in system of logic, such as is found in systems employing series wiring, or a system of logic provided by a programmed controller.

3. Final actuating devices, such as solenoid valves that open when an electric circuit is energized, an indicating light to show that the next step in starting or shutting down can be followed, or an interlocked alarm to warn of a developing malfunction.

Manual reset switches are quite often used on interlock devices in order to force the operator to check what the problem may be in the process flow. At times, interlocks require bypasses in situations in which a shutdown of a key vessel or machinery may cause extensive delay in restarting. The interlock should probably ring an alarm first. Some interlocks have a time-delay feature to avoid nuisance trips.

Hard restraints

The use of hard restraints means that positive action has been taken to prevent a serious accident. In pressure vessels, this means that a positive device has been installed to prevent a serious overpressure explosion from occurring; such devices usually consist of safety and relief valves, rupture disks, or both.

Safety and relief valves are governed by ASME code requirements, which cover such things as settings versus allowable pressure and construction and installation recommendations. Because of the peculiar nature of some chemical process plants, the threat of a rapid chemical reaction taking place within a vessel system must be carefully evaluated by responsible plant engineering people who are familiar with the process. Some processes require several approaches in dealing with a potential explosion. These start with proper system design to minimize the chance of an explosion, including careful selection of instrumentation and controls; these beginnings are backed up by process control display boards for evaluation and surveillance by trained operators. Containment is sometimes possible by designing the equipment or enclosure so that it can withstand the maximum possible pressure developed in a runaway reaction. The final defense, however, is always a means of venting in order to prevent pressure buildup.

Restraint of chemical runaway reactions

Establishing the appropriate size of relief devices for a runaway reaction requires research in thermokinetics in order to determine the speed of the reaction—a critical factor in venting considerations. The chemical parameters to consider are broken up into the thermodynamic, kinetic, and physical properties of the chemical substances under review:

1. *Thermodynamic*
   Reaction energy
   Adiabatic temperature increase
   Specific quantity of gas generated
   Maximum pressure in a closed vessel

2. *Kinetic*
   Reaction rate
   Rate of heat production
   Rate of pressure increase in a closed vessel
   Adiabatic time to maximum rate
   Apparent activation energy
   Initial temperature of detectable exothermic reaction

3. *Physical*
   Heat capacity
   Thermal conductivity

Pressure changes in a closed pressure vessel will be dependent on the reaction rate that may occur. Calculations for the reaction rate are a responsibility of process designers. When reaction kinetics for runaway reactions cannot be adequately calculated, the usual procedure is to scale up from laboratory experiments. In some cases, past practices of other plants with similar processes are used as a guide.

In addition to the use of hard restraints, other options can be considered for protection against runaway chemical reactions. Examples are:

1. Injection of a neutralizer into the reaction vessel at the first indication of lack of control

2. Quick lowering of the reaction temperature by water spray, either within or outside the vessel

3. Rapid, deliberate depressurization by channeling the contents to an empty container or flare stack if atmospheric venting may be a problem

4. Deliberate venting by lifting the test handles of relief valves

Each procedure may not be proper in all cases. For example, adding water for cooling might *increase* the chemical reaction in some chemical processes. The same may be true of depressurization. Thus, careful technical review is needed to determine the best defense that should be used to avoid a runaway chemical reaction. The process designers are usually the best qualified to determine what the strategy should be to control a chemical reaction, because of their intimate knowledge of the process variables and the reactions that may occur if design parameters are not followed. This knowledge is in many cases of a proprietary nature; however, plant safety specialists should seek any information necessary to avoid runaway chemical reactions.

Pressure relief valves

Pressure relief valves are classified as "direct-acting" and "pilot-operated." The terms "relief valves" and "safety valves" also need defining. A *relief valve* is an automatic pressure-relieving device that is actuated by pressure that pushes against a spring-loaded seat at the inlet of the valve; it is primarily intended for liquid service. A relief valve opens in proportion to the increase in pressure above the opening pressure. A *safety valve* is primarily used for gas or vapor relief but also has the distinguishing feature of opening rapidly by a pop-type action. A safety relief valve, if so designated by the valve manufacturer, and so certified, is thus suitable for liquid, vapor, or gas service.

Code safety valves.   The code requires all safety, safety relief, and relief valves to be of the direct, spring-loaded type. Pilot-operated valves may be used on pressure vessels, but the pilot valve must be self-actuated, and the main pressure relief valve must open automatically at its set pressure, not at a greater pressure; it must be able to discharge its full rated capacity if a part of the pilot should fail.

Safety relief valves should follow code-approved construction. Figure 6.7a shows an identifying symbol indicating code certification. Figure 6.7b shows a safety valve with a huddling chamber to provide the pop action for instant lifting of the valve when the set pressure is reached. Figure 6.7c shows a valve used for high-temperature service (usually over 450°F), in which the spring is outside the seat area so that the spring is not in contact with the high-temperature gas or fluid, thus isolating it from the effect of the gas or liquid.

When a relief valve bears the ASME and National Board stamping, it is the manufacturer's guarantee that the rules of the ASME code have been followed in the construction of the product.

In brief, the major constructional requirements are that the disk and seat be of noncorrosive material, the seat being fastened to the body so that it cannot lift with the valve disk. All parts should be constructed so that no failure of any part will interfere with full discharge capacity of the valve. The seat may be inclined at any angle between 45 and 90°.

The relief valve must be of the direct, spring-loaded type. Code states do not allow the installation of weight and lever types or deadweight safety valves, for the adjustment of such valves is too easily tampered with. Their use in non-code states is not recommended and is considered dangerous.

The ASME unfired pressure vessel code requirements for overpressure

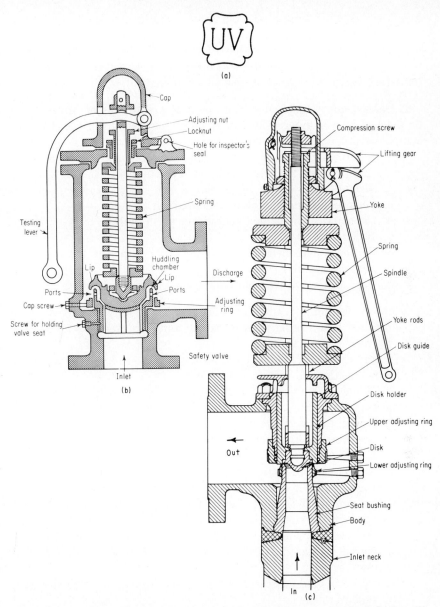

Figure 6.7 (a) Official ASME stamp for standard, unfired pressure vessel relief valve; (b) pop safety valve with huddling chamber to provide additional lift to quickly open the valve; (c) safety valve for high-temperature service, with springs outside the valve casing so that their properties are not affected by temperature.

relief devices differ somewhat from those listed for fired boilers. The reason for this is that there is a broad range of overpressure devices for unfired pressure vessels that does not exist for boilers. The following requirements should emphasize this difference between the requirements for safety valves in unfired vessels versus boilers.

1. Safety, safety relief, and pilot-operated pressure relief valves should be marked in a manner that cannot be obliterated in service; the code symbol used should convey the following information:
   a. Identifying name or trademark of manufacturer of the valve, including manufacturer's design or type number
   b. Size in inches of the valve inlet
   c. Set pressure in pounds per square inch
   d. Capacity of the valve in cubic feet per minute of air at 60°F and 14.7 psia
   e. For code-classified unfired boilers, the capacity (in steam) in pounds per hour
   f. Year the valve was built, or some coding to identify the year

The above stamping requirements are applicable only to valves on $\frac{1}{2}$-in pipe or larger. For smaller valves, the stamping requirements for size and capacity may be omitted.

2. Liquid relief valves must be at least of $\frac{1}{2}$-in iron pipe size and require the following stamped information:
   a. Name or trademark of the manufacturer, including design or type number
   b. Size in inches of the valve inlet
   c. Set pressure in pounds per square inch
   d. Relieving capacity in gallons of water per minute at 70°F

3. Pressure relief valves in combination with rupture disks require that the markings be placed on the valve or on a plate securely fastened to the valve. The combination should be marked for the following:
   a. Capacity of the combination, in pounds of saturated steam per hour or in cubic feet of air per minute at 60°F and 14.7 psia
   b. Name of manufacturer of the *valve,* with the design or type number
   c. Name of manufacturer of the *rupture disk,* with the design or type number
   d. The safety valve set pressure (on the valve) and the bursting pressure at coincident disk temperature (on the rupture disk)

The code requires that the space between a rupture disk and a relief valve be provided with a pressure gage, try cock, free vent, or other suitable telltale indicator to show that the rupture disk may be cracked and leaking. See Fig. 6.8.

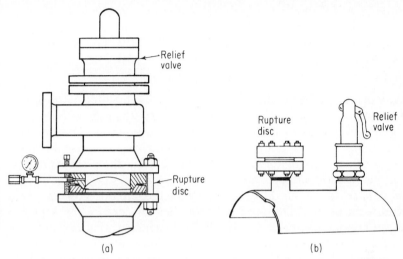

Figure 6.8 Relief valve and rupture disk protection may be needed in certain services. (*a*) Rupture disk is used in corrosive service to provide primary relief, with the relief valve being secondary. This arrangement protects the relief valve. (*b*) Relief valve opens first, and if pressure rises rapidly, rupture disk opens to provide additional capacity.

Rupture disks if installed ahead of the inlet to a relief valve, must burst so that there is no chance of interfering with the proper functioning of the relief valve downstream, including restricting the flow to the inlet of the relief valve.

If the rupture disk is installed on the outlet side of a relief valve, it is necessary that the back pressure on the relief valve not affect the proper opening of the relief valve at set pressure or restrict the capacity of the valve such that pressure can rise above code limits.

The contents of the vessel must be clean to avoid any gumming or clogging of the space between the rupture disk and the relief valve.

Rupture disks must be stamped for:

1. The name of the manufacturer or trademark, with the design or type number and rupture disk lot number.

2. Size in inches.

3. Bursting pressure, in pounds per square inch, and coincident temperature, °F.

4. Capacity in pounds of saturated steam per hour or cubic feet of air per minute at 60°F and 14.7 psia. The code permits stamping the capacity rating for other fluids or gases.

The use of relief valves in series with rupture disks is required in many process applications. Carbon-type disks can be placed ahead of a relief valve to protect the valve from corrosive contents (see Fig. 6.8a). Rupture disks are used *after* a relief valve to minimize the loss by leakage through the relief valve of valuable contents or lethal or other obnoxious or hazardous contents.

Rupture disks. Rupture disks are classified as permissible, nonreclosing pressure relief devices that may be used to satisfy the code relief requirements for overpressure protection of a pressure vessel. They are not spring-loaded devices but are really rupture diaphragms or membranes of metal or other sheet material that will burst when the pressure reaches the set-to-operate point.

Set to operate is defined by the code to be the bursting pressure of the rupture disk device. The code requires that all rupture disks have a stamped burst pressure that falls within a manufacturing design range. To understand how the stamped burst pressure relates to the set-to-operate requirements of the code, it is first necessary to understand the concept of "manufacturing design range." The ASME code states, "The manufacturing design range is a range of pressure within which the average burst pressure of test discs must fall to be acceptable for a particular requirement as agreed upon between the rupture disc manufacturer and the user or his agent."

Since rupture disks are manufactured from many different materials, in many different diameters, having the possibility of many specified burst pressures, across a large range of coincident temperatures, it would be impossible to maintain an inventory of all materials in an infinite number of thicknesses to meet all these requirements. In order to cope with these conditions and to provide rupture disks economically, manufacturers have established various ranges or tolerances around the nominal burst pressure, and refer to them as the "manufacturing design range."

Establishing burst pressure for rupture disks. The ASME code provides for the stamped burst pressure, within the agreed upon manufacturing design range, to be established by one of the following three methods:

1. At least two sample rupture disks from each lot are ruptured at the coincident temperature and one disk at room temperature. The stamped burst pressure is the average of those tests performed at coincident temperature, and this average should not fall outside the agreed upon manufacturing design range.

2. At least four sample rupture disks, but not less than 5 percent from each lot, are burst at four different temperatures. From these tests, a curve can be plotted of burst pressure versus temperature. The stamped burst pressure at the coincident temperature for this lot of rupture disks

is interpolated from this curve and must not fall outside the agreed upon manufacturing design range.

3. For prebulged, solid metal, and graphite rupture disks, a curve of percentage change of burst pressure at temperatures other than ambient may be established for a particular lot of material. It is necessary to burst at least four disks of one size at four different temperatures. From these tests, a curve can be plotted of percentage change of burst pressure versus temperature.

For a given lot of rupture disks made of the same material, at least two disks must be ruptured at room temperature. The percentage change of burst pressure for the coincident temperature taken from the curve established for the material shall be applied to the burst test to establish the stamped burst pressure at the coincident temperature. Again, the stamped burst pressure shall not fall outside the agreed upon manufacturing design range.

The code also requires the manufacturer to guarantee that the rupture disk will burst within ±5 percent of its stamped burst pressure at its coincident temperature. This means that the burst tests performed in accordance with the above three methods must not spread from their average more than ±5 percent, regardless of the manufacturing design range. This tolerance is commonly referred to as the "rupture tolerance."

As can be noted, a rupture disk has to be replaced once it functions. It is necessary either to select a disk that is not too close to the operating pressure or to operate pressure vessels sufficiently far from the relieving pressure of the safety devices. A 20 percent margin between operating pressure and set relief pressures is generally recommended to avoid leakage from pressure relief devices. The following conditions may affect the fatigue or creep life of rupture disks:

Rupture disk materials. Certain materials will resist creep and fatigue better than others. However, the choice of materials is dictated by operating conditions of temperature, pressure, corrosion, etc. Sometimes a wide choice is available, at other times only one or two. The operating life will be affected by the choice of disk material.

Disk thickness. Thin disks will not stand pressure cycling as well as thicker materials; therefore, more margin is required; i.e., the higher the rupture pressure for a given size and type of material, the longer the disk will be expected to perform satisfactorily at a given margin.

Temperature. Since materials tend to creep at elevated temperatures, more margin is needed at elevated temperatures and less at lower operating temperatures.

Pulsating or cyclic service. This is a major factor affecting operating life. Probably the most severe service and the one requiring the greatest margin is alternating pressure and vacuum. Next in severity is pulsating pressure. Steady pressure is the least severe service. All are influenced by operating temperature.

Rupture disk discharge.   Rupture disk discharge requires special consideration if venting is to the atmosphere. Because fragments from rupture disks are released upon rupture, because material vented may be toxic or flammable, and because blowdown may be at very high velocity, discharge from the vent opening can be hazardous. Provisions must be made to prevent possible injury to personnel and damage to equipment.

Discharge piping should be braced or fastened to withstand the loading or pressure during discharge. When a disk ruptures and the exhaust is to the atmosphere, there is a rocket-type thrust against the piping. This exhaust thrust must be taken into account when designing the equipment and the discharge piping.

Discharge piping should be of ample size to handle the expected capacity. It is always best to run individual lines from each disk. If this is impractical, the common line should be one or more sizes larger than the discharge line from the largest disk. This will minimize the effect of pressure buildup in the common header and reduce the back pressure to which other disks might be exposed. If disks are discharged to a common header, all connected to that header should have vacuum supports. Vacuum supports are designed to stand full vacuum continuously with some safety factor. If differential pressures are expected to get much above 15 psi, special back-pressure supports should be used.

Relief device settings

All pressure vessels require positive relief devices to guard against overpressure. The pressure setting of the relief device cannot exceed the allowable working pressure (AWP) of the vessel it protects. When more than one pressure relief device is used, one must be set at the AWP and the others must not be set at more than 105 percent of the maximum allowable working pressure. The set-pressure tolerance normally permitted by the code is 2 psi for pressures up to and including 70 psi, and 3 percent for pressure above 70 psi.

Rupture disks must be guaranteed by the manufacturer to burst within plus or minus 5 percent of the disks' stamped bursting pressure at the coincident temperature.

Relief device capacity

The capacity of the pressure relief devices, other than for unfired steam boilers, must be sufficient to prevent the pressure from rising more than 10 percent above the maximum allowable working pressure of the pressure vessel, except under the following conditions:

1. When there are multiple pressure-relieving devices, they shall act so that the pressure does not rise more than 16 percent above the maximum allowable working pressure.

2. In addition to overpressure that may occur from process control failures, the code recognizes that a fire in the area of pressure vessels could also cause a pressure rise in the vessel. The code requires that this additional fire or other external heat hazard be considered and that additional relief valve capacity be provided so that, in aggregate, the pressure buildup cannot be more than 21 percent of the allowable working pressure of the vessel.

Relief device installation

All pressure relief devices must be installed so that they are readily accessible for inspection and repair. Pressure gages should be installed near the pressure relief devices in order to note operating conditions at all times. The gages should have an upper dial reading of at least 1.25 times the set pressure of the relieving devices or twice the maximum allowable working pressure of the vessel.

Pipes and fittings between the pressure vessel and the pressure-relieving device must have an internal area at least equal to the inlet area of the pressure-relieving device so that its capacity is not restricted. Safety valves, safety relief valves, and pilot-operated valves must be connected to the vessel in the vapor space above any contained liquid, while liquid relief valves must be connected below the normal liquid line of the vessel.

Intervening stop valves

No intervening stop valves are permitted between the vessel and its protective devices, except:

1. When there is more than one pressure-relieving device and the block or stop valves are so constructed (interlock mechanism), or positively controlled, that closing of the maximum number of block valves at one time will not reduce the pressure-relieving capacity of the valves still protecting the vessel to below the code-required relieving capacity.

2. When inspection and repairs may often be needed, provided the block valve can be locked or sealed open when the vessel is under normal operation; it shall not be closed except by an authorized person, and then only if the person stays by the block valve and watches pressures and temperatures, while the pressure-relieving valve is repaired. When the repairs to the relieving device are completed, this authorized person must seal and lock open the stop valve before leaving the stop valve site.

Discharge lines from pressure-relieving devices must be of a size adequate to prevent any serious back pressure from developing that could reduce the capacity of the pressure-relieving device. If there is a chance of a liquid's accumulating on the discharge side of the pressure-relieving device, a suitable drain should be installed to prevent this.

Test levers

Pressure relief valves for air, hot-water, and steam service must have a test lever so that the valve disk can be lifted for testing purposes when the pressure is at least 75 percent of the set pressure of the valve. The design must permit the valve and lever to snap closed when the exterior lifting force is removed. Testing by the hand lever should be done periodically to prevent the valve from seizing and thus not working when needed. Rust, scale, and simmering can cause a safety valve to become stuck in the closed position, unless the crud is blown free by testing. The code permits pressure relief valves not to have hand testing devices for other types of services but recommends that testing levers be installed, unless the escaping vapors may create another dangerous hazard to the surroundings.

Seals

Seals are required on all external adjustments to prevent unauthorized tampering with the overpressure device. Adjustments to valves should be performed only by the valve manufacturer or its authorized agent, or by recognized pressure relief valve repairers. The National Board now has an accreditation system for repairers of pressure relief valves, and those who are qualified may use the repair symbol.

Conversion of Section I steam safety-valve capacity to capacity for other fluids

Steam safety and safety relief valves that are permitted by the high-pressure boiler section of the ASME code, if properly certified for pressure and capacity and bearing the official ASME-approved safety valve symbol, may also be used on pressure vessels handling other fluids or gases. The code provides the following equations for converting the steam capacity to the capacity of other fluids or gases.

The general equation for the capacity of fluids other than steam is

$$W = CKAP \sqrt{\frac{M}{T}}$$

where W = rated capacity of any gas or vapor, lb/h
C = a constant, dependent on the ratio of specific heats of the fluid or gas (see also Chap. 1)

K = valve coefficient of discharge
A = actual discharge area of the safety valve, in²
P = 1.1 times the set pressure of the valve plus atmospheric pressure, psia
M = molecular weight of gas
T = absolute temperature at inlet, or °F + 460

The rated capacity for steam W_s (in pounds per hour) is given by the code as

$$W_s = 51.5 \, KAP$$

The code also provides Table 6.3 to determine the constant C from the ratio of the specific heats k. Some molecular weights for common gases are shown in Table 6.4, based on the ASME Section VIII, Division 1, code (see also Chap. 1).

There are many practical applications of these capacity conversion equations. Two examples will be cited.

Example The displacement of an air compressor requires an air safety-valve capacity on the air tank of 12,500 ft³/min of air at 60°F and atmospheric pressure at sea level. A steam-rated safety valve is available with the required pressure setting, inlet pipe size, etc., but it is stamped for 30,000 lb/h for steam service. If air has a density of 0.0766 lb/ft³ and a k of 1.40 (specific heat ratio), would the steam-rated valve meet code requirements for capacity?

answer It is obvious that the KAP factor in the flow equations is a constant for a particular valve size. Therefore, by using the known facts in the air equation, solving for KAP, and then substituting in the steam equation, we should be able to solve the problem.

For air, from Table 6.4, $M = 28.97$; $T = 60 + 460 = 520$°F abs. The cubic feet of air per minute (W_a) must be converted to pounds per hour.

$$W_a = 12{,}500 \times 60 \times 0.0766 = 57{,}450 \text{ lb/h}$$

From Table 6.3, with $k = 1.40$, $C = 356$. Substituting in the equation and solving for the KAP factor

$$W_a = CKAP \sqrt{\frac{M}{T}}$$

$$57{,}450 = 356KAP \sqrt{\frac{28.97}{520}}$$

$$161.38 = 0.236KAP$$
$$KAP = 684$$

Substituting this KAP value in the steam equation

$$W_s = 51.5KAP \quad \text{or} \quad W_s = 51.5 \times 684 = 35{,}226 \text{ lb/h}$$

Therefore, the steam-rated safety valve would meet code requirements for the required air capacity.

TABLE 6.3   ASME Code for Determining Constant C from Ratio of Specific Heats k

k	Constant C	k	Constant C	k	Constant C
1.00	315	1.26	343	1.52	366
1.02	318	1.28	345	1.54	368
1.04	320	1.30	347	1.56	369
1.06	322	1.32	349	1.58	371
1.08	324	1.34	351	1.60	372
1.10	327	1.36	352	1.62	374
1.12	329	1.38	354	1.64	376
1.14	331	1.40	356	1.66	377
1.16	333	1.42	358	1.68	379
1.18	335	1.44	359	1.70	380
1.20	337	1.46	361	2.00	400
1.22	339	1.48	363	2.20	412
1.24	341	1.50	364		

Example   What would be the capacity of the previous example's safety valve if the gas were butane, with a $k = 1.31$, and the inlet temperature were 70°F?

answer   For butane

$M = 58.12$, from Table 6.4
$C = 348$, from Table 6.3
$T = 70 + 460 = 530°F$ abs
$KAP = 684$, from previous example

Using

$$W = CKAP \sqrt{\frac{M}{T}}$$

and substituting

$$W = 348(684) \sqrt{\frac{58.12}{530}} = 78,824 \text{ lb/h}$$

TABLE 6.4   Molecular Weights of Common Gases

Gas	Molecular weight	Gas	Molecular weight
Air	28.97	Freon 22	86.48
Acetylene	26.04	Freon 114	170.90
Ammonia	17.03	Hydrogen	2.02
Butane	58.12	Hydrogen sulfide	34.08
Carbon dioxide	44.01	Methane	16.04
Chlorine	70.91	Methyl chloride	50.48
Ethane	30.07	Nitrogen	28.02
Ethylene	28.05	Oxygen	32.00
Freon 11	137.371	Propane	44.09
Freon 12	120.9	Sulfur dioxide	64.06

SOURCE: ASME code, Section VIII, Division 1.

Steam Safety Valves

Many plants use waste-heat boilers in the part of the process involving the pressure vessel system. These types of boilers must have safety relief valves that follow the power boiler rules from Section I of the *ASME Boiler and Pressure Vessel Code*. For plant operating, inspection, and maintenance personnel involved with these waste-heat boilers, the following safety relief valve recommendations of the code might be useful. The applicable boiler code should be consulted for further details, as should local jurisdictional requirements.

Safety valves should be connected directly to an independent nozzle on the boiler without any intervening valves of any description. Threaded connections may be used up to and including 3 in in diameter. For boilers operating at over 15 psi, all safety valves over 3 in in diameter should have flanged inlet connections.

Safety valves discharging steam over 450°F from superheaters should have a flanged or welded inlet connection for all sizes. Also, such valves should be constructed of steel or alloy steel throughout, suitable for heat resistance at maximum steam temperatures. The spring in superheater safety valves should be fully exposed (Fig. 6.7c) so that it will not be in contact with high-temperature steam.

It is important that the nozzle opening to and the escape piping from the safety valve be at least as large as the safety-valve connection. If two or more safety valves are connected on a common nozzle or fitting, the area of this nozzle or fitting should at least equal the combined areas of all safety valves served.

Safety-valve springs

The safety-valve spring is usually of round or square stock, for maximum clearance between the coils. If the coils come in contact, the valve cannot lift. It is for this reason principally that the maximum range of adjustment permitted with a spring is 10 percent of its rated setting. This rule is for safety valves set at up to 250 psi. For higher pressures, the allowable range of adjustment is 5 percent of the spring rating. If the setting is changed to a greater deviation, a new spring and nameplate should be installed by the manufacturer's representative.

A lifting lever is required, in order to lift the valve from its seat when there is 75 percent of the popping pressure in the boiler. Lifting levers that can lock the valve in raised position are not approved.

Blowback

Blowback, or *blowdown,* is the number of pounds of drop in boiler pressure from the value at which a safety valve pops to the point where the valve closes. For pressure up to 100 psi, the blowback should be not over 4 percent, but not less than 2 lb. Higher pressures call for a minimum blowback of 2 percent of the popping pressure. Safety valves used on forced-circulation boilers of the once-through type may be set and adjusted to close after blowing down not more than 10 percent of the set pressure. The valve for this special use must be so adjusted and marked, and the blowdown adjustment must be made and sealed by the manufacturer. A lesser blowback may result in a destructive chattering (rapid popping and seating) action. Too great a blowback wastes steam and fuel. Although the code is silent regarding maximum blowback, it is usually good practice to adhere to the minimum allowed.

Escape pipes

Escape pipes should be used if the discharge is located where workers might be scalded. A proper escape pipe is as essential to the safety of plant personnel as the safety valve is to the boiler. Too often a worker has been opening a stop valve when a safety valve, having no escape pipe and pointing directly at the person, pops. To be standing in the path of a high-pressure, 3- or 4-in jet of steam is usually fatal.

Every escape pipe should be at least 6 ft high. If headroom makes it impossible to terminate the escape pipe within a reasonable distance from the ceiling, it should extend out through the building wall or roof. If the roof is a flat roof where workers may be, the escape pipe should extend at least 6 ft above it. If a horizontal escape pipe is more practical, it should discharge at a safe location.

It is essential that the escape pipe diameter be at least equal to the size of the safety valve. If a length of over 12 ft is necessary, it is better to use a diameter $\frac{1}{2}$ in larger for each 12 ft in length. A long line with no increase in diameter will cause a back pressure because of flow friction and may cause serious chattering of the safety valve. All 90° bends should be avoided if possible.

The escape pipe should be supported independently of the safety valve. Serious stresses may be set up in the safety-valve body, connection, or boiler nozzle by the weight of a heavy, unsupported escape pipe.

After a safety valve has blown many times, it is not uncommon for slight leakage to develop. Condensation of this leakage may gradually fill an undrained escape pipe with water. This condition alone prevents the safety valve from blowing at its set pressure. The popping point will be increased 1 lb for every 2.3-ft elevation of water in the escape pipe. Also, in an outdoor escape pipe exposed to severe winters, ice may form and seriously interfere with proper safety-valve operation. Every escape pipe should have a $\frac{3}{8}$- or $\frac{1}{2}$-in open drain at its lowest point. This drain should be conducted off the boiler top in order to prevent external corrosion induced by dampness.

Code rules on safety valves

The number and capacity of safety valves required on boilers are governed by code rules. The following rules on boilers must be followed.

Capacity of steam safety valves.   The safety-valve capacity on a boiler must be such that the safety valve (or valves) will discharge all the steam that can be generated by the boiler (this is assumed to be the maximum firing rate) without allowing the pressure to rise more than 6 percent above the highest pressure at which any valve is set, and in no case more than 6 percent above the maximum allowable pressure.

Popping-point tolerance.   A valve must meet the following popping-point tolerances (on a plus or minus basis):

1. For pressures up to and including 70 psi—2 psi

2. For pressures 71 to 300 psi—3 percent

3. For pressures 301 to 1000 psi—10 psi

4. For pressures over 1000 psi—1 percent

Number of safety valves.   One or more safety valves must be set at or below the maximum allowable pressure. The highest pressure setting of any safety valve cannot exceed the maximum allowable working pressure by more than 3 percent. The range of pressure settings of all the saturated-steam safety valves on the boiler cannot exceed 10 percent of the highest pressure setting to which any valve is set.

Each boiler requires at least one safety valve, but if the heating surface exceeds 500 ft^2 or the boiler is electric with a power input over 500 kW, the boiler must have two or more safety valves. When not more than two valves of different sizes are mounted singly on the boiler, the smaller valve must be not less than 50 percent in relieving capacity of the larger valve.

Every superheater attached to a boiler with no intervening valves between the superheater and boiler requires one or more safety valves on the superheater outlet header. With no intervening stop valves between the superheater and the boiler, the capacity of the safety valves on the superheater may be included in the total required for the boiler, provided the safety-valve capacity in the boiler is at least 75 percent of the aggregate safety-valve capacity required for the boiler.

The superheater safety valves should always be set at a lower pressure than the drum safety valves so as to ensure steam flow through the superheater at all times. If the drum safety valves blow first, the superheater could be starved of cooling steam, leading to possible superheater tube overheating and rupture.

Pilot-Operated Pressure Relief Valves

Pilot-operated valves are completely different in design and operating characteristics from the standard relief valves previously described.

Types of pilot-operated valves

Though there are a number of different types of pilots designed to work in conjunction with pilot-operated valves, they all fall into two general categories: (1) snap-action, flowing and nonflowing types and (2) modulating, flowing and nonflowing types.

Snap-action, flowing pilots reduce dome pressure of the main valve rapidly when set pressure is reached. This type of pilot flows a portion of the process material continuously through the pilot exhaust while the main valve is open. In most cases, flow is relatively small. Design of a snap-action, nonflowing pilot is quite different from that of flowing types. Flow to atmosphere through a nonflowing pilot occurs only when the pilot depressurizes the dome for main valve opening. At reseat pressure, flow occurs through the pilot to repressurize the dome for main valve closing, with no flow to atmosphere.

Back pressure on pilot-operated valves

A pilot-operated valve, unlike conventional or balanced-bellows valves, may open as a result of back pressure greater than inlet pressure. For example, where a pilot-operated pressure relief valve discharges into a pressurized header, and inlet pressure at the valve is lower than header pressure (perhaps because another higher-set pressure relief valve is relieving into the same header), this pressure difference can cause the main valve of the pilot-operated valve to open and thus cause backflow through the main valve owing to the higher header pressure. A similar condition could occur during upstream maintenance of equipment of the pilot-operated pressure relief valve. This problem can be avoided by installing suitable check valves on the discharge side of the pilot or main valve to prevent backflow. For pilot-operated pressure relief valves with nonflowing pilots, a double or "shuttle" check valve can prevent back pressure from discharging through the pilot vent when the main valve is open and relieving. This may also eliminate erratic blowdown or chatter of the main valve and will keep a nonflowing pilot from becoming a flowing one.

Proper selection of a pressure relief valve for use where back pressure can occur is very important. The following valves, in order of preference, may be used to provide the required relief capacity and ensure stable, chatter-free operation:

1. Pilot-operated safety relief valves

2. Balanced spring-loaded valves (bellows or balanced-piston type)

3. Conventional spring-loaded valves

Pressure and Vacuum Relief Valves

Many process pressure vessels operate under possible vacuum conditions when emptying contents or when vapor pressure in a tank declines rapidly as steam is lost in a heating coil within the tank. In cold climates, heat is needed to pump oils, fats, and similar substances. This generates some vapor pressure within the tank so that there is positive pressure within the tank that stores the substance to be pumped. Most of these storage tanks are designed for low *internal* pressure. When the vapor pressure inside the tank collapses with the loss of heat, a vacuum is created inside the tank that usually results in an inwardly collapsed tank unless the unit has a vacuum relief valve. Jacketed steam kettles have a similar exposure when the chamber holding the contents is subjected to a vacuum. The inner jacket will collapse if it is too thin and the mixing chamber is not equipped with a vacuum breaker.

Combination pressure and vacuum relief valves are available to guard vessels against overpressure as well as vacuum. Sometimes these valves are called "breather and vacuum valves." See Fig. 6.9.

The pressure-vacuum (PV) valve may consist of two disks or pallets (Fig. 6.9a), one for pressure and one for vacuum. The weight of the pallet determines the pressure range. Failure of these valves is caused by the disk or pallet's sticking to the seat. The reason for sticking ranges from frozen condensate to sticky substances in the atmosphere or in the tank. The maintenance required on these valves varies with the conditions. Liquids in the tank and a dust-laden atmosphere outside the tank would require a high frequency of inspection to ensure proper operation. An exposure to below freezing temperatures requires taking proper precautions to keep the valve from freezing up.

Several basic types of PV vents are currently available. They are described by vent manufacturers as falling into three main categories:

1. Solid pallet (hard pallet to hard seat) valve.

2. Diaphragm pallet valve.

3. Liquid-seal valve.

Solid and diaphragm pallet vent valves generally require an overpressure or overvacuum beyond the initial opening or set point before maximum flow capacity is achieved. The extent of opening is dependent on the size and design of the unit and on the design and loading of the pallets.

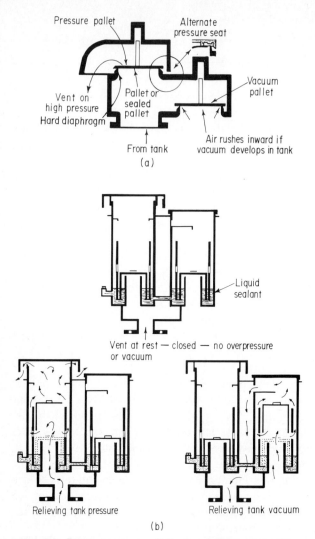

Figure 6.9 Pressure-vacuum relief valves. (*a*) Pressure-vacuum relief valve has two pallets to be lifted for relief—one for vacuum and one for pressure; (*b*) some relief valves use liquid as a seal.

Between the point of the initial opening and the fully opened condition, the capacity of the vent is affected by the change in lift of the pallet, the change in orifice discharge coefficient, and the change in pressure. At and beyond the point of full opening, the capacity is affected only by the change in pressure. After the pressure is relieved, the vent valve closes at, or slightly below, the set point. Since it is not possible to achieve maximum capacity without some overpressure or overvacuum, vents must be selected to provide the desired venting capacity without exceeding the maximum allowable pressure for which the storage tank was designed. The usual designs of pallet vent valves make field adjustment of set points difficult. In some designs, field changes in set points may be accomplished by replacement of pallets or pallet assemblies.

In liquid seal valves (Fig. 6.9b), a liquid provides the closure against flow through the vent until overpressure or overvacuum occurs. The physical properties of the sealing liquid usually are such that it has a very low freezing point as well as a high boiling point. In addition, the sealing liquid should not contaminate the storage tank contents nor, in turn, be contaminated by the tank contents.

Volatile-Substance Storage Tanks

In the petrochemical industry, large storage tanks generate vapors from ambient temperature changes or evaporation from filling and emptying of the tanks. Where the tanks are operated under slight vapor pressure, these breather vacuum relief valves may serve to limit this pressure. Storage tanks in the petroleum industry may also be designed to operate as (1) fixed-roof tanks, (2) floating-roof tanks, (3) variable-vapor-space tanks, and (4) pressure tanks. The type of tank to be used depends on the substance to be stored. In many instances, the most economical tank can be selected only after a detailed study comparing loss from, and the cost of, different tanks. For stocks having a low true vapor pressure, less than 2 psia, the fixed-roof tank generally will be the most economical selection. For stocks of the motor-gasoline range of volatility at high throughputs, the floating-roof tank generally will be the best choice, but at lower throughputs the variable-vapor-space tank generally will be better. For stocks that boil at atmospheric pressure and storage temperature, pressure tanks are best generally; however, in some cases use of the fixed-roof tanks in conjunction with a vapor-recovery system may offer more advantages.

Boiling points in storage

The tendency to boil in storage is a function of vapor pressure, altitude, barometric pressure, and liquid-surface temperature. Maximum liquid-surface temperatures vary throughout the United States. Table 6.5 shows, for various temperatures, the maximum Reid vapor pressures (RVP) of stocks that can be stored at atmospheric pressure without general boiling (but with high loss rates).

Breather valves for storage tanks

The fixed-roof tank has several openings in the roof for venting, gaging, and sampling. To maintain a gas-tight roof, accessory equipment of a gas-tight design must be provided for these openings.

TABLE 6.5   Maximum Reid Vapor Pressure of Stocks Stored at Atmospheric Pressure

Area	Maximum liquid-surface temperature, °F*	Maximum Reid vapor pressure, lb*
West coast (tempered by Pacific Ocean)	80	18
	90	15.5
Gulf coast, Atlantic seaboard, and	100	13.5
northern middle west	110	11.5
Midcontinent area and arid southwest	115	11
	120	10

* The limits must be reduced for locations at higher altitudes to account for lower barometric pressure.

The accessory for the vent opening is called a "breather valve," "pressure-vacuum relief valve," or "conservation vent." When operating properly, this device prevents either the inflow of air or the escape of vapors until some preset vacuum or pressure is developed. Most breather valves, especially the metal-to-metal types, allow some leakage below the pressure or vacuum setting. A tight breather valve is important in reducing evaporation loss.

The pressure and vacuum settings of a breather valve are dictated by the structural characteristics of the tank and should be within safe operating pressure limits. A certain amount of pressure and vacuum beyond these settings is necessary to overcome pressure drop in order to obtain required flow. Proper size and settings can best be determined by reference to API *Standards for Petroleum Products.*

Breather valves should be designed to give:

1. High-flow capacity at relatively small pressure or vacuum above the setting

2. A gas-tight seal

3. Freedom from sticking or freezing

4. Easy access to all parts for inspection and maintenance

Diaphragm and liquid-seal valves have less leakage than metal-to-metal types. For dependable service, diaphragms should be resistant to tank vapors.

Open vents of the mushroom or return-bend type should not be used on fixed-roof tanks storing volatile oils, as they permit high loss. These vents are merely hooded openings equipped with protective screens. The opening is turned down to prevent any blockage by ice or snow.

Venting accessories sometimes used are flame arresters, flame snuffers, and flash screens. They usually have little effect on vapor loss except when they are installed between the tank and vent valve and must be removed for cleaning.

Flare stacks

Discharges from vent valves in the petrochemical industry are usually directed to flare stacks, where the gas is burned in order to prevent inundating the surroundings with flammable or obnoxious gas. The height of the flare stack must be sufficient to prevent gas inversion and also to protect personnel and property from the radiant heat of the burning gas at the tip of the flare stack. Usually, continuous pilot-light burners are used on the flare tip in order to ensure discharge gas ignition.

QUESTIONS AND ANSWERS

6-1   What is meant by "mode of control"?

ANSWER: "Mode of control" means the manner in which the automatic controller acts and reacts to restore a variable quantity in a process vessel, such as pressure, flow, or temperature, to a designed control or desired value. The three controller systems used to control a process are pneumatic, electric, and electronic.

6-2   Why are manual-reset controls useful on pressure or temperature high-limit controls?

ANSWER: The manual-reset mechanism on the high-limit control emphasizes the fact that the operating control has malfunctioned, thus prohibiting further vessel operation until corrected. At times, this malfunction may be caused by fused contacts, a leaking gas valve, a shorted wire, etc.

6-3   What is meant by "blowback"?

ANSWER: It is the number of pounds per square inch of steam pressure drop from the point at which a safety valve pops to the pressure at which it reseats.

6-4   What controls the amount of blowback?

ANSWER: The blowback adjusting ring.

6-5   What is a huddling chamber?

ANSWER: It is a chamber exposing the underside of the valve disk in a safety valve to increased pressure area on its primary lift. Pressure acting on the increased area results in the pop, or secondary, lift.

6-6 What principle do some safety valves use in place of the increased area exposed by the huddling chamber?

ANSWER: The reaction principle.

6-7 Why go to the expense of installing huddling chambers or reaction flow to pop a safety valve?

ANSWER: Without such installation, a gradual lifting and seating of the valve would rapidly ruin the valve seat by cutting or wire-drawing action of the steam, or gases with similar characteristics.

6-8 On what principle do the majority of pressure gages operate?

ANSWER: On the bourdon tube principle; that is, a curved tube tends to straighten when subjected to internal pressure.

6-9 Where and why is a siphon required in pressure gage lines?

ANSWER: The siphon is located as a pigtail or drop leg in the piping to the gage; it is designed to trap condensate and to prevent live steam from entering the bourdon tube. It prevents the tube, springs, and other delicate parts from being subjected to high temperatures.

6-10 What is a compound gage?

ANSWER: A gage reading positive pressure in pounds per square inch on one side of the zero reading and negative pressure (vacuum) in inches of mercury on the other side.

6-11 If two pressure-relieving devices protect a pressure vessel stamped for 200 psi AWP, what is the maximum pressure setting permitted for the second pressure-relieving device if the first is set at the AWP per ASME rules?

ANSWER: 105 percent of 200 = 210 psi.

6-12 What is the minimum capacity required for the two pressure-relieving devices in Prob. 6.11?

ANSWER: There must be sufficient capacity so that the pressure in the tank cannot rise above 116 percent of 200 = 232 psi with the two safety devices relieving.

6-13 Differentiate between a pressure relief valve and a nonclosing relief device.

ANSWER: The pressure relief device is designed to reclose after the blowdown pressure is reached and to prevent any further fluid or gas flow. A nonreclosing relief device does not reseat after pressure bursts it open.

6-14 If a tank holds (1) highly corrosive gas; (2) lethal gas, which relief device would you place closer to the tank when a combination of safety relief valve and rupture disk is to be used?

ANSWER: (1) The rupture disk should be placed first to protect the relief from being affected by the corrosive gas to such an extent that it might corrode and bind the valve shut. (2) The relief valve should be placed nearer to the tank to prevent the lethal gas from escaping from a leaking safety valve.

6-15   A jacketed mixing tank with an AWP of 50 psi operates under pressurized mixing conditions and is protected by a rupture disk. Over what pressure setting range should the manufacturer guarantee the disk to burst if it is stamped to burst at 50 psi?

ANSWER: Plus or minus 5 percent of 50, or a minimum of 47.5 psi and a maximum of 52.5 psi.

6-16   Can steam-rated safety relief valves used on boilers be used on pressure vessels (unfired)?

ANSWER: Yes, if the steam-rated valve is rated in capacity according to Section I rules and is stamped properly, including the official ASME code symbol for an accredited valve.

6-17   What capacity would a steam-rated safety valve have to have in order to protect a pressure vessel that requires 2500 lb/h relieving capacity of a gas with a $k = 1.34$ and with a molecular weight of 17.03? Maximum gas temperature is 175°F.

ANSWER: Use

$$W = CKAP \sqrt{\frac{M}{T}}$$

where $C = 351$ (see text)
 $\quad M = 17.03$
 $\quad T = 460 + 175$
 $\quad W = 2500$

and solve for KAP in the equation

$$KAP = \frac{2500}{57.48} = 43.49$$

Now use

$$W_s = 51.5 \times KAP = 51.5 \times 43.49 = 2240 \text{ lb/h}$$

6-18   Define "coefficient of discharge" as it affects safety relief valves.

ANSWER: This is the ratio of actual flow to theoretical flow and must be determined by test.

6-19   May cast-iron seats and disks be used in pressure-relieving devices?

ANSWER: No, the use of cast iron is prohibited because of the risk of a cracking failure from rapid thermal changes and similar considerations.

6-20   What is the set-pressure tolerance for a 350-psi pressure relief valve that is to be used on an unfired pressure vessel?

ANSWER: Minimum set pressure, 339.5 psi; maximum set pressure, 360.5 psi (3 percent range).

6-21   When may a stop valve be placed between the vessel being protected and the pressure-relieving device?

ANSWER: On a pressure vessel system with multiple safety relief valves, block valves may be installed on a few safety relief valves, *provided* the remaining safety valves without block valves meet the code requirements for capacity needed in the pressure vessel system. Block valves may be used in a multiple system in which all the safety valves discharge into a common header and it is necessary to prevent backflow into a vessel. The block valve must have a locked or sealed mechanism that only an authorized person is permitted to use. This authorized person must stay near the valve when it is closed and must lock and seal it in the open position at other times.

6-22   Define the term "pilot valve."

ANSWER: A pilot valve is an auxiliary pressure-relieving valve that serves to actuate a major relieving device when it is itself actuated.

6-23   Is a vacuum relief valve a nonreclosing pressure relief device?

ANSWER: A vacuum relief valve is designed to admit air or other hooked-up gas to a pressure vessel that develops above-design internal negative pressure, and is designed to reclose and prevent further flow of air or other gas once the excess vacuum is eliminated.

6-24   Define "chatter" of a pressure-relieving device.

ANSWER: *Chatter* is the abnormal, rapid, reciprocating motion of the movable parts of a pressure relief valve in which the disk contacts the seat. It is usually caused by a pulsating pressure system in which the set pressure is too close to the operating pressure. Chatter will cause excessive seat wear and eventual leakage.

6-25   Why is proper selection of rupture disk material important?

ANSWER: Service conditions may affect the membrane and cause disk failure. For example, corrosion may thin the membrane, cause pinholes to develop, and thus cause the disk to rupture well below the marked burst pressure.

6-26   What is the effect of an exothermic reaction's reaching thermal instability?

ANSWER: When held in a closed vessel or container, every chemical combination that reacts exothermically either will reach a condition of thermal equilibrium, in which the heat generated equals the cooling rate supplied, or will thermally spiral upward uncontrolled, with subsequent pressure increase in the closed vessel. This is the reason a hard restraint on pressure increase is required.

6-27   How can cooling rates be increased in a reactor with an exothermic mixing operation?

ANSWER: In lieu of adding additional heat transfer surface, it is frequently possible to use a refrigerated coolant in order to increase the temperature difference for heat transfer.

6-28   What criteria must be considered in choosing valve material?

ANSWER: The expected service temperatures and pressures, the corrosiveness of the fluids or gases to be handled, and the relative costs of valves made of different materials.

6-29   What does the term "trim" mean in regard to valves?

ANSWER: *Trim* refers to the wear-prone parts of valves, such as seats, stems, and packing. These parts require wear-resistant materials; otherwise, excessive replacement costs may be developed.

6-30   Differentiate between valves with butt-weld end connections and those with socket-weld end connections.

ANSWER: Valves with butt-weld end connections are machined on the end connections with an inside diameter to match the bore of the pipe before butt-welding the valve to the pipe. Socket-weld end connections are available on small forged steel valves through the 2-in size. The pipe to be welded to the valve slips into the socket and is supported and aligned by the socket. There must be at least $\frac{1}{16}$-in clearance between the pipe and the bottom of the socket in order to avoid overstressing the connected pipe.

6-31   How can a pressure switch be used on a pressure vessel system?

ANSWER: A pressure switch is used to activate an electrical circuit when the process pressure exceeds or falls below a set point. The switch can be used to trigger an alarm, activate or deactivate a control mechanism, or do both. For example, vents and valves may be opened or shut, pumps and compressors started or stopped, and alarm horns sounded by a pressure switch. There are four sensors used with pressure switches—bourdon tubes, bellows, diaphragms, and pistons. The sensing element must be protected with a sealing unit when the service is corrosive.

6-32   How are vent valves used?

ANSWER: These are generally placed at the high point of a pipeline, vessel, or other piece of equipment to allow the bleed-off of vapor or gas, and may be in the open position under normal operation. Most systems require at least one vent valve for removal of air at plant start-up and for removal of process vapors before inspection and maintenance.

7

Maintenance, Inspection, and Repair

Maintenance

All plants must be properly maintained, and how this vital function is carried out will have a significant impact on a plant's profit. Breakdown maintenance, if practiced, can be expensive. The unexpected shutdown can cause process lines to produce no income. Labor to repair is unplanned and thus may involve premium overtime pay. Parts may not be available from normal sources and thus may be more expensive in the rush to restore production. It is important for a plant to have an effective organization to avoid the pitfalls of breakdown maintenance so that proper maintenance is carried out in a planned manner and within budget requirements.

Planned or preventive maintenance

There are many terms used to define maintenance work other than the breakdown type. *Planned* or *preventive maintenance* (PM) is maintenance carried out on a scheduled basis in order to catch wear and tear of components before a failure occurs. A modification of a regularly planned and scheduled maintenance program is to combine this scheduled maintenance, based on review of past inspections, testing, and wear rates, with a program of diagnostic readings. For example, the diagnostic procedures might suggest that increased pressure drop through a heat exchanger means tube fouling is occurring; therefore, it is time to remove the exchanger from service for tube cleaning. A similar diagnostic approach is possible to analyze increased use of steam, the longer need for batch treatment, etc., all of which serve to provide some form of feedback on the urgency of the maintenance that may be needed.

Predictive maintenance

"Predictive maintenance" is another term for a controlled method of instituting corrective action, as a result of periodic measurements that point out problems that may be developing on the equipment. An effective predictive maintenance program can reduce costly, unexpected downtimes by recognizing deteriorating conditions through the use of monitoring measurements. One result of a monitoring program may be that overhauls are not necessary, because the variables subject to periodic measurement prove to be constant and within acceptable limits. Measurements can be made with installed sensors and recorded for continuous monitoring. Supplementary measurements can be made with portable equipment to verify sensor readings or to obtain even more accuracy. Each pressure vessel system has some measurable item that can be used to schedule maintenance work. However, some inspections and maintenance work are dictated by jurisdictional requirements, and thus overhauls are scheduled to coincide with these jurisdictional inspections, especially internal inspections.

Performance monitoring

Performance monitoring is a real adjunct of maintenance and thus requires close coordination and cooperation between operating people, testing people, and the maintenance department. In utilities it is common to establish benchmarks of equipment performance when the equipment is new, or after a major overhaul. Performance is then subsequently monitored in order to track efficiency of performance. When the efficiency of performance drops to a calculated level that justifies the expense of shutting down for cleaning, repairs, or other maintenance work, operating and maintenance personnel work with the test group in an effort to restore performance efficiency. Therefore, maintenance is involved not only with the tracking of physical deterioration of production equipment but also with the loss in output due to scale, plugging, wear on controls, and similar items that affect the output of the pressure vessel system.

Contract maintenance

In large pressure vessel complexes, maintenance may be farmed out to contract maintenance firms, especially for the so-called turnaround inspection and maintenance work that is scheduled usually on an annual basis. The advantage of this program is that a plant does not need to employ the large in-house maintenance staff that would be necessary for the annual maintenance period; rather, it can retain a smaller staff, for checking and repairing routine wear and tear items such as packing on valves.

The ultimate responsibility for all maintenance work is with plant management and their technical staff. They should work closely with contract maintenance people in order to note plant conditions and also to make the final decisions on whether to repair or replace.

Reasons for maintenance

Maintenance on pressure vessel systems is necessary because many vessels are subjected to severe service conditions and environments, often harsher or more destructive than planned for in the original design. Some of the more common causes for service failures are as follows: excessive stresses, overpressure, external loading, thermal or mechanical fatigue or shock, overheating, corrosion, and hydrogen embrittlement. Steps should be taken to minimize these causes. For this reason, pressure vessels and auxiliary equipment should be inspected frequently by competent personnel in accordance with codes; standards; federal, state, and local regulations; and insurance company recommendations. These are only *minimum requirements,* however.

Loss in performance output is now recognized as a major reason for preventive maintenance, in addition to the traditional purpose of preventing breakdowns. In a highly competitive world, efficiency of operation is becoming a critical item from a profit-and-loss point of view. For this reason, while contract maintenance might prove beneficial for major planned shutdowns, the daily routine of maintaining equipment for good performance and reliability is still an essential part of good plant management programs. Maintenance departments must look at pressure vessel systems on at least a daily basis for signs of wear and tear, deterioration, leaks, noises, and high temperatures and pressures that will be signals of the need for maintenance if these are in any way abnormal.

Establishing benchmark readings

The nuclear industry has established an inspection and maintenance program that is now being followed by most large pressure vessel system operators. It consists essentially of obtaining code-constructed and -designed pressure vessels, establishing benchmark readings after the equipment is installed, and then monitoring the benchmarks in future inspections to note changes, and thus determine if repairs may be needed.

The same program can be adopted both for conventional pressure vessels in the large complexes and for individual vessels. There are many items that begin with the ordering of a vessel. Items to be considered are:

1. *Material.*   Is it suitable for service? Is the corrosion allowance over the thickness required by the code? Is the material allowed by the code?

2. *Welding.*   Is the welding procedure adequate for the joint? Can it be repeated in the plant during repairs? Is full, spot, or no radiography to be ordered? The weld efficiency can be affected by these choices, as can calculations of required thickness. Efficiency can also be influenced by whether the welds are ground flush or left in the as-welded condition.

3. *Postweld heat treatment.*   The code does not require all welds to be postweld heat-treated. Depending on the service, postweld heat treatment can prevent "locked-up" welding stresses, which may require added maintenance and repairs in the future.

Part of establishing benchmarks for future inspection and maintenance is checking any new installation or addition to a pressure vessel system at the plant site. This would include such items as:

1. Are hangers, guides, anchors, and supports installed according to specifications?

2. Is connected piping installed so that no misalignment exists that may impose additional stress on the piping system and the connected pressure vessel?

3. Are welding procedures in place and being followed during the erection, and are only qualified welders being used for the pressure vessel connections?

4. Is fit-up of welds being followed, and are surfaces properly prepared for welding?

5. Are the necessary nondestructive tests performed per specifications or code requirements?

6. Are all safety and relief valves installed properly, and are they of the correct pressure and capacity setting?

7. Have the necessary field hydrostatic tests been performed within the specification range for the design pressure and temperature, or as per code requirements?

8. Have the necessary documents for the field erection been filed for jurisdictional review, National Board registration, and permanent plant records?

Vendor surveillance

"Vendor surveillance" is the name usually applied when owner's or user's representatives visit a fabricator's shop in order to note the quality of the work. Another objective of the visits is to make sure the fabrication is on schedule, so that there is no delay in placing a process line into service.

Code construction is usually specified; however, some maintenance organizations monitor the code construction as part of their quality control program by visiting the fabricator's shop at so-called checkpoints to make certain that code requirements are met on the following items:

1. Rolling of plate or bending of pipe

2. Fit-up of welded joints

3. Proper use of welding procedure

4. Cleanliness of weld prep areas before and during welding

5. Proper preheat and temperature control

6. Removal of all fabrication attachments

7. Proper postweld heat treatment

8. Correct inspection techniques

9. Correct completion and signing of data reports by the fabricator and its inspection agency

Equipment inventory

One of the first requirements in a preventive maintenance program is to identify the pressure vessels that require periodic surveillance and inspection. The frequency of this surveillance and inspection should be determined on the basis of hazard, importance to production, expected rate of deterioration, and extent of sensor and control installation for the vessel system. Determining this frequency is most conveniently accomplished by establishing an equipment card system. The cards should show:

1. Name of vessel and manufacturer, date of manufacture, and date of installation.

2. ASME and National Board numbers with data reports attached for the vessel. These should show AWP and other relevant code data for the vessel.

3. In-plant use of the vessel and its location in the plant.

4. Prior inspection findings, or past test results on output performance.

5. Past repairs made and the reasons for these repairs. Included in the repairs should be a description of materials, welding procedures used,

preweld and postweld heat treatment that may have been applied, and the NDT used to prove the soundness of the repair.

6. A note showing the potential effect on the process line if the vessel has to be taken out of service.

7. Any bypasses to hooked-up spares. The same applies if a spare is available but needs to be hooked up in order to restore production. If there are no spares, the name of the nearest supplier should appear on the card with phone numbers and contacts shown.

8. Blueprints of the vessel as shown by the vessel manufacturer, along with blueprints of associated piping and control hookups to the vessel.

9. Any other relevant data on contents, precautions to be used before entering the vessel, and similar inspection problems.

Work order and checklist system

Some companies use a work order system to manage and control the expenses of maintenance work. Checklists on what to do to make the equipment available for inspection and maintenance are also a management tool to ensure that an organized effort is being applied in preventive maintenance. Management must exercise control of maintenance, because maintenance can become a significant part of operating costs. There is an increased emphasis on life extension of equipment, and the reliability and availability of equipment for a long run without problems are promoted by planned preventive maintenance. It has been found that a preventive maintenance program combined with interpretation of performance efficiency and diagnostic readings can reduce maintenance costs in the long run, and thus increase output at a lower cost.

Computerized maintenance programs

Computerized maintenance programs are becoming more prevalent. With these, management can compare maintenance work achieved or proposed with expected expenditures much more quickly and accurately. Poor equipment performance requires more maintenance, and one result of monitoring this performance on the computer will be that poor equipment will not be reordered.

With advances in computer technology and the subsequent reduction in system costs, computers are now being used in more preventive maintenance management systems. Multiple CRT terminals allow the operations of the various departments to be integrated more easily. When setting up any computer maintenance program, it is necessary to include a history of each piece of equipment, details of maintenance to be performed, frequency of maintenance, automatic issuing of work orders, and inventory control of spare parts. Obviously, there are many other functions that can be added, but the cost benefits may not always be there.

Part of all programs is the control and recording of labor costs and the materials used in carrying out the work. Work orders, which can be issued automatically by the computer on a weekly basis, are used to control this function. In addition, work orders should be manually issued to cover work due on unscheduled breakdowns or additional work revealed during normal maintenance and inspection. This allows a complete history of the equipment to be maintained. All costs associated with labor and materials used should be entered on the work order for entry into the computer at a later time. This allows historical data to be maintained on all equipment for use in preparing department budgets and for analysis of equipment problems. Inventory control of parts and material should be tied into the work order procedure so that stock supply can be optimized.

A feature of some PM programs is the capability of the computer to supply a list of parts and their numbers that historically have been required for the same type of maintenance. This list is automatically supplied by the computer at the time the work order is issued. By integrating the materials management function into the same computer, it is possible to ensure that parts and material are available when maintenance is scheduled. No plant should be without some type of maintenance management system, which, if designed and operated correctly, will more than pay for itself within a couple of years. When management, operations, and maintenance all work as one cooperative group, process plants will function to their maximum efficiency and unscheduled outages will be minimized.

Maintenance organizations

In large process plants there are several organizational formats for maintenance programs. The organization may be of the area, central, or modified area-central type, using either contract or owner maintenance personnel, depending on many factors peculiar to the specific plant. The following points should prove helpful in organizing the maintenance function:

1. Maintenance and operations should be separate functions, with the common objective of assuring safe, continuous, and profitable operation of the plant, but with primary attention focused differently—maintenance, on the equipment; operations, on the process.

2. It is essential in high-pressure plants that continuous training be applied; mechanics and supervisors must have the continuing experience of working on and learning about specialized equipment. This precludes the type of central organization in which different mechanics are sent in to perform maintenance on the equipment each time it is shut down.

3. Some plants have had good experience using a minimum number of assigned personnel for maintaining specialized equipment, and supplementing these with additional people as the work and work load requires.

4. The organization should include technical personnel organized so they can focus their efforts on the technical and problem-solving aspects of maintenance.

As pressure levels increase and plants become larger and more complex, the use of first-class engineering talent in maintenance becomes essential. Engineers can find both challenge and reward in a well-organized department oriented toward the goals of optimum maintenance; engineers can be effectively used as maintenance supervisors, planners, technical supervisors, maintenance specialists, or as equipment problem solvers. In-depth training in maintenance methods, equipment design, and equipment operation is essential for engineers in high-pressure maintenance. This training may be obtained in many ways, some of which are cooperation with equipment manufacturers by means of factory training programs or visits and discussions with equipment designers; study of equipment, via manufacturers' drawings and literature; study of equipment during construction and maintenance; on-the-job training; and visits to other plants.

Preventive maintenance in high-pressure plants

The objective of preventing major failures is especially important for high-pressure plants, such as in the chemical and petrochemical industries (see Fig. 7.1). The pressure vessels are usually of special importance in the process stream and have high energy stored in them. In addition, there are the hazards of fire and even death that may exist with any penetration of the pressure-containing parts. It has been stated that preventive maintenance means periodic inspections, adjustments, testing, performance monitoring, and, if needed, repairs to prevent breakdowns of the forced outage type. The trend is away from extensive dismantling and toward diagnostic monitoring and testing to note if any changes are taking place, either in performance or physical condition. Extensive use of NDT is part of the preventive maintenance effort.

Safety in maintenance, inspection, and testing

As process plants become larger and more complex, the continuous need for stressing safe procedures in performing maintenance work is essential. General safety rules include the following:

1. Open flames and welding should not be allowed in the vicinity of vessels that are being inspected.

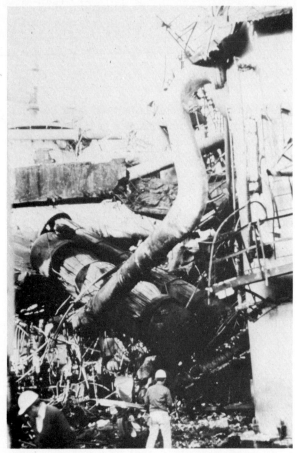

Figure 7.1 Failure of this ethylene oxide plant pressure vessel was caused by control failure. (*Courtesy of the Hartford Steam Boiler Inspection and Insurance Co.*)

2. Loose objects around all openings should be removed or properly secured.

3. Loose and torn clothing should not be worn by persons making inspections.

4. Protective hats should be worn as a safeguard where there is exposure to falling objects.

5. Nonsparking tools should be used on vessels where there is the possibility of fire or explosion. Tools should be kept clean to prevent them from *becoming* sparking tools.

6. Suitable protective clothing, including gloves, headgear, eye protection, respirators, and shoes, should be provided and worn as required.

The importance of adequate maintenance of all such equipment cannot be too strongly stressed.

7. Electrical equipment in the vicinity of pressure vessels should be de-energized before inspection. This is a safeguard against electric shock in the event that equipment is grounded to piping connected to the vessels.

Precautions during welding and cutting. Welding and cutting during maintenance work require extra attention in areas where flammable gases may be present. The safety rules of the jurisdiction should be followed, as well as those of fire and compensation insurance companies. Many fires have been started during weld repairs when sparks from welding apparatus have ignited combustible material.

For welding work, a special permit should be issued for a specific duration. During the work, the area should be continuously checked with a combustible-gas detector to ensure that it is free from contamination by any flammable mixtures. Wetted canvas or tarpaulin screens should be erected around the area to prevent sparks from flying into adjacent areas, which might contain combustible material. A fire hose and extinguishers should be on hand as an additional precaution.

The American Welding Society has published guidelines for welders doing maintenance work in the field. Among the recommendations to welders are the following precautions against a fire's erupting from the welding repairs.

Welding or cutting in atmospheres containing dangerously reactive or flammable gases, vapors, or liquids, or dust should be avoided.

Heat should not be applied to a container that has held an unknown substance or a combustible material that when heated can produce flammable or explosive vapors.

Heat should not be applied to a work piece covered by an unknown substance or whose coating can produce flammable, toxic, or reactive vapors when heated.

Adequate procedures should be developed and proper equipment used to do the job safely.

Adequate ventilation should be provided in work areas to prevent accumulation of flammable gases, vapors, or dusts.

Containers should be cleaned and purged before heat is applied.

Closed containers, including castings, should be vented before preheating, welding, or cutting. Venting will prevent the buildup of pressure and possible explosion due to the heating and the resultant expansion of gases.

The following sources provide more detailed information on fire hazards from welding and cutting operations:

1. ANSI Z49.1, *Safety in Welding and Cutting,* published by the American Welding Society, P.O. Box 351040, Miami, Florida 33135.

2. NFPA Standard 51B, *Cutting and Welding Processes,* National Fire Protection Association, Batterymarch Park, Quincy, Massachusetts 02269.

3. *Code of Federal Regulations,* Title 29, "Labor," Chapter XVII, Part 1910, OSHA General Industry Standards, available from the U.S. Government Printing Office, Washington, D.C. 20402.

4. AWS F4.1, *Recommended Safe Practices for the Preparation for Welding and Cutting Containers That Have Held Hazardous Substances,* American Welding Society, P.O. Box 351040, Miami, Florida 33135.

Protective clothing.   Federal safety standards require the use of personal protective equipment when making inspections or performing maintenance work in hazardous areas. Even if employees buy their own protective devices, the ultimate responsibility for the devices and clothing used is still with the employer. Not to be forgotten is the right of an employee to complain directly to the U.S. Department of Labor, an action that could bring a surprise visit by a compliance officer to the plant.

Proper safety clothes can help cut injuries and reduce time lost (and also reduce compensation payments), resulting in increased efficiency and productivity.

Use of masks.   Fumes or gases that originate from beneath dry sludge accumulations that have not been removed from pressure vessels may present hazards of explosion or toxicity. Breaking the crust of the sludge while it is in the tank may release these poisonous gases. When there is some question as to the thoroughness of cleaning, the inspector should wear an approved mask, preferably an air-line mask equipped with a hand blower. Canister-type masks are suitable for periods of short exposure in locations *where there is no lack of oxygen* and the concentration of fumes, vapors, or gas is not above the Bureau of Mines–approved rating for the particular canister or organic filter in use. All masks should be checked for defective parts before wearing and should not be used beyond the time recommended by the maker.

The chemical filters used in the mask should be checked to make certain that they are suitable for the chemical exposure at hand and should not be used beyond the time recommended by the maker. All masks should be checked for defective parts before wearing.

When an air-line mask equipped with a hand pump is used, a reliable person should be assigned the full responsibility of operating the hand pump. Air-line masks using compressed air should be equipped with Bureau of Mines–approved filters and reducing valves (to limit air pressure to a maximum of 5 pounds) to prevent contamination of air with oil and moisture. The suction opening to the compressor should be located in an area free of carbon monoxide fumes or other contaminants. Carbon monoxide (under certain circumstances) may be produced in compressed air equipment if any part of the equipment becomes considerably overheated.

No person should be allowed to enter a vessel without proper protection unless it is definitely established that there is no oxygen deficiency and that toxic and flammable vapors are not present.

When vessels are located in a pit, added hazard exists in the potential presence of fumes—vapor or gas—that are heavier than air, lying in "pockets" *outside* the vessel. Pits usually retard the free circulation of air, limiting the effectiveness of normal circulation and making the use of blowers essential. The pit should have a stationary ladder or stairway, well maintained and positioned to allow easy exit in case of emergency.

Use of lifelines.   A vessel *or pit* that has contained a hazardous substance should not be entered unattended. A lifeline securely attached to a safety harness, with the free end attended by a second person, should be used. The lifeline tender should give full attention to the person inside the vessel or in the pit. The free end of the lifeline should be snubbed to a stationary object so that it cannot accidentally fall into the pit.

In case of an emergency, the lifeline tender or the pump tender for the air-line mask should not enter the vessel or pit without first summoning other assistance and donning an air-line mask and lifeline, nor without being relieved by a fully competent attendant capable of handling the lifeline and maintaining adequate air supply to the mask. This is of vital importance, as many uninformed people rush to aid others but, instead, seriously impede progress.

Supervision of maintenance

As process plants become more complex, there is a definite need for written maintenance procedures and supervision. Detailed maintenance procedures should be in writing on each major pressure vessel in the process stream. The instructions and procedures should consider all possible hazards and should include:

1. Instructions for isolating, blanking off, draining, venting, and purging the vessels to be worked on. Locking out of connected electrical equip-

ment is another important consideration. Reference should be made to original drawings, as well as to supplementary sketches made of work that may have to be done on the vessel system.

2. A work permit outlining the scope of activity.

3. A list of the type of testing and maintenance equipment to be used.

4. A list of the protective clothing and other protective equipment that may be required.

5. Emergency steps or equipment that may be required during the course of the testing and maintenance work.

6. Complete instructions on testing limits, as in hydrostatic testing, and the permissible limits for any contaminants.

7. Instructions on special hazards that may be present and how these can be avoided (purging, cleaning, etc.).

8. Names and responsibilities of each member of the testing, inspection, and maintenance staff in any major pressure vessel work.

Inspections

Pressure testing and external inspection

The following precautions should be followed in the pressure testing and external inspection of unfired pressure vessels.

1. Body contact with hot surfaces or projecting objects should be avoided. (This is also applicable in making internal inspections, especially when in cramped quarters.)

2. Persons should not stand in line with pressure-relieving devices.

3. Staging should have guardrails to a height of 42 in, with midrails and toeboards, and should otherwise be in good condition.

4. Defective ladders should not be used.

5. Fittings or other appurtenances on a vessel should not be used as means of support.

6. Extra caution should be observed on wet and slippery surfaces.

7. If there is danger of being splashed with strong acid, acid-proof goggles and face masks should be worn.

8. Protective equipment, such as rubber clothing and rubber boots, should be worn when the need is indicated.

9. When subjecting a vessel to a hydrostatic test, precautions should be taken to ensure that all air has been vented. Air trapped in pockets presents an extremely dangerous situation.

Rotating vessels—block out and tag controls

Safety precautions observed about rotating machinery also apply to vessels having rotating parts. During a static inspection, the controls operating the rotating mechanism should be locked out and properly tagged.

Safety in performing internal inspections

Vessel entry. Vessel entry for internal inspections requires a safety system approach to avoid asphyxiation or contact with harmful residues. A vessel previously filled with harmful contents should be steamed and flushed until its contents have been reduced to a tolerable limit.

No person should be permitted to perform hazardous work, or enter a vessel or confined space, until a work permit signed by an authorized person has been issued. The permit should have as a part of it a checklist that indicates the following:

1. The atmosphere inside the vessel or space has been tested and found safe for entry, and all the necessary lines have been blanked or disconnected, and the lines so identified and listed. Where lines cannot be blanked off because of interference with production, valves to the vessel should be closed and locked. Warning signs that people are working inside should be placed close to the valves. If the valves on a vessel containing hazardous substances fail to shut tight, the vessel *should not be entered* until conditions have been corrected and rendered safe.

2. The vessel has an access hole large enough to allow a person wearing safety harness, lifeline, and emergency respirator to enter and leave easily.

3. All the relevant safety equipment is present, including a list of it and confirmation that it is in working order.

4. A ventilator has been set up and is in working order.

5. Specific instructions have been issued for safely making the repairs or inspection.

The permit is issued to the authorized person of the maintenance crew, who signs it, indicating that it has been read and that the equipment has been accepted for work. A permit should be valid only for the shift on which it was issued. If the job requires more time than that, the permit should be signed back to the issuer and a new one written after the shift change.

Checks should be made frequently with an explosimeter or chromatograph, or both, to ensure safe flammability limits. Vessels should be purged with air both to cool it and to eliminate any oxygen deficiency. A test for oxygen adequacy by means of an oxygen probe, Orsat, or chromatograph analyzer (preferably by two different sampling and testing methods) should be made.

A vessel that has contained an acid or alkali should have air circulated through it, after it has been copiously flushed with water (which tested neutral), then drained. A test for oxygen adequacy should be made with the aforementioned analyzers.

A vessel that has held a gas should be purged with nitrogen to within acceptable limits of flammability and toxicity, then purged with air to provide adequate oxygen, as indicated by analyses.

Purging and internal inspections. Nitrogen is extensively used to purge a vessel, or to blanket a tank's contents with inert gas to prevent chemical reactions with air. Nitrogen evaporation and presence in the air are not normally considered to be dangerous or to constitute a health hazard. This may not be true if the nitrogen concentration in a pressure vessel is too high. The atmosphere consists of approximately 78 percent nitrogen and 21 percent oxygen along with other miscellaneous gases. Because of the risk of asphyxiation from lack of oxygen, the Occupational Safety and Health Administration minimum limit for oxygen in any occupied space is 19 percent. Therefore, where it is known that nitrogen has been used for purging or inerting, the following facts should be considered:

1. It is impossible to inert a tank until all the stored liquid has been removed.

2. Never assume a vessel is safe after purging simply because the specified volume of purge gas has been used. Always check the vessel's atmosphere for oxygen concentration or flammability. Oxygen concentration should be at least 19 percent.

3. Purged tanks containing a solid residue may be unsafe. Disturbing the solid may release new flammables. In this case, continuous atmospheric monitoring is necessary.

Skin contact. Skin contact with internal surfaces of vessels should be avoided as far as possible. The necessary emergency steps to be taken in event of contact with harmful chemicals should be known *before* entering vessels that have contained such substances. It is good practice to take a bath and change clothing after inspecting a vessel, especially when it has contained chemicals that generate dangerous fumes. This will minimize the danger of burns that may result from the corrosive action of chemicals or fumes on the skin.

Interior illumination. For tanks generally, but especially those that contain explosive or flammable substances, it is recommended that only explosion-proof flashlights be used. If extension lights must be used, they should have heavily insulated cords and be equipped with explosion-proof fixtures. Extension lamps should be properly grounded. Lamps operating on a 32-volt circuit or less are preferred to those operating on a higher-voltage circuit.

Persons should avoid touching the extension cord when clothing is wet, when they are perspiring, or when they are standing on a wet surface. Even low-voltage shock under such conditions may be dangerous.

Never *assume* that an extension light is in good condition without proper examination.

Inspection of internal and external physical conditions

There are numerous reasons for making inspections of pressure vessel systems, but the paramount one is to assure safe and reliable service. Pressure vessels become affected by service conditions. Corrosion, cracking, thinning, fouling, grooving, and other metal deterioration continuously occur from service effects.

Each type of pressure vessel has its own peculiar area to be watched, either because of process effects or because of the nature of design and application. Some vessels may have an inherent hazard related to material properties. For example, a cast-iron roll has a cracking hazard if, in operation, too wide a temperature swing is experienced.

Inspections are made to detect the conditions that cause pressure vessel failures, among which are the following:

1. *Corrosion* is still the most prevalent reason for vessel failures. It can be controlled by better selection of material or by process changes. Failure prevention is by periodic inspection, supplemented by thickness tests to determine wear. The most likely places for corrosion to occur is in the liquid-vapor interfaces of a vessel, vapor zones, and zones of high fluid velocity. With proper visual monitoring, supplemented by thickness tests and calculation for allowable pressure, the vessel's condition can be known well enough to prevent the unexpected shutdown from a failure. As seen earlier, some people call this approach "predictive maintenance."

Localized corrosion at crevices, elbows, and sharp corners also requires inspection in order to detect possible cracking. Visual inspections can be supplemented by magnetic-particle and dye-penetrant examinations.

2. *Erosion* is another cause for metal deterioration, and it usually occurs at points with high entry flow of fluids or reversal of fluid flow at high velocities. Heat exchanger tubes can be eroded by high-velocity flows at entry points. Wear plates, impact baffles, and a more even flow pattern are the methods used to reduce erosion damage. Internal inspections should determine if these devices are in suitable condition to prevent erosion. Tubes can be checked by eddy-current tests to note if thinning due to erosion is occurring.

3. *Sharp corners* and abrupt changes in sections of shells or heads

require examination for repetitive cycling cracks, usually referred to as "fatigue cracks." If such cracks are found or suspected, NDT methods can be used to determine the length and depth of the cracks.

4. *Stress-corrosion cracking* should be watched for where well-known corrodents such as chlorine or sodium might attack susceptible material.

5. *Weld-seam deterioration* is a common problem in some pressure vessels such as digesters, creosoting cylinders, and vulcanizers. Cracking in heat-affected zones at welded seams may appear in service and requires careful inspection during internal inspections.

6. *Creep* occurs in vessels operating at high temperatures but usually is allowed for in design and is not a problem unless design conditions are exceeded. Creep is a slow plastic deformation under constant stress of material which results in permanent deformation of material, even when the load or stress is removed. Inspection for creep is thus one of looking for changes from original dimensions.

7. Entryways, openings, and nozzle connections should be examined for cracks and *evidence of distortion*. Gasket seating faces should be checked for possible wire cutting due to leakage.

The above provides only a sample of defects that may be found during inspections.

Deciding on the extent of inspection programs

Inspection programs are usually patterned on plant needs and the physical, chemical, and economic factors that can influence output or plant reliability. From a broad safety perspective, an inspection program can consist of the following activities:

1. Establish benchmark conditions when equipment is new or has had an extensive overhaul. Equipment record cards and code construction data will assist in this initial evaluation.

2. Set up an operating inspection program to note if the vessel is operating within process parameters and within the pressure and temperature ratings. This would include a check on the alarms and shutdown and pressure-relieving devices that provide protection against abnormal conditions.

3. Review with operation and test personnel the efficiency of performance and whether the pressure vessel may require internal cleaning to restore output efficiency. A review of pressures, temperatures, flows, and similar indicators of normal and abnormal performance will assist in determining if an internal inspection may be required.

4. Review past incidents of failure, abnormal condition, or crack discovery in order to develop schedules of inspection for fatigue cracks after

a given number of cycles. This will require coordination with the operating personnel and careful record keeping.

5. Perform similar reviews on such items as water treatment, concentrations of mix, etc., where past analysis of failures indicates a need for controlling these items in order to avoid stress-corrosion cracks.

6. When calculations of wear rate and records indicate that corrosion allowances may have been used up, conduct more frequent internal inspections to track this thinning action.

7. Determine the condition of connected valves during operating inspections in order to note if leakage is occurring through wearable items such as packing and valve seats.

8. Follow manufacturers' instructions; most manufacturers have explicit instructions on what inspections may be required on their equipment.

9. Cooperate with the jurisdictional inspector when mandated inspection is due.

Types of inspections

Inspections can be broadly divided into the following types:

1. Jurisdictional inspections
2. Internal inspections
3. External or operating inspections
4. Nondestructive testing and inspection

Jurisdictional inspections. States and municipalities have installation and inspection regulations for pressure vessel systems. The main purpose of these inspection regulations is to protect the lives and limbs of employees and the public, as well as to avoid serious property damage. When a state or municipality has adopted laws on pressure vessels, the vessels require ASME design and construction and compliance with National Board rules for reinspection and repairs. Most pressure vessel inspections for the jurisdiction are made by commissioned inspectors of licensed insurance companies or by qualified jurisdictional inspectors.

The jurisdictional inspections may be of only the external type in some jurisdictions, where in the opinion of the commissioned inspector there is evidence that internal corrosion is not serious enough to warrant an internal inspection.

Frequency of jurisdictional and plant inspections. The frequency and types of inspections to be made will depend on the vessel and its service. Age and location plus physical condition will determine when the next inspection should be performed. Some vessels with noncorrosive contents, built and installed according to code requirements, with proper and functional safety devices, may by jurisdictional rules require only an external inspection at two- to three-year intervals. Most inspectors making jurisdictional inspections, however, consider the plant maintenance that is being provided before allowing lengthy intervals between jurisdictional inspections. For example, on such objects as air tanks, they will check whether the plant has a program for draining the condensate periodically out of the tank and for checking the operation of the pressure control switch and the safety valve. The burden of inspecting and maintaining is still primarily with the plant staff, and jurisdictional inspections must be considered primarily as a *verification inspection* that the plant program is satisfactory. Plant personnel must also track efficiency of performance of pressure vessel systems used in processes, something with which jurisdictional inspections are only indirectly involved. Insurance companies making pressure vessel inspections stress more the implementation by the plant of inspection programs, with the insurance and jurisdictional inspectors acting as monitors of the program. Most pressure vessel insurers have the staff to help implement an inspection program to be carried out by qualified and trained plant personnel.

Management responsibility. This policy of placing the burden for inspections of pressure vessel systems on plant management was born out of necessity. Large process plants have an extreme assortment of high-pressure vessels in their process streams and thus have the potential of having an accident occur not only from the usual overpressure failures, but also from explosions due to leakage of flammable contents and to rapid chemical reactions. Such conditions require engineering controls, safeguards, and protection plans beyond the normal exposure analysis involved with checking only for physical deterioration.

Internal inspections. Internal inspections have as their purpose the cleaning out of scale and other deposits that may be affecting output. Even more important, a visual inspection can be made to determine if corrosion thinning is taking place, cracks are developing, welds are still sound, and connections to controls and safety relief valves are unobstructed and, in general, to make sure no serious deterioration is taking place.

Some vessels in the process industries require the removal of internal components before inspections can be made, as in the case of a typical packed distillation column. Vapor enters through a feed pipe located below the support plate or grid of the lower bed. Random or structured packing lies on top of the plate and is, in turn, covered by a hold-down grid. Above this, a liquid distributor, which receives a mixture of feed and liquid from the upper bed, redistributes the mixed liquid to the lower bed.

A liquid collector set on a ringed channel provides this mixture; it collects the liquid from the upper bed and mixes the liquid with liquid feed piped into the channel. The upper bed of packing, like the lower one, is held up by a support plate and secured by a hold-down grid. Reflux from the condenser returns to the top of the bed through liquid-distribution internals.

In practice, entry ports are provided for easy access to the liquid collection and redistribution system. This is extremely useful for servicing the liquid distributor during a turnaround or shutdown. As can be noted, these types of pressure vessels require the removal of internal packing and screen in order to expose shell surfaces to visual examination.

Cleaning before inspection. The cleaning of vessels containing petroleum products before inspection should follow these guidelines:

1. The cleaning operation should be under the supervision of qualified personnel.

2. Supply piping should be blanked or the control valves locked in an approved manner.

3. The vessel should be emptied by pumping out and floating solids for manual removal.

4. Entry ports and handholes should be opened and the vessel vented.

5. Internal surfaces should be washed down with a hose and steamed.

6. Sludge deposits remaining should be removed by the use of suitable nonsparking tools. The washing should then be repeated on the portions of the vessel where the sludge has been removed.

7. Tests should be made under the supervision of qualified personnel to determine whether or not explosive or toxic vapors are present. Only equipment that is maintained in good operating order should be used for making these tests.

Inspection intervals. Intervals between internal inspections may be determined by any of the following:

1. Jurisdictional requirements

2. Company policy based on past experience

3. Past readings and conditions found

4. Manufacturer's recommendations

5. Special hazards due to content leakage, such as the release of toxic or flammable substances

6. Wear rates and thickness tests that indicate a vessel may be nearing its minimum required thickness and thus may require more frequent inspections

A good rule of thumb to follow on internal inspection frequency, when only the wear rate is used as a criterion and when the vessel is approaching the minimum thickness required, is to schedule internals at half the indicated remaining life expectancy.

External inspections. Most external physical inspections must of necessity concentrate on checking operating pressures versus allowable pressure; if possible, testing the safety relief valve by means of the test lever; and making sure the safety valve is properly stamped for allowable pressure and capacity and comparing this with the allowable pressure on the pressure vessel. Evidence of any form of leakage requires careful attention, because this could be due to corrosion thinning, cracks, or even melting of a piece because of high temperature. Leakage coming from behind insulation, supports, or other obstruction to a clear sighting of metal surfaces requires removal of the obstruction for a closer visual and perhaps NDT examination using such techniques as magnetic-particle or dye-penetrant tests. Entry ports and handholes require gaskets—a favorite place for leakage to occur. Leaks, of course, can deteriorate the metal to the point that it may require repairs if the minimum thickness required for the vessel has been reached.

Where it is impractical to test the safety relief valve in service because of hazardous or costly liquid and gas contents, or because the process may be seriously disturbed, it is common practice to have the safety relief valves removed during the internal inspection of the vessel and to have the valve tested with air pressure by an authorized safety-valve repair shop or by the safety-valve manufacturer who has the ASME and NB stamp for safety relief valves.

Generally, the intent of the external inspection is to ensure that the installation still conforms to design or code requirements and that there is no evidence of physical deterioration that might jeopardize the continued safe operation of the vessel.

Nondestructive testing and inspection. Where NDT examinations are made to supplement a visual inspection, the readings or findings of the NDT examination should be included in the report. Wearable pressure vessels such as carbon-steel digesters are usually subjected to thickness testing with ultrasonic instruments. The grid mapping of thickness readings is usually done on a form similar to that shown in Fig. 7.2. By using a laid-out map of the vessel, it is possible to repeat readings in the same area as in previous years in order to note changes. This procedure also

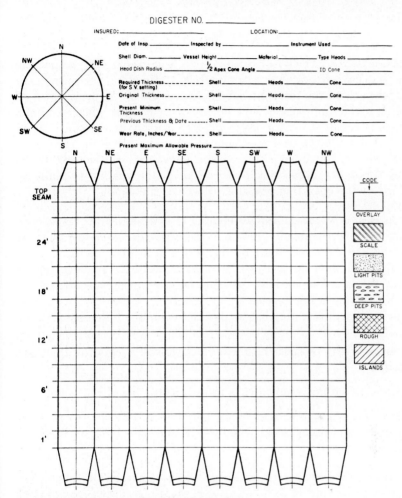

Figure 7.2 Digester thickness and wear report.

permits establishing a corrosion rate for the vessel, if the thinning is due to unavoidable process conditions causing corrosion.

A report of findings is an important element of any inspection program. This can be on a check-off form or in narrative outline form supplemented by mapped-out diagrams as shown in Fig. 7.2. One map can be used to record the physical conditions found. The second map could show the thickness readings obtained on the vessel. The map shown in Fig. 7.2 is for a digester, as used in the paper industry.

If corrosion or any other deteriorating condition is found in a vessel, it should be described (pitting, grooving, general corrosion, etc.). The type of corrosion can be shown on a grid map, using the symbols designated by the plant for the various defects. Where thickness readings are taken, these can be inserted in the proper location on the grid map.

Thickness testing. Thickness testing is an important NDT activity in plants that have wearable pressure vessels. Quite often, the instrument personnel perform the tests and submit the results to the pressure vessel group. A better approach would be first to calculate what minimum thickness is required for the given pressure setting of the safety relief valve so that when the tests are made and readings below the acceptable level are reached, qualified pressure vessel personnel can be alerted to the condition found. A decision can then be made on whether to make repairs, adjust the operating pressure, or replace the vessel.

Interpretation of inspection and test results. It is essential for responsible personnel, knowledgeable in pressure vessel strength requirements, to interpret all the data generated by inspectors and NDT personnel. The conditions can then be analyzed so that a decision can be made on whether the vessel is suitable for more service or requires corrections before being returned to service.

Periodic thickness testing has as its chief value the following:

1. Determines the present thickness of the vessel versus the thickness required

2. Helps to determine wear rates, and thus future life of the vessel

3. Permits decisions to be made on whether to replace the vessel, make repairs of the thinned area, or reduce the allowable pressure if process conditions permit

Calculating minimum thickness and wear rates. Minimum thickness and wear rates can be calculated by using code equations. Two corrosion rates (a long-term one and a current one) should be calculated, because, at times, corrosion may accelerate as a plate or head nears the minimum thickness required for the operating pressure.

$$\text{Long-term corrosion rate, inches per year} = \frac{\text{original thickness} - \text{present thickness}}{\text{number of years in service}}$$

$$\text{Present corrosion rate, inches per year} = \frac{\text{last thickness reading} - \text{present reading}}{\text{number of years between tests}}$$

Example A digester was installed in 1954 with an original thickness of 1.75 in. In 1975, the vessel was checked and had a minimum thickness of 1.57 in, while in 1978, the thickness was 1.31 in. Calculate long- and short-term wear rates to 1978.

answer

$$\text{Long-term wear rate} = \frac{1.75 - 1.31}{24} = 0.018 \text{ in/year}$$

$$\text{Short-term wear rate} = \frac{1.57 - 1.31}{3} = 0.086 \text{ in/year}$$

Example If the above vessel were required to be operated at 150 psi and the above minimum readings occurred on the shell, with an original weld efficiency of 80 percent, an inside diameter of 10 ft, and plain carbon-steel construction with an allowable stress of 13,750 lb/in², what would be the remaining life of the vessel before the pressure would have to be reduced or a stainless-steel weld overlay applied for corrosion resistance?

answer In order to determine remaining life to code minimum thickness, it is necessary to use the shell equation from Chap. 5.

$$t = \frac{PR}{SE - 0.6P}$$

where $P = 150$ psi
$R = 60$ in
$E = 0.80$
$S = 13{,}750$ lb/in²

$$t = \frac{150(60)}{13{,}750(0.8) - 0.6(150)} = 0.825 \text{ in required}$$

The remaining allowable wear before the minimum required thickness is reached = $1.31 - 0.825 = 0.485$ in. The remaining life, using the short-term wear rate = $0.485/0.086 = 5.6$ years.

Thickness testing permits plants to project their future plans for their vessels.

NDT inspections for flaws. Besides thickness testing, NDT is used for flaw detection on vessels that have seen service and are periodically inspected for deteriorating conditions. The methods used include the following:

1. *Visual* inspection is the most widely used and can be quite successful in finding propagating surface cracks, especially in components of a vessel that are prone to stress concentration or fatigue. Visual inspection requires that the inspector have a basic knowledge of causes of failures so that the inspection can be concentrated on the areas of the vessel most likely to have cracks, wear, grooving, pitting, corrosion, and similar evidence of distress. Good visual inspections can detect damage and thus trigger repair procedures at an early stage before any catastrophic failure is experienced.

2. *Dye-penetrant* and *magnetic-particle* inspections are quite often used to better identify surface defects or to develop more detailed data about a suspected area of a vessel.

3. *Radiography* can be used to inspect for fatigue cracking. However, radiography requires careful alignment, heavy screening, and a substan-

tial investment of money, time, and skilled personnel, which makes it less attractive than other NDT methods in this application.

4. *Ultrasonic* inspection can detect flaws that cannot be found using visual observation, dye-penetrant inspection, or magnetic-particle testing. Discontinuities can be detected in regions that are not accessible for visual observation. Drawbacks include (1) relatively high cost, (2) requirement for skilled operators, and (3) a need for trained technicians to interpret the data.

5. *Eddy-current testing* (ET) is extensively used in heat exchanger tube testing for determining tube wall thinning and for identifying cracks. ET works on the principle of creating a uniform magnetic field around a tube and simultaneously measuring the disruption in that field that is caused by certain conditions, such as tube wall thinning or other defects. Identification of the types of defects present and quantification of the amount of wall loss are determined by comparison of the type and intensity of instrumental signals obtained from a given tube versus those obtained with a standard containing known, artificially created defects. ET is as much an art as a science. Extremely poor results can be obtained if the operator is not familiar with typical failure modes.

Tube wall thickness and material can also affect ET accuracy. For example, a 0.0005-in-deep pit may be identified in an 18-gage admiralty tube, but only a pit 0.010 in deep or greater may be identified in a 22-gage, high-alloy, pit-resistant, austenitic stainless-steel tube. This is a 10 percent resolution for the 18-gage admiralty tube and a 36 percent resolution for the 22-gage stainless-steel tube. Ferritic stainless-steel tubes present additional difficulties for accurate eddy-current analysis.

Standard *leak testing,* such as hydrostatic testing, or a gas test can also be used to determine the condition of tubes in heat exchangers.

Supplementing the visual inspections with an appropriate NDT inspection should help the preventive maintenance program minimize the chance of a serious unexpected equipment failure.

Hydrostatic testing. Hydrostatic testing is used to check if the construction of a vessel is satisfactory by noting if any leakage is occurring in welded joints, flanged joints, and other pressure vessel components. This test is also used after repairs for vessels in service, and when there is any doubt as to the extent of a defect or detrimental condition after a vessel has seen service. The standard ASME hydrostatic test pressure for new construction is $1\frac{1}{2}$ times the stamped allowable working pressure that is to be applied to the vessel in service. The National Board recognizes a test to the setting of the safety relief valve as being satisfactory for checking vessel tightness. The pressure test should not exceed $1\frac{1}{2}$ times the maximum allowable working pressure stamped on the vessel, with adjustments allowed for the allowable stress of the vessel at operating temperature versus ambient or water temperature at the time of testing. Water temperature for testing on pressure vessels should not be below 60°F to avoid brittle failures due to transition-temperature problems with some steel. This minimum test temperature will also prevent vapor from the air from condensing on vessel surfaces and thus masking leaks.

The ASME and the NB recommend a maximum test fluid temperature of 120°F. To avoid brittle fractures, large process plants use the following additional guidelines: 60°F for vessels with wall thicknesses up to a maximum of 1 in; 80°F, up to 2 in; 100°F, up to 3 in; and 120°F for vessels with thicknesses of 4 in and over.

Hydrostatic tests should be maintained at the specified test pressure so that a thorough examination for leaks can be made. Some plants use the rule of 1 h under test pressure per inch of maximum wall thickness in the vessel. There is some stress relieving taking place during hydrostatic tests; however, yielding is a sign of overstressing. Pressure gages and temperature gages should be carefully calibrated before any testing is done.

There are certain precautions to be followed in conducting hydrostatic tests:

1. The equipment foundation and supporting structure must be checked for their ability to support the combined weight of the equipment and the liquid required to fill it.

2. Serious damage can occur to any closed vessel or system through extreme temperature changes, and care must be exercised to avoid freezing or thermal expansion. It is recommended that the temperature of the test medium be at least 60°F or as close to the ambient temperature as practicable if it is above 60°F. The risks of failure by brittle fracture due to the progressive lack of ductility with decreasing temperatures should be considered at the design stages and suitable precautions taken.

3. An internal examination before testing is advisable when a vessel has been operating under corrosive conditions to ensure that its strength has not been adversely affected. Thus, in determining the test pressure, due allowance can be made for the condition of the equipment under consideration.

4. Drainage facilities should be adequate in the test area to remove all the test fluid when the vessel that has been tested is being emptied.

5. All vents and pressure gages should be properly sited. Vents should be situated so that all the air can be expelled and should be of adequate size to avoid the possibility of creating vacuum conditions in the pressure shell when the vessel is being drained on completion of the test. Pressure gages should be properly connected and calibrated.

6. When the test is applied, all air should have been expelled and the test pressure gradually increased. The test should be maintained for a sufficient length of time to allow:

    a. The authorized inspector to examine all parts of the equipment under test

    b. Any weaknesses or defects in the equipment to manifest themselves and be observed.

7. The test fluid will normally be water, and when austenitic materials are being tested, it should preferably be demineralized and contain less than 30 ppm of chloride ions. Other fluids may be used when water is precluded for process reasons; when this alternative fluid is flammable, its flash point should not be less than 113°F (45°C).

8. Vessels in the open should not be tested during inclement weather that may conceal evidence of leakage.

9. Any pressure vessel system subjected to its maximum test pressure should not be approached for close examination until after a reasonable time has elapsed. It is recommended that the test pressure be reduced by 10 percent before carrying out a close visual examination. Special consideration should be given to high pressures and large volumes where the contained energy is high.

10. An internal examination after testing is advisable when a pressure vessel has a complex or novel design, in order to ascertain whether the test pressure stresses have caused damage, or when an examination before testing has revealed any dubious feature such as surface defects or corrosion that may have propagated during the test.

Pneumatic testing.   Pneumatic tests are permitted on vessels that (1) are designed or supported in such a way that they cannot be safely filled with water or (2) cannot be readily dried and where traces of any liquid used in testing could present operating problems for the vessel or process. Pneumatic tests are conducted usually with air and must be at least equal to $1\frac{1}{4}$ times the maximum allowable working pressure stamped on the vessel. The code prefers that components first be tested by normal hydrostatic tests. Welds and similar joints should be nondestructively tested before any pneumatic test, in order to detect any deficiencies before subjecting the vessel to this potentially dangerous test.

Precautions in conducting a pneumatic test include the following:

1. The test area should be cordoned off, because pneumatic tests involve confining a gas under pressure that can expand in volume if the vessel should fail unexpectedly.

2. Reference should be given to the resistance of the vessel to fast fracture. Extra precautions may be necessary for brittle materials.

3. The procedure should be carefully monitored to prevent local chilling during filling and emptying of the vessel; also the formation of condensation should be avoided.

4. Steps should be taken to reduce the internal volume of the vessel where practicable. This has the effect of reducing the energy stored in the vessel while under test.

5. When the source of test pressure is capable of producing pressures higher than the pressure desired for the test, precautions against overpressure of the vessel or system under test should be taken by the use of reducing valves and safety valves of adequate size.

6. The vessel under test should not be subjected to shock loads such as hammer testing.

Leak tests.   The hydrostatic and pneumatic tests are considered strength tests. These are tests in which a pressure is applied greater than the maximum pressure generated in service but less than that which would cause yielding and permanent deformation. The strength test provides demonstrable proof that the pressure vessel system can safely withstand the normal service pressure in operation.

Leak tests are used to apply a pressure differential across components of a pressure vessel in order to detect leakage paths or leakage rates; they would be used, for example, in tube testing inside a shell in a heat exchanger. The pressures applied in leak testing may be less than the maximum service pressure, and the main purpose of the leak test is to find evidence of leaks in such places as expanded, riveted, bolted, or welded joints and welded seams, or to find evidence of defects in material, such as cracks, porous areas that have been thinned, and similar material defects.

Other tests for leaks include vacuum tests, search gas tests using a neutral gas, air tests, and service fluid tests.

Each plant must become familiar with the types of pressure vessels that must be inspected and the problems that may be expected in the vessel on the basis of service conditions, method of operation, design, history, or knowledge of a problem on a similar vessel at another site. Many pressure vessels require special attention owing to wear rates and thus require a thickness testing program. Digesters in the paper industry were the first vessels to draw attention to thinning problems and the need to make periodic thickness tests.

Inspection and stress-corrosion cracking

Causes of stress-corrosion cracking.   A recent continuous digester explosion also called attention to the need for postweld heat treatment of welds in plates with thicknesses under $1\frac{1}{2}$ in (see Fig. 7.3), even though the code does not always require postweld heat treatment below this thickness unless the purchaser of the vessel specifies it. It has been concluded by some investigators that cracking in pressure vessels occurs as a result

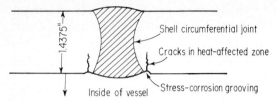

Figure 7.3 Stress-corrosion cracking in welded joints that were not postweld heat-treated. The contents of the vessel tended to cause corrosion in stressed areas.

of stress-corrosion cracking (SCC). SCC is caused by the action of a tensile stress during exposure to a specific corrosive environment.

The tensile stress is provided primarily by residual stresses left in the welds after fabrication, although contributions from the internal operating pressure and thermal gradients in the vessel are also important. For example, the corrosive environment in digesters is engendered by kraft cooking liquor. Alkaline solutions are well known to cause SCC of carbon steels, although not generally at the low NaOH concentrations found in kraft liquors.

The presence of tensile stress in a shell has been shown to correlate with vessel cracking. Vessels that have been fully stress-relieved are less susceptible to cracking than vessels that were only partially stress-relieved. Stress relief redistributes, rather than eliminates, residual welding stresses in a shell. In many cases, these redistributed stresses, together with operating stresses, are still sufficient to cause cracking in a vessel. The results of fracture mechanics experiments show that large changes in crack propagation rates can occur with small differences in the crack-tip stress intensity. On this basis, any reduction in tensile stress on the shell should be beneficial in lowering crack propagation rates in the vessel, as should reduction in corrosive concentrations.

Inspection procedure.   Where stress-corrosion cracking in welds is suspected, a procedure recommended by NDT specialists can be used. First, the weld surface should be prepared so that it is free of corrosion, pitting, and surface abnormalities. Any weld undercutting should be removed by smoothing the surfaces of the welds so that they blend together with a 3 : 1 taper.

After the weld surface has been prepared, a certified level 2 examiner, using a dc yoke, should magnetize the area to be tested. Direct current provides enough magnetic force to penetrate the weld surface $\frac{1}{4}$ in, and the yoke assembly eliminates the possibility of arc striking, which could set up an inner-granular stress condition.

Once the area has been magnetized, a solution of suspended fluorescent iron particles is sprayed on the magnetized area and illuminated by near ultraviolet light (black light). Because surface cracking creates flux-leakage fields in the magnetized area, the suspended iron particles outline the leakage. The black light illuminates the fluorescent particles, making the leakage visible and revealing the extent of cracking.

For stainless-steel welds that are nonferrous, or not able to be magnetized, liquid-dye-penetrant testing is used in these areas to reveal cracking. The weld surface to be checked is wire-brushed to remove any surface corrosion, sprayed with a cleaner, and then wiped dry to remove dirt and grease. A liquid penetrating dye is applied, which seeps into invisible surface irregularities over a 10- to 30-min dwell time. After the excess penetrant is removed, a developer is applied. Through capillary action, the developer draws the high-contrast dye from the flaw, revealing the defect.

The key to an effective program is using a sufficiently sensitive testing discipline to reveal the characteristics of SCC. Wet, fluorescent magnetic-particle and liquid-dye-penetrant tests are both extremely sensitive techniques that have proved effective in revealing the fine, granular cracking that is characteristic of SCC. Before the development of magnetic-particle testing (MT) and liquid-dye-penetrant examination, some inspectors thought that radiographic and visual examinations were sufficient. While these techniques have their merits, they are not sensitive enough to reveal SCC characteristics as economically as do the MT and liquid-dye-penetrant tests.

Use of nondestructive testing to discover cracks, and the repair of these cracks, should not be thought of as a cure for the effects of caustic embrittlement. Instead, such testing and repair should be considered as a means of determining the condition of a carbon-steel vessel.

Inspection of stainless-steel overlaid and lined vessels

Stainless-steel weld-overlaid vessels or stainless-steel lined vessels also require a planned inspection program. Part of this program should be internal inspections in order to make sure that the corrodent fluids have not penetrated the stainless-steel barrier.

Streaks or spots of the fluid are usually a positive indication of a break or pinhole in overlay or seam and are evidence of concentrations behind the surface and of rapid attack and thinning of the plate at this point. Puncture of the shell or thinning to dangerous proportions can occur in a relatively short time. Defects in the overlay or lining may be the following:

- Checks and cracks
- Slag inclusions

- Thin beads in overlay
- Grooves or deep valleys between beads
- Pitting
- Pattern of corrosion
- Washing
- Attack of beads in overlay or lining welds
- Wide beads of overlay with less than 50 percent lap

Visual inspection of clad plates follows the same pattern as that used for the lined ones. Special attention should be given to the weld seams and to any bulged areas or blisters. This may indicate separation of the stainless clad from the base steel. Care must be exercised in the examination to prevent any damage to the surface of the lining such as scratching, gouging, or marking with any material that might react with the stainless steel. No wax or grease marking crayons or pencils should ever be used on any plate surface, as they may contain sulfides or other materials that could attack the plate or sheet.

Inspection of rotating rolls

Rotating rolls in all types of industries require inspection for deterioration and overpressure. Dryer rolls are often supplied steam through a pressure-reducing valve. It is essential to install a safety valve *after* the reducing station and to set it at a pressure not higher than the allowable pressure for the dryer rolls being supplied the steam. The capacity of the safety valve should match all the steam that can be passed through the reducing valve, or bypass, in order to protect the rolls properly. (See Fig. 1.1.)

Some pressure vessels are periodically machined to restore smooth surfaces and tapers for quality control reasons. It is necessary to track the thinning that may reach the required thickness for the required pressure. Yankee dryer rolls are an example of vessels requiring periodic machining.

The machining of the shell of a Yankee dryer roll is a very intricate operation inasmuch as the roll not only has to be finished to a very special polish but also must be carefully crowned so that in operation the pressure between it and the press roll is uniform for the full length of the press roll.

The press roll pushes against the Yankee roll with a pressure of from 200 to 400 lb/linear in, and some attempts have been made to go to still higher pressures. This means that on a 12-ft-long roll the total pressure is in the neighborhood of 70,000 lb, which is just about equivalent to the weight of the Yankee roll. This tremendous pressure requires crowning of both the Yankee roll and the press roll in order to take care of the calculated deflection.

In actual practice crowning in excess of that required by calculation is used. Careful investigation has shown that this is required because of the difference of expansion between the shell and heads of the Yankee roll. There is a difference of temperature in the shell between the area covered by the paper and the ends of the rolls. In the center of the roll the temperature may be as low as 190 to 200°F, whereas the rest of the roll may be at a temperature corresponding to the steam pressure, less a slight loss due to radiation. Calculations made on a dryer roll using 30 lb/in² steam pressure show that on an ordinary 12-ft-diameter Yankee dryer the face is concave to an extent of 0.025 in because of the difference in expansion between the center of the shell and the ends. This alone requires 0.05 in of crown, either in the press roll or in both the press roll and the Yankee roll, to correct for the differential expansion.

These types of steam vessels also require internal inspections to determine if any cracking is occurring at head and shell bolted connections and at any sharp-cornered areas of the vessel. Condensate removal in these vessels is also important; therefore, siphons should be examined to see if they are functional. External surfaces as well as internal surfaces require inspections for gouged marks where other components of a paper machine ride on the dryer.

It is necessary to establish an early program of determining how much stock can be removed from a vessel requiring periodic machining so that the minimum allowable thickness required is not exceeded.

Example Measurements show that a cast-iron Yankee roll with Class 60 material has an OD of 16 ft 4.6 in and an ID of 16 ft. If the safety valve is set for 110 psi, how much more metal is available for machining without having to reduce pressure?

answer Use

$$t = \frac{PR}{SE - 0.6P}$$

where $P = 110$ psi
  $E = 1$
  $S = 6000$ lb/in² (SF = 10 for CI)
  $R = 96$ in

$$t = \frac{110(96)}{6000 - 0.6(110)} = 1.780 \text{ in required}$$

The metal left for machining is $2.3 - 1.780 = 0.52$ in.

Rotating rolls with holes drilled in them for moisture removal under vacuum also require inspections for hole plugging and cracking between holes due to the combined action of large bending stresses and a corrosive environment. Penetrant testing can be used to detect incipient failure in nonferrous rolls. Ultrasonic testing and magnetic-particle testing are most often used on ferrous material.

Suction rolls operate in hot environments containing many potentially corrosive chemicals and deposits that, combined with operating stresses and residual stresses incorporated at the time the roll was made, can cause short-term corrosion fatigue failures.

The most common mode of failure for suction-roll shells appears to be corrosion fatigue. Corrosion erosion, pitting corrosion, stress-corrosion cracking, and overloading have also been observed failure mechanisms that can initiate or be secondary factors in suction-roll shell failures.

A program of hole cleaning with high-pressure showers is often used to remove possible corrodent deposits. Periodic hole-diameter testing with a special gage has also been used in order to track deterioration of the metal between the holes that, thus, reduces ligament efficiency. There is also a concentrated effort being made to find an economical stainless steel for rolls operating in hostile environments. One such effort is on the suction rolls found in the paper industry.

Inspection of low-pressure vessels

Low-pressure vessels require internal inspections at periodic intervals, because corrosion and service may affect the welds and surrounding plate. A rash of deaerator cracking incidents, and even ruptures, recently has drawn attention to the fact that low-pressure equipment can also fail unexpectedly.

One of the main functions of the deaerating heater is to drive the free oxygen out of water. While performing that task, it is subjected to the greatest amount of corrosion in the entire boiler system. The condensate is relatively pure and, as such, has a low hydroxyl-ion concentration, with a low pH value. Therefore, the steel in the deaerator is attacked faster, as it contains condensate with a concentration of hydrogen and hydroxyl ion slightly below the neutral pH of 7. In order to reduce or slow the corrosive attack, it becomes important to maintain a pH value between 8 and 9.

Failures due to corrosion have occurred in both welds and base metal owing to wasting away of the through thickness to unsafe levels that could not support the loadings imposed. This occurs at a slow rate and can be detected by proper inspections.

Heat exchanger maintenance and inspection

Heat exchanger problems are usually problems of lack of performance or problems of leakage, generally in the tube area. They usually are taken for granted unless performance suffers or operating problems develop.

This vast array of vessels requires preventive maintenance and inspection to avoid problems not only in the heat exchanger but also quite often in process streams. For example, a simple tube failure in which the pressure in the tube is higher than that of the heating medium can cause high-pressure backflow into such a vital and expensive piece of equipment as an extraction turbogenerator, when the heat exchanger is used as a staged high-pressure feedwater heater. Therefore, one of the first inspections required for heat exchangers is a review of the protective scheme not only for the heat exchanger itself but also for the equipment that may be connected to the heat exchanger. This also involves a review of the overpressure protection that is provided to the shell in the event that a tube should fail. Some heat exchangers, such as feedwater heaters, require high- and low-level alarms and motorized valves so that the heater can be quickly isolated from a ruptured tube failure. These controls and protective devices to prevent backflow or high pressure from a tube failure require at least a monthly check for proper operation, so they don't become bound with sludge and noncondensable gases on the connections to the sensing elements. Many a plant has suffered lengthy repairs and loss of income because the consequences of a tube failure in a heat exchanger were not properly diagnosed for the protection needed, or because the testing of these devices at prescribed intervals was neglected. At times, a domino effect in the process is possible as a heat exchanger with leaking tubes pollutes the entire process stream.

Performance problems. Performance deterioration is usually due to tube fouling. Internal leakage at tube ends, between passes, and at intermediate tube supports or flow baffles is another obvious reason for nonperformance. Air or gas binding can affect flows or create erratic pressure conditions. Performance monitoring consists of taking benchmark readings of pressure, temperature, and in-and-out flows when the exchanger is new, and then comparing performance test results at later intervals to note if any deterioration or changes are taking place.

Fouling of heat exchanger surfaces depends on many variables and is taken for granted in some applications. The heat exchanger is periodically removed from service in order to clean the surfaces and thus restore efficient output. At this outage, internal inspections are possible to look for mechanical defects in the exchanger.

Fouling is the deposition of undesirable materials on the heat exchanger surface, which increases resistance to heat transmission. Fouling is a complex phenomenon and may be due to sedimentation, crystallization, chemical reaction, polymerization, coking, growth of organic material such as algae, and corrosion. These fouling mechanisms may operate in isolation or may occur together. The rate of fouling is controlled by physical and chemical relationships that, in turn, are affected by the operating conditions. The operating variables that have important effects on fouling processes are:

1. *Flow velocity.* Very strong to moderate effect on a majority of fouling processes.

2. *Surface temperature.* Affects most fouling processes, and in particular crystallization and chemical reaction.

3. *Bulk-fluid temperature.* Affects rate of reaction and crystallization.

4. *Materials of construction.* Possible catalytic action and corrosion.

5. *Surface.* Roughness, size, and density of cavities will affect crystalline nucleation, sedimentation, and the adherence tendency of deposits.

The materials of construction and the surface nature have the greatest effect in initiating fouling rather than in continuing and sustaining it.

As already noted, deposits on the heat transfer surface due to fouling increase the overall thermal resistance; in addition, they lower the overall heat transfer coefficient of the exchanger. For heat exchangers to maintain satisfactory performance in normal operation, and a reasonable service time between cleanings, it is important during design to provide a sufficient surface via a fouling allowance appropriate to the expected operating and maintenance conditions.

Appropriate values for fouling resistances involve physical and economic considerations, which vary from user to user—even for identical services. The user should specify design fouling resistances on the basis of past experience and current, or projected, costs. In the absence of such information, the user may be guided by TEMA standards.

Mechanical problems. Mechanical problems develop in heat exchangers and require that the unit be monitored for repairs or adjustment. Among these problems are:

1. Tube leakage in the region of the tube sheets or intermediate tube supports. Visual inspections during hydrostatic tests may pinpoint the tubes that are leaking. Eddy-current tests are also used to detect tube deterioration. The causes of this leakage are many.

2. Corrosion erosion at tube inlets. This is caused by flow maldistribution or excessive turbulence. Thinning of tube ends on one side is evidence that corrosion erosion is taking place.

3. Impingement attack on tubes. This is characterized by sharply defined pits in otherwise smooth tube walls. The pits can be elongated in the direction of flow.

4. Tube-hole wire drawing or wormholing due to joint leakage. This may also erode the tube sheet in the area where the holes are worn. The cause of the leakage must be found and corrected.

5. Tube-sheet ligament cracking. This is usually caused by excessive thermal gradients between the tube sheet and the tube-hole area. In other words, most cracking is caused by an expansion and contraction problem around the ligament. Corrosion on the shell side could affect heat transfer and cause ligament cracking. Visual, dye-penetrant, magnetic-particle, and hydrostatic tests are used to detect the cracks so that repairs can be made.

6. Tube end fatigue. Failures because of this are usually caused by the repetitive stresses generated by tube vibration. Breaks or cracks due to fatigue are usually found where the tube emerges from the tube sheet or at intermediate supports, where large flexing may occur from bending on long tubes. Where tubes are welded to the tube sheet, cracking may occur in the welds.

7. Crevice corrosion. This may appear in the tube-to-tube-sheet interface if the tube rolling did not assure full contact of tube to tube sheet.

8. Stress-corrosion cracking. This can occur in tubes when the tube metal is under high tension in service and a corrodent that attacks the metal is present in the heat exchanger flows. Material selection is still the best way to minimize stress-corrosion cracking. Good joint design to avoid thinning of tube metal from rolling is another method used to combat stress-corrosion cracking. Visual inspections, appropriate NDT examinations, and hydrostatic or pneumatic tests are used to detect this type of cracking problem.

9. Vibration damage. This is usually in the form of excessive tube wear or cracks appearing in the areas of tube supports. Long heat exchangers are very vulnerable to this problem. The tubes can be excited by the velocity of flow across them, and forms of sympathetic vibration may emerge. If there are indications of vibration damage (rubbing and erosion), it may be necessary to review the speed of flows *at all load conditions* and compare vibrations at these speeds with the natural frequency of the heat exchanger components, assembly, and connected piping.

Some remedies are (*a*) installing deresonating baffles (when the vibration-inducing mechanism is acoustical coupling), (*b*) lacing the tubes with wire between the baffles, (*c*) sliding flat bars between the tubes at locations between the baffles, (*d*) changing the flow pattern to divided-, split-, or cross-flow when thermally possible, (*e*) altering the character of the shell-side stream, (*f*) retubing with tubes that have a natural frequency in a safe range, and (*g*) replacing the bundle with one designed to be free of vibration damage.

10. Improper tube-to-tube-sheet fabrication. This problem may include incomplete fusion between tubes and tube sheet, inclusions in the

weld, heat-affected-zone cracks, porosity, and similar welded-joint problems. Welded tube-to-tube-sheet joints require visual inspections to detect cracking, weld washout, undercutting, and similar defects that, if ignored, can lead to leakage problems in service.

Preventive maintenance inspections.   Preventive maintenance inspections take four forms on heat exchangers:

1. Periodic visual inspection of internal components
2. Visual inspections supplemented by NDT (eddy-current, dye-penetrant, magnetic-particle tests)
3. Hydrostatic or pneumatic tests to check the integrity of tubes
4. External inspections to check for changes in performance and to monitor controls, safety devices, external problems such as leaks, and similar operating variables to note if any changes or any abnormal operating conditions are occurring.

Tube plugging.   Tube plugging is a common method of expeditiously stopping a leak in order to return a heat exchanger to service. Repairs are postponed in this way until (1) so many tubes become plugged that output is affected or (2) tube replacements are made at the next maintenance shutdown of the plant. It is usually at this time that an investigation is made on why the tubes started to leak.

Records of tube plugging in shell-and-tube heat exchangers provide information regarding the present condition and the rate of change of the condition. These records should include a drawing of each tube sheet, with plugged tubes shown and date of plugging indicated. These data should be recorded on a tube plugging diagram.

The records should be supplemented with notes regarding the apparent cause of failure, the mode of unit operation before failure, and, in the case of U-tubed heat exchangers, the pass in which the failure occurred. An example of such a note would be "600 admiralty condenser tubes failed following a 6-month outage."

Typically, such records are not maintained in plants. If a life extension study is to be performed on such a unit, the information will have to be obtained by other means. This type of record keeping should be performed in order to give an early warning of a problem even if the component is not completely overhauled.

Visual examination of shell-and-tube heat exchangers is not usually fruitful, since most failures are hidden by the pressure vessel shell. Observation of the tube plugging pattern may reveal some useful information. Borescope inspections may be possible with some heat exchangers.

Repairs to Pressure Vessel Systems

National Board R stamp

As more jurisdictions in the United States pass laws on installation of unfired pressure vessels in conformity to ASME Section VIII, Division 1, standards and also pass requirements for reinspection at periodic intervals, a program of registering qualified repair organizations has evolved which in some respects duplicates the issuance of ASME new-construction stamps to boiler and pressure vessel manufacturers. The aim of registering repair organizations is to promote uniform repair procedures. This involves using code-approved material, establishing a welding procedure, using welders that have been qualified for the repair or alteration that may be required, and having an inspection agency verify that the repairs or alterations meet code and NB requirements. The National Board of Boiler and Pressure Vessel Inspectors has developed an R stamp issuance program to any organization, manufacturer, or owner-user group that desires to be certified as a registered repair organization. Most jurisdictions have accepted this program of *nationally* certifying repair organizations, because it promotes uniform repair procedures and safe practices throughout the United States.

The National Board publishes a document entitled *The National Board Inspection Code,* which is used as a guide by inspectors and others to help maintain the integrity of boilers and pressure vessels. It contains rules and guidelines for inspection after installation, repair, alteration, and rerating, thus helping to ensure that this machinery continues to be used safely. When a weld repair or alteration is completed, the organization that carried out the work is required to complete a record-of-weld-repair form that details the type of repair or alteration made by that organization. The record-of-weld-repair form is required to be signed by the organization carrying out the work, verifying that it was carried out in accordance with the *National Board Inspection Code* and the jurisdictional requirements. In addition, this form is to be signed by the authorized inspector who accepted the repair or alteration; the form verifies that the work was carried out in accordance with the *National Board Inspection Code.* The weld-repair form is forwarded to the enforcement authorities for boilers and pressure vessels in a jurisdiction. In the event of a repair or an alteration to a boiler or pressure vessel, the authorized inspector of the jurisdiction should be contacted for guidance in complying with the National Board rules or those existing in a jurisdiction on repairs or alterations (usually similar to the nationally recognized NB rules).

Requirements and guidelines concerning the NB R stamp include the

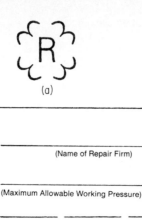

(a)

(b)

(c)

Figure 7.4 National Board repair stamps. (*a*) National Board R stamp to show repairs were made to a code vessel; (*b*) nameplate showing the name of the repair organization and date of repairs; (*c*) alteration nameplate.

following, but the NB rules should be consulted, as these can change with time:

Authorization to use the stamp bearing the official National Board repair symbol as shown in Fig. 7.4 will be granted by the National Board pursuant to compliance with its rules and requirements.

Repair organizations, manufacturers, contractors, or owner-users that make repairs to boilers and pressure vessels classified under the *National Board Inspection Code* may apply to the National Board of Boiler and Pressure Vessel Inspectors by completing the application forms for the loan of a repair stamp and for authorization for its use.

Each applicant must agree, if authorization to use the stamp is granted, that the stamp is at all times the property of the National Board and will be promptly returned upon demand. If the applicant discontinues the repair of such vessels or if the Repair Certificate of Authorization issued to such applicant has expired and no new certificate has been issued, the stamp will be returned to the National Board.

Each repair symbol stamp must be serialized and used only by the repair organization to which it was issued.

The holder of the National Board repair symbol stamp or holder of an applicable ASME code symbol stamp must submit to the authorized inspection agency that accepts the work and to the state or city of the United States or province of Canada (jurisdictional authority) a National Board Form R-1, Report of Welded Repair, in accordance with the *National Board Inspection Code* (see Fig. 7.5).

Alterations of a boiler or pressure vessel must be made in accordance with requirements of the *National Board Inspection Code*.

A nameplate may be used, or where permitted, the repair symbol may be impressed directly below the original stamping on the vessel. If a nameplate is used, it must be welded or permanently attached either below or adjacent to the original stamping. Each repair organization, manufacturer, contractor, or owner-user that completes the repairs of a boiler or vessel shall hold a valid Certificate of Authorization for use of the National Board repair symbol stamp when required by the jurisdictional authorities of the states of the United States or provinces of Canada.

Before issuance or renewal of a National Board Repair Certificate of Authorization, all requirements, including a written quality control system, material control, repair design, fabrication, examination, and authorized inspection, must be met. In addition, it is required that the applicant successfully demonstrate the implementation of its written quality control system. In areas where there is no National Board member jurisdiction or where a jurisdiction elects not to review a manufacturer's facilities, the review will be performed by a representative of the National Board of Boiler and Pressure Vessel Inspectors.

The quality control system must describe and explain what documents and procedures the repair organization will use to validate a repair that would restore the boiler or pressure vessel to a safe condition and maintain the validity of the original vessel.

Subject to the acceptance of the National Board member jurisdiction, the authorized repair organization may make changes to the quality control system. Where there is no National Board member jurisdiction, or where the jurisdiction elects not to perform the function, changes to the quality control system must be acceptable to the National Board.

FORM R-1, REPORT OF WELDED ☐ REPAIR OR ☐ ALTERATION
as required by the provisions of the National Board Inspection Code

1. Work performed by _____
               (name of repair or alteration organization)   (P.O. no., job no. etc.)

_____
                           (address)

2. Owner _____
                        (name)

_____
                         (address)

3. Location of installation _____
                         (name)

_____
                         (address)

4. Unit identification _____ Name of original manufacturer _____
            (boiler, pressure vessel)

5. Identifying nos.: _____
        (mfr's serial no.) (original National Board no.) (jurisdiction no.) (other) (year built)

6. Description of work: _____
           (use back, separate sheet, or sketch if necessary)

_____

_____

_____

_____ Pressure test, if applied _____ psi

7. **Remarks:** Attached are Manufacturers' Partial Data Reports properly identified and signed by Authorized Inspectors for the following items of report: _____

_____

_____

_____

_____

_____
           (name of part, item number, mfr's name, and identifying stamp)

CERTIFICATE OF COMPLIANCE

The undersigned certifies that the statements made in this report are correct and that all design, material, construction, and workmanship of this _____ conform to the National Board Inspection Code.
                         (repair or alteration)

Certificate of Authorization no. _____ to use the _____ symbol expires _____, 19 ___

Date _____ 19 __ _____ Signed _____
                 (repair or alteration organization)   (authorized representative)

CERTIFICATE OF INSPECTION

The undersigned, holding a valid Commission issued by The National Board of Boiler and Pressure Vessel Inspectors and certificate of competency issued by the state or province of _____ and employed by _____ of _____ has inspected the work described in this data report on _____, 19 ___ and state that to the best of my knowledge and belief this work has been done in accordance with the National Board Inspection Code.

By signing this certificate, neither the undersigned nor my employer makes any warranty, expressed or implied, concerning the work described in this report. Furthermore, neither the undersigned nor my employer shall be liable in any manner for any personal injury, property damage or loss of any kind arising from or connected with this inspection, except such liability as may be provided in a policy of insurance which the undersigned's insurance company may issue upon said object and then only in accordance with the terms of said policy.

Date _____, 19 ___ Signed _____ Commissions _____
                    (Authorized Inspector)   (National Board (incl. endorsements), state, prov. and no.)

This form may be obtained from the National Board of Boiler and Pressure Vessel Inspectors, 1055 Crupper Ave., Columbus, OH 43229   NB-66 Rev 4

Figure 7.5 The National Board welding repair and alteration form is now required by most jurisdictions with boiler and pressure vessel inspection laws.

When required by the jurisdictional authority, the holder of a National Board repair symbol must have in force at all times a valid inspection contract or agreement with an authorized inspection agency that employs authorized inspectors as defined in the National Board bylaws. The authorized inspection agency participates in the review of the quality control system for issuance or renewal of the repair authorization certificate, and any revisions to the quality control system shall be subject to the acceptance of the authorized inspection agency.

Further details on the securance of an NB repair stamp by repair organizations, owner-users, etc., desiring to perform their own repairs according to NB rules should contact:

The National Board of Boiler and Pressure Vessel Inspectors
1055 Crupper Avenue
Columbus, Ohio 43229

Another requirement of the NB for repairers is to have a *written* quality control manual covering the following areas:

1. An organizational chart of people responsible for supervising repairs and alterations on boilers and pressure vessels.
2. The scope of repair and alteration work that the organization intends to perform and be qualified for.
3. Methods to be used in ensuring the use of code-approved material, rods, etc., in the repair.
4. Methods of carrying out the proposed repairs. This section should have step-by-step instructions as to how a specific repair is to be carried out. It should name the title of the person responsible for developing these instructions and should include any necessary drawings. This step-by-step repair outline should include material check, reference to items such as the welding procedure specifications (WPS), fit-up, NDT technique, and heat treatment and pressure test methods to be used. There should be a space for "sign-offs" at each operation to verify that each step has been properly performed. In addition, the authorized inspector may wish to make an inspection of a specific operation before the next step is commenced. The quality control manual should provide for this.
5. Welding, NDT, and heat treatment. The quality control manual should indicate the title of the person or persons responsible for the development and approval of the welding procedure specification and its qualification, and the qualification of welders and welding operators. It is essential that only welding procedure specifications and welders or welding operators qualified as required by the *National Board Inspection Code* be used in the repair of boilers and pressure vessels. Similarly, NDT

and heat-treating techniques must be covered in the quality control manual.

6. Pressure tests. Reference should be made in the quality control manual to required pressure tests upon completion of the repair.

7. Acceptance and inspection of repair. All repairs are subject to the acceptance of the authorized inspector responsible for the boiler or pressure vessel. The quality control manual should clearly indicate the method whereby the authorized inspector is informed of the contemplated repair, repair work while in progress, and final inspection or acceptance of the completed repair.

8. Report of Welded Repair form. The quality control manual should be specific as to who is responsible to prepare, sign, and present the Form R-1, Report of Welded Repair, to the authorized inspector for acceptance. In addition, the distribution of this form or any other necessary documentation should be described in the manual.

Repairs and alterations

Repairs. *Repairs* under NB rules are work that is necessary to restore a pressure vessel to a safe and satisfactory operating condition so that the original pressure and temperature rating of the vessel is retained. Typically, repairs would include weld repairs, such as for cracking; pressure part replacement as indicated by the original design; installation of a flush patch with the same thickness and material as on the original design; replacement of a cylindrical course of the same thickness and material as on the original design; and building up by welding wasted areas, provided the remaining thickness is equal to code-required thickness for the pressure and temperature of the original design. Practically all repairs require the approval of the qualified inspector representing the jurisdiction for the location in which the vessel is located. This could be an inspector commissioned by the jurisdiction.

Alterations. The NB rules on pressure vessels classify an *alteration* as a change in any item from the original construction. This includes making changes in order to rerate the vessel, such as changing the allowable working pressure. Alterations generally require a U stamp versus an R stamp. If the U stamp holder cannot submit revised calculations and drawings to show that the vessel still meets ASME code requirements, most jurisdictions, including the NB, require that a professional engineer experienced in pressure vessel design prepare calculations, drawings, welding procedures, and similar details and have these approved by an inspection agency, usually of the firm that will do the work, and by the owner's inspection agency if vessel insurance is involved. The jurisdiction will review the proposed alterations to ensure that all requirements of the

ASME code will be met in order for the vessel to continue to be considered a code vessel. The NB requires an alteration form to be filed with them to show that an NB-registered vessel has had code-permissible changes made. The purpose of these requirements and documented procedures is to prevent unauthorized and possibly unsafe changes from being made on code-registered vessels. Reference is again made to the *NB Inspection Code,* and the organization itself, for details on approval of repairs and alterations. Most authorized or commissioned boiler and pressure vessel inspectors of a jurisdiction can tell process plants using pressure vessels how to obtain approval for repairs or alterations through the local jurisdiction or the NB.

Pressure vessel field repairs

The best way of avoiding the need for repairs is to set high standards for design, material selection, and fabrication, but inevitably wear and tear takes place on pressure vessel systems in service that must be corrected by proper repair in order to restore the system's functional capability. NB rules for repairs usually require repair procedures and quality control of the repair that match in most cases the original code requirements. Field repairs require far greater attention to detail, and the surrounding conditions, than does fabricator shop welding.

Determining the cause of defect.   It should be standard practice by plant inspection staff first to determine the cause of the defect needing repair. This may affect the method of repair that is to be employed and may even require a modification of the original construction, for example, different preheat and postweld heat procedures.

Welding parameters.   Many factors contribute to the need for in-service pressure vessel repairs. Corrosion, erosion, and fatigue cracks are the most prevalent causes for repair. Most repairs require that some form of welding be done. The purpose of a repair is to retain a vessel's stamped allowable working pressure. This means repairs must duplicate as much as possible the original construction requirements of the ASME code under which the vessel was built. However, field weld repairs can present some problems, among which are:

1. *Unfavorable environmental conditions.*   There are factors detrimental to weld repair that are beyond the control of the operator. Access for repair may be poor compared with the access for the production weld. This may limit the choice of welding process purely from the standpoint of accessibility, and also for health reasons (processes that produce copious fume or require high levels of preheat may be unsuitable in an en-

closed area). Field weld repairs, as distinct from repairs in the fabrication shop, will usually need to be made in a fixed position, and environmental effects such as wind, rain, and cold may hamper the welder by their inconvenience and the personal discomfort caused. On the whole, the above comments about access and environment are also applicable to the inspection staff.

2. *Incomplete removal of the defect being repaired.* It is not uncommon for the defect under repair to be only partially removed, particularly when it is a cracklike planar defect. The remainder of the unrepaired defect may extend beyond the ends of the repair, or deeper into the material when a partial thickness repair is made. This is, in part, because of errors or poor techniques in the NDT methods used to locate the defect, together with the fact that the material in the middle of the weld is usually in compression, forcing the crack faces tightly together. It is a requirement of most standards that, during gouging out of defects, the cavity should extend at least 1 in (25 mm) beyond the ends of the detected defects; yet, incomplete removal can still occur. For this reason, it is good practice to inspect the excavated area before repair, using either magnetic-particle or dye-penetrant tests to ensure that all traces of the defect have been removed. To do this obviously requires a cavity of inspectable shape. Note that the use of an unsuitable metal removal technique (particularly chipping) can flow metal over a crack and make it undetectable by dye-penetrant inspection.

3. *Introduction of further discontinuities associated with the repair.* The types of discontinuities that may form in a repair weld are, on the whole, the same as those that might be found in an original weld made by the same process. Hydrogen-induced cracking is probably the most likely discontinuity to occur in repair welds, as a result of inadequate control of levels of hydrogen in the weld metal, and to a lesser extent inadequate levels of weld preheat and postheat. With gas-shielded- or flux-cored-arc welding repairs, solidification cracking may occur because of excessive dilution or poor bead profile, but is more commonly a result of too fast a travel speed.

Open root butt joints welded with covered electrodes and the shielded-metal-arc (SMAW) process are selected for most repair and field welds in refineries and chemical plants. They require uniform fit-up and a high degree of welder skill. They also require careful selection of welding electrodes.

For carbon-steel joints, the welder is trained to achieve radiographic quality using narrow root openings and deep-penetrating E6010 or E6011 electrodes.

Slag and porosity can generally be avoided by good welding practice, although some slight porosity is more or less inevitable at start and stop positions in manual welding.

4. *Increased residual stress and distortion.* Residual stress levels in-

fluence the structural significance of discontinuities in relation to the likelihood of brittle fracture. Residual stresses due to repair are likely to be high when a large degree of restraint is present, and may be higher than those present in parts of the original weld that were also subjected to high restraint. Distortion as a result of residual stress may also present problems, but it can be adequately controlled in many cases by careful manipulation of repair weld shape, size, sequence, and heat input to establish a balanced welding procedure. Preheat and postweld heat treatment of welded repairs is recognized as a necessary procedure in preventing residual stress cracking due to welding.

5. *Slag inclusions.* Slag inclusions (metallic matter trapped in the weld) can result from poor interpass cleaning, rust and scale on the base metal, and particles trapped along undercut and overlapped passes. When welding is being done with covered electrodes or flux-cored wire, slag rises in the weld pool to form a protective coating on the surface. This slag must be thoroughly removed before welding additional passes. In the case of overlapped and undercut beads, cleaning may be difficult and, therefore, incomplete and may result in slag trapped within the completed weld. As with porosity, where the trapped slag is of a sufficient amount, the resulting reduction in weld strength may lead to failure at the joint.

6. *Undercut, overlap, incomplete joint penetration.* Undercut, overlap, and incomplete joint penetration each represent a notch or sudden change in material continuity. When found at the surface of a weld (root or face), the notch becomes a nucleus for service failure as corrosion and stresses concentrate at it. Corrosion may result from chemicals and gases such as are transferred in process flows or from a hostile environment.

Most welded assemblies that require inspection for the protection of the public and the environment will be subjected to stresses such as load, tension, and vibration. When these stresses act upon a notch, the end result could be cracking.

7. *Porosity in welds.* Porosity is another defect attributed to moisture contamination. Moisture from the atmosphere, oil, grease, or a damp electrode creates gas vapors that form bubbles or voids within the solidifying weld. Where a significant amount of porosity exists, the effect of this reduction in sound weld metal will be to reduce the fatigue strength of the connection at the weld.

Where significant porosity has formed at the surface, a notch will be created that can promote mechanical or metallurgical failure.

8. *Cracks in welds.* Cracks represent a serious threat to a weld's integrity. They are prone to spreading and can bring about total failure of the joint. Material verification, proper electrode selection, preheating, postheating, and sufficient root size are all areas that quality control inspection will concentrate on to control crack formation.

Proper matching of the electrode with the material to be welded will prevent unequal contraction of the base metal and weld metal, the stresses of which could lead to cracking.

Rapid cooling of a weld in high-strength alloyed steels can lead to cracking in the heat-affected zone. The application of preheat and post-heat will slow the cooling rate down and prevent the formation of a brittle and crack-prone microstructure.

In a highly restrained joint of heavy wall thickness, a root pass of inadequate size can become overstressed and crack.

Visual and NDT examinations are required on welded repairs in order to detect some of the errors in field welding technique. When a weld defect is identified as not being acceptable by NB or ASME code construction or repair rules, in most cases, corrective action must be taken to rectify the condition. This usually consists of removing the discontinuity or defect according to code requirements, to be followed by an approved welding procedure to redo the weld repair.

At times, new welding procedures may have to be developed, and welders qualified for this procedure, before a satisfactory repair can be attained.

Physical condition repairs

Pitting repair. The pitting repair required will depend on the extent or spread of the pitted area and the depth of the pits. Pits are a manifestation of localized corrosion. If shallow and isolated, so that the vessel's corrosion allowance has not been exceeded, pits can be ground smooth and blended into the vessel's wall to eliminate any stress concentration. Deeper pits up to the minimum thickness required for the maximum allowable pressure can be ground out to sound metal, checked by magnetic-particle or dye-penetrant inspection to make sure no cracks exist at the bottom of the pit, and then welded in to restore a smooth surface of the vessel. Pitting that is closely grouped within an 8-in circle must be treated as a wasted-away area. The minimum thickness for the vessel will have to be determined for the allowable working pressure. Thickness measurements can be made in an unaffected area and the depth of the pits subtracted in order to obtain the thickness of sound metal. If the thickness is still adequate, the pits can be cleaned by grinding and filler-metal welding can be used to restore the surfaces. In the paper industry, when wearable vessels such as digesters reach the minimum thickness required for the allowable pressure on the vessel, a stainless-steel, overlaid weld is applied. The surfaces must be properly prepared before the weld overlay is applied. The surfaces in this area must be thoroughly cleaned of all scale, oil, rust, grease, and content deposits before any welding is applied. No credit can be given for any pit fill-up welding or weld overlay in calculating allowable pressure. The allowable pressure is based on the minimum *sound* plate thickness that is left on the vessel in the affected pitted area. Another way of viewing this restriction is to consider any welding to restore a pitted surface to the original thickness to be a means of restoring the corrosion allowance, not the strength of the vessel.

Pitting that has reduced the thickness within an 8-in circle to below that needed for the maximum allowable pressure may necessitate a reduction in the allowable working pressure, or the welding in of a replacement plate to restore the thickness to the original design.

Wasted or general corroded areas. Buildup by welding is permitted on general corroded areas, provided the remaining thickness is adequate for the allowable working pressure of the vessel. No credit can be taken for the welded, built-up area. There are certain precautions to be taken in any stayed area: Stay bolts must be removed in the wasted area before welding; the stay bolt holes must be rethreaded after welding; and the built-up welding applied to restore a wasted or corroded area cannot be applied over stay bolts.

Tube sheets are treated as stayed areas. The tubes in any wasted area must be removed before welding. The tube holes in this area should be checked for cracks to make sure the corrosion has not affected them. After the built-up welding, the tube holes should be reamed before new tubes are installed, to assure a good rolled tube-to-tube-sheet joint.

In any wasted area, whether stayed or unstayed, if the deterioration has reduced the thickness below that required for the allowable pressure, two choices are available: (1) reduce the pressure to correspond to the reduced thickness or (2) weld in a replacement piece, following code welding requirements, in order to restore the original thickness of the vessel. In the latter case, flush-welded patches should be used. The welds on high-pressure equipment should be heat-treated or stress-relieved and field-radiographed to make sure a good joint was made between the patch and the original plate.

Wasted areas around entry holes and handholes can be built up by welding if the remaining thickness satisfies or is equal to the required code thickness for the allowable pressure. If thinning has occurred beyond this required thickness, a preformed access hole ring can be fitted to the shell or head portion and the ring flush-welded to the plate.

Flush-welded inserts. Flush-welded inserts or patches require that full-penetration welds be made in order to assure a sound joint. Code original-construction requirements must be met, including fit-up, grinding of the weld surfaces flush, and NDT inspections. Generally, radiographic weld examination will be required by most jurisdictions and authorized inspectors on any welds over 8 in long. Gasket leakage is the main reason for the wasting away of metal around these access openings. Proper gasketed joint maintenance will avoid the need for this type of repair.

Cracks.   Cracks should trigger an intense investigation on why they have occurred; otherwise, they may occur again. Cracking can come from a slow, progressive physical cause, such as repetitive stress, or from fatigue. This latter type of cracking will be initiated at a surface and propagate in a direction perpendicular to an applied tensile load. Physical and fatigue cracks can be detected by visual and NDT examinations before they penetrate into the required-thickness area of the plate. *Cracks* are classi-fied as "relevant discontinuities," which means that the physical struc-ture of the material has been affected and that if the discontinuity is left uncorrected, it could continue to grow in service until a failure occurs. Cracks require careful analysis so that their causes can be corrected. This could involve a change in design, such as avoiding sharp corners; it could require a change in operation to prevent thermal cycling. Many cracks, however, appear to be due to some form of discontinuity that was intro-duced during fabrication. Factors leading to such cracks include porosity, improper heat treatment of heat-affected weld zones, slag inclusions, and similar variables that are found in fabricating or installing pressure ves-sels. Other cracks occur because of environmental factors that attack the metal; these cause stress-corrosion cracks to appear.

The NB rules for crack repairs emphasize the need for determining the cause of the cracks and for carefully examining the crack area by visual inspection and NDT to make sure that all discovered cracks have been removed before making any repairs. If a crack has penetrated the full thickness of the plate, all welding repairs require full-penetration welds to be applied in order to assure a good welded joint. Complete NDT examination after welding is required for this type of crack repair, follow-ing original code construction requirements. Postweld heat treatment is usually needed, even if it is only of a localized nature.

Weld cracks.   Modern construction of boilers and pressure vessels has been possible with the advancement of welding knowledge. However, many variables must be considered in obtaining a proven, good welded joint. It is necessary to consider the following variable factors: weldabil-ity of the base metal, shape of the joint, welding process to be used, procedure to be followed in performing the weld, size and type of elec-trode, current (or temperature of weld) to be applied, preheat and post-heat treatment to be used, NDT to be applied to check the joint, and qualification of welders. Final acceptance considerations may also involve a hydrostatic test.

Defects in welding techniques or procedures can produce weld cracks. Small cracks can usually be veed out and rewelded. Methods of NDT such as dye-penetrant testing are used to determine if the end of a crack has been reached in the veeing-out process. Preheat and postheat treatment may be required on the weld repair following ASME Section IX rules. The repair must usually be approved after completion and then signed off (witnessed) on a form by an authorized inspector.

Defects in welds considered unacceptable are cracks, areas with incomplete fusion or lack of penetration, and slag inclusions of $\frac{1}{4}$ in for plate thickness up to $\frac{3}{4}$ in, with the maximum slag inclusion length being $\frac{3}{4}$ in for plate thickness over $2\frac{1}{4}$ in. The ASME publishes porosity charts that detail the number of occurrences and size of permitted porosity in a given length per weld thickness. Welds can fail from the application of repeated stress, especially where a discontinuity may be present.

Heat-affected-zone cracking has been the subject of much research. Brittle fractures in the heat-affected zone may be initiated by the presence of hydrogen, reheat cracks, fatigue cracks, or lack of root or sidewall fusion. Postweld heat treatment will generally improve the HAZ toughness of most grades of carbon steels, and this will help in preventing brittle failures in the HAZ of a weld. As a quality control device in ensuring a sound weld, NDT inspections of welds have been materially beneficial in reducing weld crack failures.

Repairs to cracks. It is necessary to distinguish the depth of cracking according to whether or not the cracks enter the minimum required code wall thickness for the part under consideration. Minimum code-required thickness of a component can be established by calculation or by referring to the data sheet for the vessel as prepared by the manufacturer. This information should be reviewed before repair.

Cracks with depths that do not violate minimum code-required thickness can be blended in by grinding with a 3 : 1 taper. The bottom of any groove should be ground with a generous radius to avoid any stress concentration. Cracks whose removal by grinding or air-arc gouging involves violation of the minimum required wall thickness as specified by Section VIII of the ASME code must be repaired by welding. Weld repair is necessary to restore the minimum required wall thickness and avoid derating of the pressure vessel.

Ground areas should be finish-ground and examined with wet, fluorescent magnetic-particle inspection techniques to ensure that all the cracks have been removed before any repair welding is commenced. All completed repair welds should be nondestructively examined to ensure that no flaws are present.

Postweld heat treatment is generally recommended for carbon-steel plate. All repairs must be approved by an authorized inspector before they are begun and carried out by NB repair rules. Welding procedure specifications must detail the welding parameters to be used in the repair, and welders must be qualified for the welding that is to be applied in the repair in accordance with the ASME code, Section IX.

Lap cracks. Vessels with longitudinal lap joints are prone to cracking failures. Probably the most dangerous of all fatigue cracks is the "lap crack" developing unseen between rivet holes of the longitudinal lap-riveted seam of old lap-riveted shells. This defect is induced by the fact that a lap-seam shell is not rolled into a true circle, and a bending stress is concentrated at the offset of the lap by the breathing action of the shell. Many serious explosions have resulted from such cracks. Usually, leakage from the seam appears as a warning of a lap crack. If any leakage exists or suspicion of this defect develops, the plate should be radiographed along the lap seam. It is often necessary to remove several rivets in suspected regions so that the inside of the rivet holes may be examined. No repairs are permitted on lap cracks. The vessel should be condemned immediately.

Heat exchanger repairs

Heat exchanger repairs can present problems, especially if the tubes are of corrosion-resistant material, such as titanium, that are welded to a steel tube sheet. Welding together dissimilar metals requires having a proven welding procedure. Repairs must follow this proven welding procedure, which generally can be obtained from the original-equipment manufacturer. Some contract maintenance and repair organizations also have developed welding procedures for repairing dissimilar metal welds.

Tubes that leak in heat exchangers are often expeditiously plugged and the vessel returned to service. Modern practice includes eddy-current and borescope inspections in order to determine the cause for the tube failure. Wear due to aging is the usual reason tubes are plugged, because it is known that retubing will eventually be required in order to restore the tube life of the exchanger. However, if a tube should fail on a relatively new heat exchanger, a thorough investigation as to reason is needed, because it is a symptom of possible design or operating problems that need to be resolved. The defective tube should be removed for a complete metallurgical analysis of the reason for the failure.

The question of how many tubes can be plugged before retubing is needed involves economic considerations, process scheduling, and the effect on output performance. There is evidence, however, that some tube plugging may increase flow velocities in the remaining tubes in service, and this may cause tube vibration, erosion of tube ends, and increased pressure drop through the exchanger. Thus, accelerated wear may take place not only on the tubes but also on the tube sheets and intermediate supports. Periodic tube testing by modern NDT methods for wear, erosion, and cracks, and then replacement of the tubes that are potential sources of in-service failures, is perhaps the best way of maintaining the integrity of a heat exchanger in service.

Sleeving. Sleeving has been used on heat exchangers where crevice corrosion and cracking of tubes has occurred in the tube–tube-sheet joints. Where tubes are only partially rolled into the tube sheet, thus leaving a crevice between the tube and the tube sheet, caustic and other corrodents can settle in this crevice and attack the *outside* diameter of the tube in the unrolled portion of the tube–tube-sheet joint. Accumulation of the corrodent in the crevice, combined with the hydraulic and pressure stress on the tube, can cause intergranular attack of the tube material. Corrosion-resistant sleeving techniques have been developed in order to prevent this type of crevice corrosion and cracking. Most heat exchanger manufacturers can provide details for this type of repair.

Seal welding. Seal welding of tubes may be used to prevent corrodents from penetrating a rolled joint. However, since tubes are relatively thin in comparison with tube sheets, the ends of the tubes must have sufficient thickness to prevent the welding heat from melting or burning through the tube. The welding procedure must also consider any problems of dissimilar metals that may exist between the tubes and the tube sheet.

Welding repairs to stainless steel

Stainless-steel welding repairs require consideration of the metallurgical changes that may occur in some stainless steels. The austenitic grades of stainless steel are the most easily welded. There are two problems in welding of these alloys. The first relates to the precipitation of grain boundary carbides in the heat-affected zone adjacent to the welds. Carbide precipitation can be minimized by using titanium- or columbium-stabilized electrode grades in the stabilized, annealed condition. Following welding, a high-temperature-solution anneal can be performed on the component to redissolve any grain boundary–precipitated carbide. The other problem relates to the high-temperature strength of the weld metal itself. An all-austenitic weld deposit has a fine grain structure, which has poorer creep properties than a coarse grain structure. To counteract this effect, it is desirable to have a small amount of delta ferrite present in the weld. Typically, a suitable 4 or 5 percent ferrite stainless-steel electrode is an effective means of offsetting this grain boundary weakness. However, an excessive amount of ferrite will lead to a lowering of the high-temperature strength and to poorer corrosion resistance. A second danger of excessive ferrite is the promotion of sigma-phase formation at high temperatures.

Sigma phase. At temperatures below 1500°F (815°C) for a 50-50 chromium-iron alloy, a hard, brittle, intermetallic compound called "sigma" forms. Sigma forms slowly, first developing at grain boundaries, and is readily identifiable by microscopy and x-ray diffraction. Its formation is favored by high chromium levels. The presence of the sigma phase increases hardness slightly and decreases ductility—however, not enough to be of serious consequence in most pressure vessel applications. It should be noted that sigma formation is possible between about 1050 and 1700°F (565 and 925°C) in austenitic stainless steels containing more than 16 percent chromium and less than 32 percent nickel.

Sensitization. The heat of repair-welding austenitic stainless steels can cause sensitization. When austenitic stainless steels are heated between 800 and 1500°F (425 and 815°C), intergranular attack will occur in corrosive environments. The most critical temperature is about 1200°F (650°C). Sensitization occurs through chromium depletion in a region immediately adjacent to grain boundaries. Both chromium and carbon are distributed throughout the austenitic structure and can combine to form chromium carbide. The temperature at which this occurs most rapidly is close to 1200°F (650°C). At lower temperatures, diffusion rates of the atoms are too slow to cause the formation of chromium carbide, and at higher temperatures, decomposition of chromium carbide occurs. The precipitation of the carbide occurs most readily at grain boundaries, because the diffusion rates of carbon and chromium are greatest along these boundaries.

The carbon atoms can migrate to the grain boundary from all parts of the crystal more easily than the chromium atoms can; therefore, chromium is depleted from more localized regions near the grain boundary, forming an envelope of chromium-depleted material. This region, therefore, is susceptible to corrosion.

Stainless steels lose their excellent corrosion resistance below 11 percent chromium. The depletion of chromium by the formation of chromium carbide forms a thin layer, adjacent to the grain boundaries, of material containing less than 11 percent chromium, and therefore, a sensitized alloy is not corrosion-resistant.

Postweld heat treatment is not usually performed on austenitic stainless alloys except to relieve residual stresses. Because sensitization occurs between 800 and 1500°F (425 and 815°C), the stress-relief temperature should be higher than 1500°F (815°C), except in the stabilized, annealed condition. For type 321 and 347 alloys that have been given a stabilization anneal, stress relief can be achieved in the 1300 to 1400°F (700 to 760°C) range without fear of detrimental chromium carbide precipitation at grain boundaries.

Need for metallurgical review of stainless-steel repairs. Repairs to stainless steels, including repairs to stainless-steel overlaid vessels and stainless-steel cladding and lining, requires a careful review and control of the welding repair; otherwise, some of the possible changes to stainless steel from the welding heat may result. It is important to develop a welding procedure for these metals that can be used in field repairs. In most cases, the manufacturer of stainless-steel vessels should be consulted on the best methods to use in repairing them.

Repairs involving dissimilar metals. The electrodes selected and the procedure developed require special attention when dissimilar metals are to be repaired. Among the items to be considered are the following, as detailed in general terms for a stainless-steel clad repair:

1. When an alloy is to be welded on the base steel, an electrode having higher alloy content or one specially formulated for the purpose is used to compensate for dilution.

2. Clad plate $\frac{1}{2}$ in (13 mm) and under in thickness is usually welded entirely with an appropriate alloy electrode, while thicker plate is usually welded by a combination of alloy and steel welds. Where a combination weld is used, the edge is prepared so that the steel side is welded first without picking up alloy from the cladding. The joint is then backgouged, and the proper alloy electrode is applied on the carbon steel. Dilution of carbon-steel welds by the high alloys can result in brittle or even cracked welds.

3. Superficial repairs or attachments of similar metals may be made with electrodes matching or compatible with the cladding metals and service. Attachments welded to the cladding will not cause bond loosening.

4. In preparing joints, welding, and conditioning, particularly for thin plates, care is taken to maintain the cladding thickness and continuity. Weld undercuts and indiscriminate grinding should be avoided.

5. In welding clad plates, just as for homogeneous plates, arc strikes with electrodes or electrode holders should be avoided. Even arc strikes with a similar electrode may leave pores and fissures that under some conditions will result in progressive corrosion. Insulated electrode holders should be used. Splatter of high-alloy surfaces with carbon-steel weld metal should also be avoided.

The 300 series of stainless steels is the most prone to becoming sensitized when subjected to welding or heat treatments such as normalizing, tempering, and stress relieving. The trend is to avoid this by using columbian- or titanium-stabilized types or the extra low carbon stainless steels and thus provide greater resistance to corrosive attack in the depleted chromium regions.

Stainless-steel weld overlay.   Stainless-steel weld overlay is used to provide corrosion resistance for a carbon-steel pressure vessel. The carbon-steel vessel must have at least the minimum code-required thickness for the operating pressure. No credit can be taken for the stainless-steel weld overlay as a strength member of the pressure vessel.

Before any overlay welding is applied, all surfaces must be cleaned following good stainless-steel welding procedures. On reinspecting overlay welds, if any small fissures, cracks, pits, occlusions, or deep valleys between the stainless-steel weld beads are found, they should be repaired during the outage of the vessel. Grinding out and rewelding with the specified stainless-steel rods is the usual practice in restoring the stainless-steel corrosion protection to the carbon-steel vessel walls.

Any weld overlay must be performed by a qualified welder following a qualified procedure. All proposed overlay welding should receive the approval and subsequent inspection of a commissioned boiler and pressure vessel inspector.

There have been many new developments in applying corrosion-resistant weld overlays as new materials are developed by companies specializing in producing corrosion-resistant steels, such as Hastelloy, a trademark of Carpenter Steel Co. The reader is referred to these specialty companies for further details on the application of corrosion-resistant weld overlays.

Management Support of Pressure Vessel System Safety

Risk management programs require management to assign decision making on safety matters to a responsible person who has been given clearly defined objectives in the safety program.

The potential for accidents in large pressure vessel systems is considerable and must be addressed by a safety program. Management support will be forthcoming if a proper hazard analysis is submitted to management defining the large risks involved, both in economic terms and from the standpoint of potential legal and governmental action. The current concern about environmental impact on the part of the public and legislators requires a higher degree of maintenance and inspection than was previously the case. The large plants can and should obtain management's support for the following policies:

1. Total plant safety audits should be made at periodic intervals in order to review for potential hazards the plant's layout, operating procedures, and maintenance and testing procedures, as well as repairs. Of paramount importance is a review of the controls and safety devices that have been installed to prevent serious overpressure or temperature excursions from taking place. Environmental impacts to surrounding property and people are today a major risk-management concern, because the public and governments expect proper precautions to be taken to avoid area disasters. A plant team effort in a safety audit could include the plant manager, the safety department head, the plant principal design engineer or someone who has a thorough knowledge of the plant design, the maintenance manager, the process staff engineer, the instrument and control engineer, and possibly a union representative with safety experience.

Periodic safety audits are required because most process plants continuously are changing through modifications of equipment and operating practices, especially when automation in a plant is a management goal.

2. Maintenance management should participate at an early stage in any new plant designs or additions. Valuable input on material selection, corrosion allowance, suitability for inspection, past operating experience with certain types of vessels, stainless-steel versus carbon-steel construction, and similar considerations can be obtained from a good maintenance and inspection staff.

3. Equipment inventory and maintenance record procedures to keep track of equipment from the time it is purchased to the time it is retired are an essential requirement in tracking equipment performance.

4. Establishing and maintaining a trained maintenance and inspection staff that is familiar with codes, repair methods, new construction requirements, jurisdictional requirements, welding variables, and nondestructive testing will help in diagnosing potential problems to equipment.

5. Establishing and maintaining a training program for the maintenance and inspection staff, including procedural manuals, training in the use of NDT equipment, welding procedures, and welding operator qualification, can keep the staff current on company and industry practices.

6. Establishment and maintenance of a program of liaison with vessel manufacturers to develop good purchasing procedures and design philosophies should be encouraged by management for their maintenance, engineering, and inspection staff.

7. Periodic plant sessions on plant problems by operation and inspection personnel should receive management support. These sessions will

help the staff strive to achieve the plant's production and profit goals as members of the management team.

The need for higher-quality plant integrity was pointed out in an American Insurance Association study* of process pressure vessel failures. Among the findings of the study were these:

1. A development of the past decades, already employed in ammonia and polyethylene processes, is the huge "single-train" process plant. As this trend spreads to other types of processing, it has the effect of placing "all the eggs in one basket." Losses due to interruption of business assume huge proportions in such operations. Trained operators must be familiar with the entire process in order to react properly to a process disturbance in the single-train plant; otherwise, a total plant shutdown may occur from a minor plant disturbance.

2. The operation of chemical processes at extremely high pressures and temperatures, such as the 125,000-psi ethylene polymerization units, is becoming more prevalent. Design and equipment technology is having difficulty in maintaining the pace.

3. More hazardous chemicals are being handled each year as the industry achieves the ability to cope in some degree with the materials. Each year the number of commercial chemicals added to the hazardous chemicals list grows larger. This can pose environmental hazards never seen before.

4. There has been a general decrease in the relative number of operating personnel for large plants as a result of automation. Therefore, the number of employees available for firefighting or other emergencies is reduced and should be continuously reviewed as a safety measure.

5. Potentially explosive situations in modern chemical plants are increasing in number. Massive spills of flammable gases that form explosive vapor clouds are more common.

6. The rapid technological changes occurring in the chemical industry by way of integrated complexes, as well as new processing, are producing new fire and explosion control problems. Research and development in prevention and protection measures is hard-pressed to cope with them. Hazard analysis should assist this effort.

7. Quality control problems during the construction phases of chemical plants appear to be increasing. Welding of critical units, in particular, requires more careful workmanship, inspection, and testing. The same is true for repair or alteration of these vessels.

---

* *Hazard Survey of the Chemical and Allied Industries,* American Insurance Association, New York, N.Y.

8. Development of new high-pressure equipment without sufficient commercial experience to produce adequate performance specifications presents many serious design and fabrication problems.

9. The requirements for highly skilled operators are becoming acute. Operator training is more important as chemical plants become increasingly complex.

10. Preventive maintenance programs are becoming an absolute necessity in the modern chemical plants, some of which are programmed to operate at maximum capacity over long periods of time.

Technology plays a central role in maintaining the standard of living to which society has become accustomed and can affect the daily life of the average citizen. Some of these impacts have been negative—for example, major power blackouts, noise near jetports, and pollution of air and water resources. In response to some of these negative aspects, public attitudes toward the social value of technology have been changing, and government, as well as private groups, has become more actively involved in questioning, and even suggesting restraining, the advancement of technology. As a result, some new criteria for acceptance of technology have arisen:

1. Safety from chronic effects, such as the long-term effects of radiation or exposure to potentially carcinogenic products or toxic materials, must be evaluated.

2. The disposal of both waste and used products that may be potentially harmful to the environment requires attention.

Process plant management and the maintenance and inspection staffs can minimize the public concern for environmental effects with good operation and maintenance programs.

QUESTIONS AND ANSWERS

7-1 Explain breakdown maintenance.

ANSWER: *Breakdown maintenance* is the practice in which a plant takes no corrective action until output performance is no longer acceptable to the process or the equipment fails mechanically, as with leaks due to corrosion or cracks.

7-2 Name three types of organizational structure for plant maintenance.

ANSWER: Maintenance structures that are based on responsibility for maintaining output can be (1) centrally organized in the plant, (2) responsible for a defined area of the plant, (3) modifications of the two.

7-3   What is the requirement for purging a vessel with gas before an internal inspection?

ANSWER: The vessel should be purged with a neutral substance such as nitrogen. After the purging and testing to see that all gas has been removed, air purging should be performed to make sure there is enough oxygen in the vessel to avoid asphyxiation. The Occupational Safety and Health Administration requires that at least 19 percent oxygen by volume be present in the vessel before entry.

7-4   Name at least four possible pressure vessel system inspections.

ANSWER: Performance inspections for efficiency; external or operating inspections; internal inspections; nondestructive testing inspections; hydrostatic or pneumatic tests and inspections.

7-5   What NDT method is not suitable for nonferrous material?

ANSWER: Magnetic-particle testing requires the material to have magnetic capability, and nonferrous material does not have this.

7-6   What level of NDT capability is needed to approve inspection and test procedures to be used?

ANSWER: Level 3 capability is required.

7-7   When are pneumatic tests permitted in lieu of hydrostatic tests?

ANSWER: Pneumatic tests are permitted by the code if the vessel has insufficient support to hold the water or when traces of water would be objectionable or hazardous when introduced into the process stream.

7-8   What problem may develop in the weld if an electrode is stripped of the coating?

ANSWER: The coating on the electrode forms a gaseous shield over the molten metal while the joint is being welded. If this protective shield is eliminated because of a lack of electrode coating, hydrogen from moisture in the atmosphere may enter the molten pool and become entrapped as the metal solidifies. This can cause hydrogen embrittlement cracks to develop in the weld.

7-9   What is another term for "fatigue limit"?

ANSWER: Preferred is "endurance limit."

7-10   What criterion is important in the selection of a weld rod in welding of stainless steels?

ANSWER: In welding, it is necessary always to select a welding rod with a filler metal that has metallurgical and mechanical properties, including corrosion resistance, at least equal to the base metals being welded.

7-11   What defects are not acceptable in overlay welding?

ANSWER: The overlay should have no cracks or porosity or other discontinuity defects such as slag inclusions, which may reduce the effective thickness of the overlay or might adversely affect the service performance of the overlay. Since most overlays are stainless steel, some of the problems in welding stainless steels must be considered. These include carbide precipitation, sensitization, and sigma-phase and martensite formation. Some of these require metallographic testing and examinations of appropriate sections of the overlay samples.

7-12   What type of vessel does the code prohibit from being repaired by brazing?

ANSWER: The code prohibits brazing construction and repair for vessels to be used in lethal service, or for fired and unfired boilers.

7-13   What are arc strikes, and what detrimental effect can they have on a welded joint?

ANSWER: Arc strikes represent unintentional melting or heating outside the intended weld deposit area. They usually are caused by the welding arc, but can be produced beneath an improperly secured ground connection or magnetic-particle inspection electrical connections (prods). The result is a small remelted area that can produce undercut, hardening, or localized cracking, depending on the metal composition.

7-14   Describe slag inclusion and its detrimental effect on a welded joint.

ANSWER: This term is used to describe the oxides and other nonmetallic, solid materials that are entrapped in weld metal or between weld metal and base metal. Slag inclusions may be caused by contamination of the weld metal by the atmosphere; however, they are generally derived from electrode-covering materials or fluxes employed in arc welding operations. In multilayer welding operations, failure to remove the slag between passes will result in slag inclusions in these zones. Slag inclusions are generally linear and may occur either as short particles or as long bands.

The majority of slag inclusions may be prevented by proper preparation of the groove before each bead is deposited, using care to correct contours that will be difficult to penetrate fully with successive passes. Slag inclusions act as stress concentration points at which fatigue cracks may develop in service.

7-15   Define the term "back-gouging" in welding.

ANSWER: Most welds used in pressure work are of the double butt variety; that is, the plates are butted together and welded from both sides. The metal of the root pass as it is being deposited does not receive the protection of the slag on its underside. The underside is therefore subject to heavy oxidation where it is exposed to the air. The oxidation or any other inclusions or defects must be chipped out before any welding is done on the second side; otherwise, defects may occur in the welding. The cleaning out is termed "back-gouging" and should be carefully done. Methods used are flame gouging, chipping, or grinding. It is believed that the first method is the best, since defects are usually exaggerated by the heat and they can be readily seen by the operator and removed.

7-16   Must a mill test report on a pressure vessel be altered if new plate material is added?

ANSWER: Definitely, so that the authorized inspector can see if the addition is of an applicable code material for the vessel. All materials used for pressure parts must be identified by a satisfactory mill test report showing both physical and chemical properties. The inspector must check these properties and see that they agree with the specifications and are suitably identified with the actual materials to be used. All material for pressure parts must be legibly stamped in some permanent manner with the manufacturer's name or symbol and the minimum tensile strength of the material and also some identifying numbers to tie in the material with the mill test report.

The material specification number, size, and thickness as ordered will be on the mill test report. The report may cover any number of plates. Shown for each plate will be a heat and slab number to identify the plate in the fabricator's shop.

7-17   How does a metallurgist use the Nelson curves for vessels intended for potential hydrogen service?

ANSWER: The Nelson curve is used to select alloyed material that will prevent hydrogen embrittlement and cracking in any type of potential high-pressure hydrogen service. This is a primary tool employed by the metallurgist when selecting materials for refinery and chemical plant service. It establishes which and how much alloy is required to withstand high-temperature hydrogen service. The higher the temperature or the partial hydrogen pressure, the more alloy is required. However, not just any alloy will satisfy these metallurgical requirements. The Nelson curve indicates clearly that molybdenum and chromium are required in these environments. Thus, an E8018-B2 electrode (with Cr and Mo) can be quite effective, while an E8018-C1 (with Ni) is no better than carbon steel.

7-18   How can an electrode's moisture content be restored to an acceptable level?

ANSWER: When the moisture in the coating gets too high, but is less than a level where obvious damage to the coating integrity has occurred, the electrode can be rebaked and so be restored to the initial condition. Manufacturers have recommendations for the time, temperature, and conditions for these reconditioning treatments. The temperature for reconditioning is high enough to drive off moisture that has become entrained in the coating, but not so high as to embrittle the coating or oxidize the alloys. After such a reconditioning treatment, the electrodes are essentially the same as when removed from a hermetically sealed container. Repeated rebaking may increase the coating fragility and contribute to the start of porosity through oxidation of the manganese and silicon. This will be obvious to the user, since poorer welding characteristics (wetting, arc control, and slag stiffness) will result.

Packaging is the first important consideration in controlling electrode moisture content. Electrodes in the various forms have been shipped in boxes for years with satisfactory results. However, unless an initial rebake is performed, the most demanding applications require a hermetically sealed and impermeable container.

7-19 What leak tests are used to determine tube–tube-sheet joint tightness?

ANSWER: Depending on the service conditions, the following tests may be used:

1. Removal of the bonnets or channel covers exposes the tube ends, which may show leaks when hydrostatic pressure is applied to the shell side of the exchanger. The hydrostatic test should be maintained for a minimum of $\frac{1}{2}$ h.

When the tube pressure is higher than the shell pressure, the covers must be in place during hydrostatic testing of the tubes. Leakage at the tube joint may not be visible but can be noted through shell-side vent or drain leaks.

2. In bubble-formation tests the shell is filled with an inert gas or air at design pressure, and after 15 min, bubble solutions are placed over the joints and tube ends. If bubbles appear, the joint is not tight.

3. Halogen leak tests are used for heat exchangers handling lethal or noxious fluids. The test involves sniffing or sucking air through a tubular probe and then passing this air over a heated platinum element. The element ionizes the halogen vapor, which then flows to a cathode plate. The extent of leakage will show as a current strength on the meter. If the meter shows no current, this means the joint did not pass any halogen.

4. Helium leak testing is used when the hazard of a leak is unacceptable. Portable mass spectrometers sensitive to the presence of helium are used to determine if any helium has leaked past the joint.

7-20 Why should gaskets be replaced after dismantling a gasketed joint or connection?

ANSWER: It is recommended that when a gasketed joint is dismantled, new gaskets be used in reassembly. The reasons for this are (*a*) composition gaskets become brittle and dried out in service and do not provide an effective seal when reused and (*b*) metal or metal-jacketed gaskets in initial compression match the contact surfaces and tend to work-harden and cannot be recompressed on reuse.

7-21 How are stainless steels affected by welding?

ANSWER: Intergranular corrosion may occur for several diverse reasons, and many opportunities exist for stainless steels to be sensitized by the fabrication process or the service environment. Stainless steels of normal carbon content will usually be sensitized by welding. Therefore, these steels should be solution heat-treated and quenched following welding for service in corrosive media. If a solution annealing and quenching treatment of the component is not feasible, low-carbon grades or stabilized stainless steels should be used. However, improper welding of these steels can also cause sensitization. Significant sensitization does not normally result when typical welding procedures and material chemistry are used and when no further heating of material occurs. The welding procedures and material chemistry should be controlled to prevent undue sensitization during welding. Control should include the following: (1) maintaining low heat input to the weld joint by controlling current, voltage, and arc travel speed, (2) limiting interpass temperature, (3) using stringer bead technique and the limiting of weaving, and (4) limiting the carbon content of the material where section thickness results in increased time at sensitization temperatures.

7-22   Name two methods of reducing the possibility of stress-corrosion cracking.

ANSWER: To avoid stress-corrosion cracking in those systems where it occurs, the stress can be eliminated or it can be reduced to a level below a certain critical stress that is dependent on the exposure temperature, the solution concentration, and the composition and microstructure of the steel material. Alternatively, the responsible chemical can be inhibited, eliminated, or reduced to a level below which it has no effect. If control of the stress, control of the chemical, or both can be accomplished, then stress-corrosion cracking can be avoided. In those cases where neither control of the stress nor control of the chemical can be accomplished, a different material must be sought that is not susceptible to stress corrosion (or other types of corrosion) in the particular environment.

7-23   Define solution heat treatment of stainless steel.

ANSWER: Solution heat treatment is a remedy that is suited to shop applications. Here the welded vessel is reheated in a furnace to a temperature of about 1922°F (1050°C), causing the chromium carbide particles to dissolve back into the grain. With the chromium redistributed uniformly throughout the material, the vessel is no longer susceptible to cracking due to loss of the corrosion-resistant chromium by chromium carbide precipitation.

7-24   Define the term "trepanning."

ANSWER: This is a method of examining a section of a welded joint by cutting out a round or boat-shaped specimen of the joint. The piece removed must be large enough to give a full cross-sectional view of the welded joint. This joint must be ground smooth and then etched with an acid solution to provide a clean, bright surface that can reveal small defects by visual inspection. Cracks through a weld can be noted by this method.

The holes or boats of the sectioned welded joint are usually welded closed by required code procedures and code-qualified welders.

7-25   What may cause an austenitic steel to become magnetic?

ANSWER: Cold-working an austenitic steel can transform it into martensite and thus make it susceptible to cracking. Such a condition can be detected by noting if the austenitic steel has acquired a high degree of magnetism.

7-26   What packing material is available for valve maintenance?

ANSWER: Teflon, straight asbestos, asbestos core packing reinforced with Inconel metal wires, carbon packing, and polymer fiber packing.

7-27   What is the difference between a dump valve and a drain valve?

ANSWER: A *dump valve* is a valve provided for the purpose of rapidly discharging a fluid from a vessel or other piece of equipment. A dump valve is located at the low point of the equipment and differs from a drain valve in the urgency of removal of the fluid. For this reason, a dump valve is generally a quick-opening device; it is sized for greater flow rate than a drain valve.

7-28   What precaution must be observed before closing a new vessel or a vessel that has been opened for cleaning or repair?

ANSWER: Make sure all tools, pipes, welding rods, rags, and other such items are removed from drums, headers, and tubes. At times, a mirror and flashlight must be used to check headers that are otherwise not accessible to inspection for foreign material. Bent tubes that cannot be looked at from end to end, such as U tubes, should be thoroughly flushed, one tube at a time. On new vessels, or where work was done in a tube area, drop rubber balls, and even steel balls, to make sure the tubes are free of obstruction. Water and air can be used to push the rubber balls through.

7-29   What precaution must be observed in turbining the tubes?

ANSWER: Turbining tubes can cause local tube wear or nicking if the turbining tool is forced through a tube or held in one position too long.

7-30   Is it safe to use portable lamps and extension cords inside vessel drums, shells, or headers?

ANSWER: Only if low-voltage lamps, 32 V or less, supplied by transformers or batteries are used to avoid electric shocks in case a lamp or bulb breaks and creates a current flow through the vessel shell. Never use extension cords without proper waterproof fittings. Make all connections outside the vessel. Light bulbs should have explosion-proof guards. Fittings, sockets, and lamp guards must be grounded.

7-31   What assistance should the owner or operating engineer give the legal pressure vessel inspector during inspections?

ANSWER: It is the responsibility of the owner to prepare the vessel for the required legal internal inspection. All openings must be removed. All scale and mud must be removed so metal surfaces are exposed for inspection. Shell sides must be cleaned so tubes can be inspected for corrosion, thinning, erosion, and signs of impingement. Baffles and intermediate supports require checking for tightness.

The inspector must be given all the help she or he needs. Point out any known defects. Station someone immediately outside the vessel during the internal inspection. If the vessel is in battery with others, make sure that all steam, water, and blowoff valves are locked and cannot be opened. Make provision for a hydrostatic test if the inspector deems one advisable. In general, assist in every way to make the examination thorough and complete.

7-32   What is meant by a shielded arc? Give a reason for using this welding approach.

ANSWER: It is an electric arc produced by a coated electrode. The coating in melting back of the arc produces an inert gas that blankets the molten pool to prevent oxidation. As the arc moves on, a brittle crust is left to blanket the cooling metal.

7-33   What is the purpose of a welding procedure and welder qualification?

ANSWER: The purpose of having a welding procedure is to demonstrate that the proposed method of welding, if followed, will produce a satisfactory welded joint within acceptable standards and that the welded joint will satisfy the stipulated service requirements. The purpose of welder qualification is to confirm that the welder can produce a weld with the specified freedom from defects for the welding procedure being followed.

7-34   Is a qualified welder permitted to make welded repairs on any part of a vessel?

ANSWER: Not necessarily. That a welder is qualified to make some welds may not mean he or she is qualified for welding (1) the particular thickness of plate or (2) the type of material, nor for welding (3) in the position of welding to be used or (4) according to the method required.

7-35   On what basis are repairs allowed on vessels?

ANSWER: Repairs permitted are based on restoring the affected part or parts to as near the original strength as possible. They are governed by code requirements for new construction, or by NB rules on permissible repairs, where the state has adopted NB rules for repairs.

7-36   Must all repairs be approved by an authorized inspector?

ANSWER: Yes, if the strength of the vessel has been impaired in any way that requires repairs involving code enforcement and interpretation. Repairs not affecting the strength of the vessel or that are of a minor or routine nature may not require approval. But the inspector should be consulted on the problem if there is any doubt about the safety of the vessel. Crack repairs, welding, tube replacement, safety-valve replacement, and similar repairs or changes require approval. The best rule to follow on any structural repairs or changes is to contact an authorized boiler inspector immediately.

7-37   What is the purpose of a quality control program in fabrication of pressure vessels?

ANSWER: The manufacturer or assembler should maintain a quality control system that will ensure that all code requirements, including those for material, design, fabrication, examination (by the manufacturer), and inspection (by authorized inspector), will be met. Provided that code requirements are suitably identified, the system may include provisions for satisfying any requirement by the manufacturer or user that exceeds minimum code requirements and may include provisions for quality control of non-code work. In such systems, the manufacturer may make changes in parts of the system that do not affect the code requirements without securing acceptance by the authorized inspector.

7-38 What type of documentation must a boiler or pressure vessel manufacturer maintain on file in order for a pressure vessel to be stamped by the ASME code symbols?

ANSWER: The type of documentation needed to satisfy code requirements varies in different sections of the code. In general, the manufacturer must keep all radiograph film and must also prepare a manufacturer's data report, which must be signed by both the manufacturer's representative and the authorized inspector. The authorized inspector should not sign the data report until he or she has carefully checked it to make certain that it properly describes the boiler or vessel to which it applies, that the boiler or vessel complies with the code, and that the data report has already been signed by the manufacturer's representative. These data reports may be registered with the National Board. For nuclear vessels, more extensive documentation is required.

Metallurgical and Welding Definitions

AIR HARDENING (air quenching) A hardening process wherein steel is heated to the hardening temperature and cooled in air. Unless steel is high in carbon, alloy, or both, it will not show much increase in hardness when air-hardened.

ALLOY A material with metallic properties composed of two or more elements of which at least one is a metal.

ANNEALING The heating and controlled cooling of solid material for the purpose of removing stresses, making it softer, refining its structure, or changing its ductility, toughness, or other properties. Some heat treatments covered by the term "annealing" include full annealing, graphitizing, malleabilizing, and process annealing.

ARC BLOW The deflection of an electric arc from its normal path because of magnetic forces.

ARC CUTTING (AC) A group of cutting processes that melt metals to be cut with the heat of an arc between an electrode and the base metal. See "Carbon-arc cutting," "Metal-arc cutting," "Gas-metal-arc cutting," "Gas-tungsten-arc cutting," and "Plasma-arc cutting."

ARC GOUGING An arc cutting variation used to form a bevel or groove.

ARC STRIKE   Any inadvertent change in the contour of the finished weld or base material resulting from an arc generated by the passage of electrical energy between the surface of the finished weld or base material and a current source, such as welding electrodes or magnetic-inspection prods.

ARC TIME   The time during which an arc is maintained in making an arc weld.

ARC VOLTAGE   The voltage across the welding arc.

ARC WELDING (AW)   A group of welding processes, wherein coalescence is produced by heating with an arc or arcs, with or without the application of pressure, and with or without the use of filler metal.

ARC WELDING ELECTRODE   A component of the welding circuit through which current is conducted between the electrode holder and the arc. See "Arc welding."

AUSTEMPERING   A patented heat-treating process that consists of quenching an iron-base alloy from a temperature above the transformation range in a medium having a high rate of heat abstraction, and then maintaining the metal, until transformation is complete, at a substantially uniform temperature that is below that of pearlite formation and above that of martensite formation.

AUSTENITE   A phase in steels that consists of the gamma form of iron with carbon in solid solution. Austenite is tough and nonmagnetic and tends to work-harden rapidly when cold-worked in those steels that are austenitic at ordinary temperatures.

AUSTENITIC STEEL   Steel that has a stable austenitic structure at normal (room) temperature.

AUTOGENOUS WELD   A fusion weld made without the addition of filler metal.

AUTOMATIC WELDING   Welding with equipment that performs the welding operation without adjustment of the controls by a welding operator. The equipment may or may not perform the loading and unloading of the work.

AXIS OF A WELD   A line through the length of a weld, perpendicular to and at the geometric center of its cross section.

BACK-GOUGING   The removal of weld metal and base metal from the other side of a partially welded joint to assure complete penetration upon subsequent welding from that side.

BACKHAND WELDING   A welding technique in which the welding torch or gun is directed opposite to the progress of welding. Sometimes referred to as the "pull-gun technique" in gas-metal- and flux-cored-arc welding.

BACKING RING   Backing in the form of a ring, generally used in the welding of piping.

BASE METAL   The metal to be welded or cut.

BEND TEST A test commonly performed by bending a cold sample of specified size through a specified circular angle. Bend tests provide an indication of the ductility of a welded sample.

BILLET A semifinished, rolled ingot of rectangular cross section or nearly so. In general the term "billet" is used when the cross section ranges from 4 to 36 in², the width always being less than twice the thickness. Small sizes are usually classed as bars or "small billets." The term "bloom" is properly used when the cross section is greater than about 36 in², though this distinction is not universally observed.

BLISTER A defect in metal produced by gas bubbles formed either on the surface or beneath the surface while the metal is hot or plastic. Very fine blisters are called "pinhead" or "pepper" blisters.

BLOWHOLE See the preferred term, "Porosity."

BRINELL HARDNESS TEST A test consisting of forcing a ball of standard diameter into the specimen being tested under standard pressure and judging the hardness of the material by the amount of metal displaced.

BURNING Heating steel to a temperature sufficiently close to the melting point to cause permanent injury. Such injury may be caused by the melting of the more fusible constituents, by the penetration of gases such as oxygen into the metal with consequent reactions, or perhaps by the segregation of elements already present in the metal.

BUTTERING The deposition of one or more layers of weld metal on one or both faces of a joint before preparation of the joint for final welding for the purpose of providing a suitable transition weld deposit for the subsequent completion of the joint.

BUTT JOINT A joint between two members aligned approximately in the same plane.

CARBIDES As found in steel, compounds of carbon and one or more of the metallic elements, such as iron, chromium, or tungsten.

CARBON-ARC CUTTING (CAC) An arc cutting process in which metals are severed by melting them with the heat of an arc between a carbon electrode and the base metal.

CARBON-ARC WELDING (CAW) An arc welding process that produces coalescence of metals by heating them with an arc between a carbon electrode and the work. No shielding is used. Pressure and filler metal may or may not be used.

CARBONITRIDING A process of case hardening an iron-base alloy by the simultaneous absorption of carbon and nitrogen through heating in a gaseous atmosphere of suitable composition, followed by cooling at a rate that will produce desired properties.

CARBON STEEL Steel the major properties of which depend on its carbon content and in which other alloying elements are negligible.

CARBURIZING Adding carbon to iron-base alloys by absorption through heating the metal at a temperature below its melting point in contact with carbonaceous materials. Such treatment followed by appropriate quenching hardens the surface of the metal. The oldest method of case hardening.

CARBURIZING COMPOUNDS Mixtures containing carbonaceous solids that will give up carbon to steel in the presence of heat. Gas rich in carbon is sometimes used in the carburizing process.

CASE The surface layer of an iron-base alloy that has been made substantially harder than the interior by the process of case hardening.

CASE HARDENING Carburizing, nitriding, or cyaniding and subsequent hardening, by heat treatment, all or part of the surface portions of a piece of iron-base alloy.

CASTING STRAINS Strains produced by internal stresses resulting from the cooling of a casting.

CAST STEEL Any object made by pouring molten steel into molds.

CHARPY TEST A test made to determine the notched toughness, or impact strength, of a material. The test gives the energy required to break a standard notched specimen supported at the two ends.

CHECK ANALYSIS Analysis of metal after it has been rolled or forged into semifinished or finished forms.

CHEMICAL ANALYSIS Qualitative analysis: separating a substance into its component elements and identifying them; quantitative analysis: determining the proportions of all component elements.

CHILL RING See the preferred term, "Backing ring."

CHIPPING One method of removing surface defects such as small fissures or seams from partially worked metal. If not eliminated, the defects might carry through to the finished material. If the defects are removed by means of a gas torch, the term "deseaming" or "scarfing" is used.

CHROMIUM A hard, corrosion-resistant metal widely used as an alloying element in steel and as plating for steel products.

CLADDING A relatively thick layer [> 0.04 in (1 mm)] of material applied by surfacing for the purpose of improved corrosion resistance or other properties.

CLEAVAGE PLANE Crystals possess the property of breaking more readily in one or more directions than in others. The planes of easy rupture are called "cleavage planes."

COALESCENCE The growing together or growth into one body of materials being welded, usually by fusion.

COATING A relatively thin layer [< 0.04 in (1 mm)] of material applied by surfacing for the purpose of corrosion prevention, resistance to high-temperature scaling, wear resistance, lubrication, or other purposes.

COLD FINISHING Changing the shape of, or reducing the cross section of, steel while cold—usually accomplished by rolling, drawing through a die, or turning.

COLD HEADING   Forcing cold metal to flow into dies to form thicker sections and intricate shapes.

COLD ROLLING   See "Cold finishing."

COLD SHUT   An area in metal where two portions of the metal in either a molten or a plastic condition have come together but have failed to unite into an integral mass.

COLD WORKING   Permanent deformation of a metal below its recrystallization temperature, which hardens the metal.

COMPLETE FUSION   Fusion that has occurred over the entire base-material surfaces intended for welding and between all layers and passes.

COMPLETE JOINT PENETRATION   Joint penetration in which the weld metal completely fills the groove and is fused to the base metal throughout its total thickness.

CONSUMABLE INSERT   Preplaced filler metal that is completely fused into the root of the joint and becomes part of the weld.

COVERED ELECTRODE   A composite filler-metal electrode consisting of a core of a bare electrode or metal-cored electrode to which a covering sufficient to provide a slag layer on the weld metal has been applied. The covering may contain materials providing such functions as shielding from the atmosphere, deoxidation, and arc stabilization and can serve as a source of metallic additions to the weld.

CRACK   A fracture-type discontinuity characterized by a sharp tip and high ratio of length and width to opening displacement.

CRATER   In arc welding, a depression at the termination of a weld bead or in the molten weld pool.

CRATER CRACK   A crack in the crater of a weld bead.

CREEP   Plastic deformation or flow of metals held for long periods of time at stresses lower than normal yield strength. Especially important if temperature of stressing is near the recrystallization temperature of the metal.

CREEP STRENGTH   The maximum stress that can be applied to steel at a specified temperature without causing more than a specified percentage increase in length in a specified time.

CREEP TEST   A number of samples, each loaded to a different stress, are placed in heating coils and held at a constant predetermined temperature. Tests are conducted for periods ranging from 1000 to 2000 hours, during which time the samples stretch. The elongation is measured and recorded at regular intervals. The results show the amount of elongation that can be expected when the steel is subjected to a given stress and temperature within a given time.

CRITICAL POINTS OR TEMPERATURES   The various temperatures at which transformations occur in steel as it passes through its critical range—on either a rising or a falling temperature. (See "Transformation range.")

CRITICAL RANGE   A temperature range in passing through which steel undergoes transformation. The preferred term is "Transformation range."

CRYOGENICS   Working with materials in environments near absolute zero ($-459.69°F$).

CUP FRACTURE   A type of fracture—which looks like a cup having the exterior portion extended with the interior slightly depressed—produced in a tensile-test specimen. Usually an indication of ductility.

DECALESCENCE   The absorption of heat, due to internal changes, that occurs when steel is heated through the critical temperature range.

DECARBURIZATION   The loss of carbon from the surface of solid steel during heating, forging, hot-rolling, etc.

DEEP DRAWING   The process of working metal blanks in dies on a press into shapes that are usually more or less cuplike in character.

DEEP ETCHING   Etching, for examination at low magnification, by a reagent that attacks the metal to a much greater extent than is normal for microscopic examination; may bring out such features as abnormal grain size segregation, cracks, or grain flow.

DEOXIDIZATION   The removal of oxygen, present as iron oxide, from molten steel by adding a deoxidizing agent such as manganese, silicon, or aluminum.

DEPOSITED METAL   Filler metal that has been added during a welding operation.

DEPOSITION EFFICIENCY   In arc welding, the ratio of the weight of deposited metal to the net weight of filler metal consumed, exclusive of stubs.

DEPOSITION RATE   The weight of material deposited in a unit of time. It is usually expressed as kilograms per hour (kg/h).

DEPOSITION SEQUENCE   The order in which the increments of weld metal are deposited.

DEPTH OF FUSION   The distance that fusion extends into the base metal or previous pass from the surface melted during welding.

DIAMOND PYRAMID HARDNESS TEST   An indentation hardness test employing a 136-diamond-pyramid indenter and variable loads, making possible the use of one hardness scale for all ranges of hardness from soft lead to tungsten carbide.

DIRECT CURRENT ELECTRODE NEGATIVE   The arrangement of direct current arc welding leads in which the work is the positive pole and the electrode is the negative pole of the welding arc. See also "Straight polarity."

DIRECT CURRENT ELECTRODE POSITIVE   The arrangement of direct current arc welding leads in which the work is the negative pole and the electrode is the positive pole of the welding arc. See also "Reverse polarity."

DISCONTINUITY   An interruption of the typical structure of a weldment, such as a lack of homogeneity in the mechanical, metallurgical, or physical characteristics of the material or weldment. A discontinuity is not necessarily a defect.

DISTORTION   A change in shape (usually refers to changes of shape caused by internal stress).

DOUBLE-WELDED JOINT   In arc and oxyfuel-gas welding, any joint welded from both sides.

DRAWING   The pulling of steel through a die, as in drawing wire, or the deforming of steel in dies on a press (deep drawing).

DRAWING BACK   Reheating after hardening to a temperature below the critical for the purpose of improving the ductility or lowering the hardness of the steel or both.

DRAWING QUALITY   A grade of flat-rolled steel that can withstand extreme pressing, drawing or forming, etc., without creating defects. Produced from deep-drawing rimmed steels or extra deep-drawing aluminum-killed steels that are specially rolled and processed.

DUCTILITY   The ability to change shape without fracture. In steel, ductility is usually measured by elongation and reduction of area as determined in a tensile test.

ECCENTRICITY OF TUBING   Variation in wall thickness, when the wall is lighter on one side and heavier at a point 180° away, resulting in the ID of the tube not being concentric with the OD.

EDDY-CURRENT TEST   Nondestructive testing method in which eddy-current flow is induced in the test object. Changes in the flow caused by variations in the object are reflected into a nearby coil or coils for subsequent analysis by suitable instrumentation and techniques.

EFFECTIVE LENGTH OF WELD   The length of weld throughout which the correctly proportioned cross section exists. In a curved weld, it is measured along the axis of the weld.

EFFECTIVE THROAT   The minimum distance from the root of a weld to its face less any reinforcement.

ELASTIC LIMIT   The maximum load per unit of area (usually stated as pounds per square inch) that may be applied without producing permanent deformation. It is common practice to apply the load at a constant rate of increase and also measure the increase of length of the specimen at uniform load increments. The point at which the increase in length of the specimen ceases to bear a constant ratio to the increase in load is called the "proportional limit." The elastic limit will usually be equal to or slightly higher than the proportional limit.

ELECTRODE   A component of the welding circuit through which current is conducted to the arc, molten slag, or base metal.

ELECTRODE EXTENSION (GMAW, FCAW, SAW)   The length of unmelted electrode extending beyond the end of the contact tube during welding.

ELECTRODE HOLDER   A device used for mechanically holding the electrode while conducting current to it.

ELECTRODE LEAD   The electrical conductor between the source of arc welding current and the electrode holder.

ELECTROGAS WELDING (EGW)   An arc welding process that produces coalescence of metals by heating them with an arc between a continuous filler-metal (consumable) electrode and the work. Molding shoes are used to confine the molten weld metal for vertical-position welding. The electrodes may be either flux-cored or solid. Shielding may or may not be obtained from an externally supplied gas or mixture.

ELECTRON-BEAM CUTTING (EBC)   A cutting process that uses the heat obtained from a concentrated beam composed primarily of high-velocity electrons that impinge on the workpieces to be cut; it may or may not use an externally supplied gas.

ELECTRON-BEAM WELDING (EBW)   A welding process that produces coalescence of metals, with the heat obtained from a concentrated beam composed primarily of high-velocity electrons impinging on the surfaces to be joined.

ELECTROSLAG WELDING (ESW)   A welding process producing coalescence of metals with molten slag that melts the filler metal and the surfaces of the work to be welded. The molten weld pool is shielded by this slag, which moves along the full cross section of the joint as welding progresses. The process is initiated by an arc that heats the slag. The arc is then extinguished and the conductive slag is maintained in a molten condition by its resistance to electric current passing between the electrode and the work.

ELONGATION   The increase in length of a test specimen after rupture in a tensile test, expressed as a percentage of the original length.

ENDURANCE LIMIT   Maximum dynamic stress to which material can be submitted for an infinite number of times without causing fatigue failure.

ETCHING   Revealing structural details by preferential attack of reagents on a metal surface.

ETCH TESTS   Etching to detect inclusions in steel. A common test is dipping a sample into acid, which reacts with the inclusions to disclose their presence.

EUTECTOID STEEL   Carbon steel with a 100 percent pearlitic structure, which is the structure developed under normal conditions of hot working and cooling when the proportion of carbon is about .80 percent. Hypereutectoid steel has a greater percentage of carbon, and hypoeutectoid steel has less carbon.

EXTENSOMETER   An instrument for measuring changes caused by stress in a linear dimension of a body.

EXTRUDING   Shaping metal in continuous form by forcing it through a die.

FACE OF WELD   The exposed surface of a weld on the side from which welding was done.

FACE REINFORCEMENT   Reinforcement of a weld at the side of the joint from which welding was done.

FATIGUE   The tendency for a metal to break under conditions of repeated cyclic stressing below the ultimate tensile strength.

FATIGUE LIMIT   See "Endurance limit."

FATIGUE TEST   Highly polished samples are subjected to stress while bending, which results in a reversal of stress for every complete revolution. The stress is reduced on each succeeding sample until the maximum stress a sample will sustain for 10 million reversals has been reached. Experience justifies the assumption that if steel can withstand 10 million reversals, it can withstand such stress indefinitely. This stress is reported as the fatigue limit.

FERROALLOYS   Iron alloyed with some element such as manganese, chrome, or silicon; used in adding the element to molten steel.

FIBER   A characteristic of wrought metal manifested by a fibrous or woody appearance of fractures and indicating directional properties. Fiber is due chiefly to the extension in the direction of working of the constituents of the metal, both metallic and nonmetallic.

FIBER STRESS   Unit stress at a certain point when overall section stress is not uniform.

FILLER METAL   The metal to be added in making a welded joint.

FILLET WELD   A weld of approximately triangular cross section joining two surfaces approximately at right angles to each other in a lap joint, T joint, or corner joint.

FINISHED STEEL   Steel that is ready for the market without any further work or treatment, such as wire, bars, sheets, rails, plates. Blooms, billets, slabs, and wire rods are termed "semifinished."

FINISHING TEMPERATURE   Temperature at which hot mechanical working of metal is completed.

FIREBOX QUALITY   Quality of plates for use in pressure vessels that will be exposed to fire or heat and the resulting thermal and mechanical stresses.

FISSURE   A small cracklike discontinuity with only slight separation (opening displacement) of the fracture surfaces. The prefixes "macro" and "micro" indicate relative size.

FIXTURE   A device designed to hold parts to be joined in proper relation to each other.

FLAKES   An internal steel fracture with a bright, scaly appearance.

FLAME ANNEALING   The direct application of a high-temperature flame to a steel surface for the purpose of removing stresses and softening metal. Commonly used to remove stresses from welds.

FLAME CUTTING Production of shapes from flat steel with single- or multiple-torch setups. Torches may be guided by hand or by an electric eye, or may be guided mechanically.

FLAME HARDENING In this method of hardening, the surface layer of a medium- or high-carbon steel is heated by a high-temperature torch and then quenched.

FLANGE QUALITY Quality of plates for use in pressure vessels that are not exposed to fire or radiant heat. Special manufacturing, testing, and marking are required.

FLAT POSITION The welding position used to weld from the upper side of the joint; the face of the weld is approximately horizontal.

FLOWABILITY The ability of molten filler metal to flow or spread over a metal surface.

FLUX A fusible mineral material that is melted by the welding arc. Fluxes may be granular or solid coatings. They serve to stabilize the welding arc and shield all or part of the molten weld pool from the atmosphere; they may or may not evolve shielding gas by decomposition.

FLUX-CORED-ARC WELDING (FCAW) A gas-metal-arc welding process that produces coalescence of metals by heating them with an arc between a continuous filler-metal (consumable) electrode and the work. Shielding is provided by a flux contained within the tubular electrode. Additional shielding may or may not be obtained from an externally supplied gas or gas mixture. See "Flux-cored electrode."

FLUX-CORED ELECTRODE A composite filler-metal electrode consisting of a metal tube or other hollow configuration containing ingredients to provide such functions as shielding atmosphere, deoxidation, arc stabilization, and slag formation. Alloying materials may be included in the core. External shielding may or may not be used.

FOREHAND WELDING A welding technique in which the welding torch or gun is directed toward the progress of welding.

FORGING A piece of metal that has been shaped or formed, while hot, by forging with a hammer (hand or power), in a press or by a drop hammer.

FORGING QUALITY A quality of semifinished steel produced for applications involving forging; requires manufacturing control for chemical composition, deoxidization mold practice, pouring, rolling, discard, cooling surface preparation, testing, and inspection. Purchaser's method of fabrication and end use are vital considerations in producing steel to this broad definition.

FRACTURE The surface of a break in metal.

FRACTURE TEST Breaking metal to determine structure or physical condition by examining the fracture.

FULL ANNEALING Heating to above the critical temperature range, followed by slow cooling through the range, producing maximum softness.

FULL-FILLET WELD   A fillet weld whose size is equal to the thickness of the thinner member joined.

FUSION   The melting together of filler metal and base metal or of base metal only, which results in coalescence.

FUSION FACE   A surface of the base metal that will be melted during welding.

FUSION WELDING   Any welding process or method that uses fusion to complete the weld.

FUSION ZONE   The area of base metal melted as determined on the cross section of a weld.

GAS-METAL-ARC CUTTING (GMAC)   An arc cutting process used to sever metals by melting them with the heat of an arc between a continuous metal (consumable) electrode and the work. Shielding is obtained entirely from an externally supplied gas or gas mixture.

GAS-METAL-ARC WELDING (GMAW)   An arc welding process wherein coalescence is produced by heating with an electric arc between a filler-metal (consumable) electrode and the work. Shielding is obtained from a gas, a gas mixture (which may contain an inert gas), or a mixture of a gas and a flux. (This process has sometimes been called "MIG welding.")

GAS POCKET   See the preferred term, "Porosity."

GAS-SHIELDED-ARC WELDING   A general term used to describe gas-metal-arc welding, gas-tungsten-arc welding, and flux-cored-arc welding when gas shielding is employed.

GAS-TUNGSTEN-ARC CUTTING (GTAC)   An arc cutting process in which metals are severed by melting them with an arc between a single tungsten (nonconsumable) electrode and the work. Shielding is obtained from a gas or gas mixture.

GAS-TUNGSTEN-ARC WELDING (GTAW)   An arc welding process that produces coalescence of metals by heating them with an arc between a tungsten (nonconsumable) electrode and the work. Shielding is obtained from a gas or gas mixture. Pressure and filler metal may or may not be used. (This process has sometimes been called "TIG welding," a nonpreferred term.)

GLOBULAR TRANSFER (arc welding)   A type of metal transfer in which molten filler metal is transferred across the arc in large droplets.

GOUGING   The forming of a bevel or groove by material removal. See also "Back-gouging," "Arc gouging," and "Oxygen gouging."

GRAIN GROWTH   The increase in the size of grains making up the microstructure of steel such as may occur during heat treatments.

GRAIN REFINEMENT   Reduction of the crystalline or grain structure by heat treating or by a combination of heat treating and mechanical working.

GRAIN STRUCTURE   The type of crystalline structure of a metal, as observed by eye or under the microscope.

GRAPHITIZING   Annealing gray cast iron so that most of the carbon is transformed to the graphitic condition. Controlled by increasing silicon and by thermal treatment.

GROOVE   An opening or channel in the surface of a part or between two components that provides space to contain a weld.

GROOVE ANGLE   The total included angle of the groove between parts to be joined by a groove weld.

GROOVE WELD   A weld made in the groove between two members to be joined. The standard types of groove welds are as follows:
    Double-bevel groove weld
    Double-flare-bevel groove weld
    Double-flare-V groove weld
    Double-J groove weld
    Double-U groove weld
    Double-V groove weld
    Single-bevel groove weld
    Single-flare-bevel groove weld
    Single-flare-V groove weld
    Single-J groove weld
    Single-U groove weld
    Single-V groove weld
    Square groove weld

HARDENABILITY   In steel, the ability to harden when cooled from its hardening temperature, as measured by its surface hardness and by the depth of hardening below the surface.

HARDENING   As applied to heat treatment of steel, heating and quenching to produce increased hardness.

HARDFACING   A form of surfacing in which a welded cladding is applied to a base metal for the purpose of reducing wear or loss of material by abrasion, impact, erosion, galling, and cavitation.

HEAT-AFFECTED ZONE   That portion of the base metal that has not been melted, but whose mechanical properties or microstructure has been altered by the heat of welding or cutting.

HEAT OF STEEL   The steel produced from one charge in the furnace, and consequently practically identical in its characteristics.

HEAT-RESISTING STEELS   Those steels that are used for service at relatively high temperatures because they retain much of their strength and resist oxidation under such conditions.

HEAT TREATMENT   An operation or combination of operations involving the heating and cooling of steels in the solid state for the purpose of obtaining certain desirable mechanical, microstructural, or corrosion-resisting properties.

HIGH-STRENGTH STEEL A specific class of low-alloy steels in which increased mechanical properties and, usually, good resistance to atmospheric corrosion are obtained with moderate amounts of one or more alloying elements other than carbon. Preferably called "high-strength, low-alloy steels."

HORIZONTAL FIXED POSITION In pipe welding, the position of pipe joint in which the axis of the pipe is approximately horizontal and the pipe is not rotated during welding.

HORIZONTAL POSITION, fillet weld The position in which welding is performed on the upper side of an approximately horizontal surface and against an approximately vertical surface.

HORIZONTAL POSITION, groove weld The position of welding in which the axis of the weld lies in an approximately horizontal plane and the face of the weld lies in an approximately vertical plane.

HORIZONTAL ROLLED POSITION In pipe welding, the position of a pipe joint in which the axis of the pipe is approximately horizontal and welding is performed in the flat position by rotating the pipe.

HOT-SHORTNESS Brittleness in metal, at an elevated temperature.

HOT WORKING The mechanical working of metal above the recrystallization temperature.

IMPACT TEST Determines the energy absorbed in fracturing a test bar at high velocity. Test may be in tension or bending, or may be a notch test (Izod V-notch or Charpy keyhole) in which the notch creates multiaxial stresses.

INADEQUATE JOINT PENETRATION Joint penetration that is less than that specified.

INCLINED POSITION The position of a pipe joint in which the axis of the pipe is approximately at an angle of 45° to the horizontal and the pipe is not rotated during welding.

INCLUSIONS Particles of nonmetallic material, usually oxides sulphides, silicates, and such, that are entrapped mechanically or are formed during solidification or by subsequent reaction within the solid metal.

INCOMPLETE FUSION Fusion that is less than complete during welding.

INDUCTION HARDENING A hardening process in which the part is heated above the transformation range by electrical induction.

INERT GAS A gas that does not normally combine chemically with the base metal or filler metal. See also "Protective atmosphere."

INTERGRANULAR CORROSION Electrochemical corrosion along the grain boundaries of an alloy, usually caused because the boundary regions contain material anodic to the center of the grain.

INTERMITTENT WELD A weld in which the continuity is broken by recurring unwelded spaces.

INTERPASS TEMPERATURE The highest temperature in the weld joint immediately before welding, or in the case of multiple-pass welds, the highest temperature in the section of the previously deposited weld metal, immediately before the next pass is started.

IRON A metallic element. However, in the steel industry, iron represents the product of a blast furnace containing 92 to 94 percent iron. Blast furnace iron is also called "pig iron" or "hot metal."

IZOD TEST A test made to determine the notched toughness of a material. The test gives the energy required to break a standard notched specimen supported as a cantilever beam.

JOINT The junction of members or the edges of members that are to be joined or have been joined.

JOINT DESIGN The joint geometry together with the required dimensions of the welded joint.

JOINT EFFICIENCY The ratio of the strength of a joint to the strength of the base metal (expressed in percent).

JOINT GEOMETRY The shape and dimensions of a joint in cross section before welding.

JOMINY END-QUENCH TEST A hardenability test in which a steel sample is heated to its proper quenching temperature and subjected to a spray of water at one end, a quenching method that provides a very rapid rate of cooling at the end sprayed, with progressively slower cooling all the way up to the other end.

KILLED STEEL Steel to which sufficient deoxidizing agents have been added to prevent gas evolution during solidification.

KIP A unit of load equalling 1000 pounds, or 453.59 kilograms.

LACK OF FUSION See the preferred term, "Incomplete fusion."

LACK OF JOINT PENETRATION See the preferred term, "Inadequate joint penetration."

LADLE ANALYSIS An assay by chemical analysis for specific elements of a test ingot sample obtained from the first part or middle part of a heat or blow during the pouring of the steel from the ladle. This is the analysis reported to the purchaser.

LAMINATIONS Defects caused by blisters, seams, or foreign inclusions aligned parallel to the worked surface of a metal.

LAP A surface defect appearing as a seam caused from folding over hot metal, fins, or sharp corners and then rolling or forging, without welding them into the surface (steelmaking).

LAP JOINT A joint between two overlapping members (welding and riveting).

LEG OF A FILLET WELD The distance from the root of the joint to the toe of the fillet weld.

LIGHTLY COATED ELECTRODE   A filler-metal electrode consisting of a metal wire with a light coating applied subsequent to the drawing operation, primarily for stabilizing the arc.

LIQUIDUS   The lowest temperature at which a metal or an alloy is completely liquid.

LOCAL PREHEATING   Preheating a specific portion of a structure.

LOCAL STRESS RELIEF HEAT TREATMENT   Stress relief heat treatment of a specific portion of a structure.

LOCKED-UP STRESS   See the preferred term, "Residual stress."

LONGITUDINAL SEQUENCE   The order in which the increments of a continuous weld are deposited with respect to its length.

MACHINE WELDING   Welding with equipment that performs the welding operation under the constant observation and control of a welding operator. The equipment may or may not perform the loading and unloading of the work.

MACROETCH TEST   Deep-etching steel in a hot acid solution to evaluate soundness and homogeneity without magnification.

MAGNAFLUX TEST   A method of detecting cracks, laps, and other defects by magnetizing the steel and applying fine magnetic particles (dry or suspended in solution). Presence of a surface or subsurface defect is indicated by a particle pattern.

MAGNETIC-ANALYSIS INSPECTION   A nondestructive method of inspection for determining the existence and extent of possible defects in ferromagnetic materials. Finely divided magnetic particles, applied to the magnetized part, are attracted to and outline the pattern of any magnetic leakage fields created by discontinuities.

MALLEABILITY   The property of a metal to deform when subjected to rolling or hammering. The more malleable a metal, the more easily it can be deformed.

MALLEABILIZING   An annealing operation performed on white cast iron for the purpose of partially or wholly transforming the combined carbon to temper carbon, and in some cases to remove completely the carbon from the iron by decarburization.

MANUAL WELDING   Welding wherein the entire welding operation is performed and controlled by hand.

MARAGING STEELS   A group of high-nickel martensitic steels developed by the International Nickel Co. Their high strength and ductility evolve primarily from the aging of a martensitic matrix.

MARTENSITE   With most steels, a distinctive structure developed by cooling as rapidly as possible from the quenching temperature. In this form, the steel is at its maximum hardness.

MATRIX   The ground mass or principal substance in which a constituent is embedded.

McQUAID-EHN TEST   A test for revealing grain size of steel by heating above the critical range in a carbonaceous medium. This causes grains to be outlined sharply when polished, etched, and viewed under a microscope. Grain sizes range from No. 8 (finest) to No. 1 (coarsest).

MEAN DIMENSION   The average of minimum and maximum mill tolerances; generally used in connection with tubing on OD, ID, or wall dimension.

MECHANICAL PROPERTIES   Properties of a material that reveal the reaction when force is applied, or that involve the relationship between stress and strain, such as modulus of elasticity, tensile strength, and fatigue limit. The term "mechanical properties" is preferred to "physical properties."

MELTBACK TIME   The time interval at the end of crater-fill time to arc outage during which electrode feed is stopped. Arc voltage and arc length increase and current decreases to zero to prevent the electrode from freezing the weld deposit.

MELTING RANGE   The temperature range between solidus and liquidus.

MELTING RATE   The weight or length of electrode melted in a unit of time.

MELT-THRU   Complete joint penetration for a joint welded from one side. Visible root reinforcement is produced.

MERCHANT BAR-QUALITY STEEL   A standard steel, free from visible pipe, widely used for general production and repair work, bracing, machine parts, welding jobs, etc. It is not particularly recommended for forging, heat treating, or complicated machine operations.

METAL-ARC CUTTING (MAC)   Any of a group of arc cutting processes that sever metals by melting them with the heat of an arc between a metal electrode and the base metal. See "Shielded-metal-arc cutting" and "Gas-metal-arc cutting."

METAL-ARC WELDING   See "Shielded-metal-arc welding," "Flux-cored-arc welding," "Gas-metal-arc welding," "Gas-tungsten-arc welding," "Submerged-arc welding," and "Plasma-arc welding."

MICROSTRUCTURE   The structure of metals as revealed by examination of polished and etched samples with the microscope.

MIG WELDING   (Metal-inert gas welding.) See the preferred terms, "Gas-metal-arc welding" and "Flux-cored-arc welding."

MODULUS OF ELASTICITY   Within the proportional limit, if the stress (in pounds per square inch) is divided by the strain (stretch in inches per inch) a value will be obtained that is called the "modulus of elasticity" of the material. This value for steel is about 30,000,000.

NETWORK STRUCTURE   A structure in which the crystals of one constituent are partially or entirely surrounded by envelopes of another constituent, an arrangement that gives a network appearance to a polished and etched specimen.

NICKEL STEEL   Alloy steel containing nickel as its principal alloying element.

NITRIDING   Adding nitrogen to the solid iron-base alloys by heating at a temperature below the critical in contact with ammonia or some other nitrogenous material.

NOMINAL DIMENSION   OD, ID, or wall thickness of tubing specified by buyer, regardless of how the tolerances are expressed.

NONCORROSIVE FLUX   A soldering flux that in neither its original nor its residual form chemically attacks the base metal. It usually is composed of rosin or resin-base materials.

NORMALIZING   Heating to about 100°F above the critical temperature and cooling to room temperature in still air. Provision is often made in normalizing for controlled cooling at a slower rate, but when the cooling is prolonged, the term used is "annealing."

OIL QUENCH   A quench from the hardening temperature, in which oil is the cooling medium.

OLSEN TEST   This is a cupping test made on an Olsen machine as an aid in determining ductility and deep-drawing properties. The test simulates a deep-drawing operation. It is continued until the cup formed from the steel sample fractures. Ductility and drawing properties are judged by the depth of the cup, position of the break, condition of the surface after the break, etc.

OPEN-CIRCUIT VOLTAGE   The voltage between the output terminals of the welding machine when no current is flowing in the welding circuit.

OVERHEAD POSITION   The position in which welding is performed from the underside of the joint.

OVERLAP   The protrusion of weld metal beyond the toe, face, or root of the weld; in resistance seam welding, the area in the preceding weld remelted by the succeeding weld.

OVERLAYING   See the preferred term, "Surfacing weld."

OXY-FUEL GAS CUTTING (OFC)   A group of cutting processes used to sever metals by means of the chemical reaction of oxygen with the base metal at elevated temperatures. The necessary temperature is maintained by means of gas flames obtained from the combustion of a specified fuel gas and oxygen.

OXY-FUEL GAS WELDING (OFW)   A group of welding processes that produce coalescence by heating materials with an oxy-fuel gas flame or flames, with or without the application of pressure and with or without the use of filler metal.

OXYGEN-ARC CUTTING (AOC)   An oxygen cutting process used to sever metals by means of the chemical reaction of oxygen with the base metal at elevated temperatures. The necessary temperature is maintained by an arc between a consumable tubular electrode and the base metal.

OXYGEN CUTTING (OC)   A group of cutting processes used to sever or remove metals by means of the chemical reaction of oxygen with the base metal at elevated temperatures. In the case of oxidation-resistant metals, the reaction is facilitated by the use of a chemical flux or metal powder.

OXYGEN GOUGING An application of oxygen cutting in which a bevel or groove is formed.

PARENT METAL See the preferred term, "Base metal."

PARTIAL JOINT PENETRATION Joint penetration that is less than complete. See also "Complete joint penetration."

PASSIVATION Generally refers to a process for the surface treatment of stainless steels. Material is subjected to the action of an oxidizing solution, usually nitric acid, which augments and strengthens the normal protective oxide film enabling the material to resist corrosive attack. The passivating process also removes foreign substances from the surface that might cause local corrosion.

PEARLITE A relatively hard constituent of steel made up of alternate layers of ferrite (iron), and cementite (iron carbide, that is, a compound of iron and carbon). See "Eutectoid steel."

PEENING The mechanical working of metals using impact blows.

PERMANENT SET Permanent change in shape due to application of stress.

PHOTOMICROGRAPH A photographic reproduction of an object magnified more than 10 times. Used to study the grain structure of steel.

PHYSICAL PROPERTIES Properties exclusive of those listed under mechanical properties such as density, electrical conductivity, and coefficient of thermal expansion. The term is often used to describe mechanical properties, but such usage is not recommended.

PICKLING Immersion of steel in a dilute solution of acid for the purpose of removing the scale.

PIERCING Process of spinning and rolling a billet over a mandrel in such a way that a hole is opened in the center.

PIPE A cavity formed in metal (especially ingots) during the solidification of the last portion of liquid metal. Contraction of the metal causes this cavity, or pipe.

PIT A depression in the surface of metal occurring during its manufacture (or as a result of corrosion).

PLASMA-ARC CUTTING (PAC) An arc cutting process that severs metal by melting a localized area with a constricted arc and removing the molten material with a high-velocity jet of hot, ionized gas issuing from the orifice.

PLASMA-ARC WELDING (PAW) An arc welding process that produces coalescence of metals by heating them with a constricted arc between an electrode and the workpiece (transferred arc) or the electrode and the constricting nozzle (nontransferred arc). Shielding is obtained from the hot, ionized gas issuing from the orifice, which may be supplemented by an auxiliary source of shielding gas. Shielding gas may be an inert gas or a mixture of gases. Pressure may or may not be used, and filler metal may or may not be supplied.

PLUG WELD A circular weld made through a hole in one member of a lap or T joint joining that member to the other. The walls of the hole may or may not be parallel, and the hole may be partially or completely filled with weld metal. (A fillet-welded hole or a spot weld should not be construed as conforming to this definition.)

POROSITY A condition of cavity-type discontinuities formed by gas entrapment during solidification.

POSTHEATING The application of heat to a weld or weldment subsequent to a welding or cutting operation.

POSTWELD HEAT TREATMENT Any heat treatment subsequent to welding.

PRECIPITATION HARDENING The process of hardening an alloy by heating it for the purpose of allowing a structural constituent to precipitate from a solid solution.

PREHEATING The application of heat to the base metal immediately before a welding or cutting operation.

PREHEAT TEMPERATURE The minimum temperature in the weld-joint preparation immediately prior to the welding, or, in the case of multiple-pass welds, the minimum temperature in the section of the previously deposited weld metal, immediately before welding.

PROCEDURE The detailed elements (with prescribed values or ranges of values) of a process or method used to produce a specific result.

PROCEDURE QUALIFICATION The demonstration that welds made by a specific procedure can meet prescribed standards.

PROCEDURE QUALIFICATION RECORD (PQR) A document providing the actual welding variables used to produce an acceptable test weld and the results of tests conducted on the weld for the purpose of qualifying a welding procedure specification.

PROCESS ANNEALING Heating to a temperature below or close to the lower limit of the critical temperature range and then cooling as desired.

PROPORTIONAL LIMIT The greatest load per square inch of original cross-sectional area for which the elongation is proportional to the load.

PROTECTIVE ATMOSPHERE A gas envelope surrounding the part to be welded, or thermal-sprayed, with the gas controlled with respect to chemical composition, dew point, pressure, flow rate, etc. Examples are inert gases, combusted fuel gases, hydrogen, and vacuum.

PULSED-POWER WELDING Any arc welding method in which the power is cyclically programmed to pulse so that effective but short-duration values of a parameter can be utilized. Such short-duration values are significantly different from the average value of the parameter. Equivalent terms are "pulsed-voltage" or "pulsed-current welding"; see also "Pulsed-spray welding."

PULSED-SPRAY WELDING An arc welding process variation in which the current is pulsed to utilize the advantages of the spray mode of metal transfer at average currents equal to or less than the globular-to-spray transition current.

QUENCH HARDENING Hardening a ferrous alloy by heating within or above the transformation range and cooling at a controlled rate. This usually involves formation of martensite.

QUENCHING Cooling rapidly by immersion in oil, water, etc.

QUENCHING MEDIUM The medium used for cooling steel during heat treatment—usually oil, water, air, or salts.

QUENCHING TEMPERATURE The temperature from which steel is quenched during a heat-treating process.

RADIOGRAPHY A nondestructive method of internal examination in which metal or other objects are exposed to a beam of x-ray or gamma radiation. Differences in thickness, density, or absorption, caused by internal discontinuities, are apparent in the shadow image either on a fluorescent screen or on photographic film placed behind the object.

RECALESCENCE The liberation of heat due to internal changes, which occurs when steel is cooled through the critical temperature range.

RED-SHORTNESS See "Hot-shortness."

REDUCING ATMOSPHERE A chemically active, protective atmosphere that at elevated temperature will reduce metal oxides to their metallic state. ("Reducing atmosphere" is a relative term, and such an atmosphere may be reducing to one oxide but not to another oxide.)

REDUCTION OF AREA The difference between the original cross-sectional area of a tensile specimen and that of the smallest area at the point of rupture. It is usually stated as a percentage of the original area; also called "contraction of area."

REFINEMENT OF STRUCTURE See "Grain refinement."

REFINING TEMPERATURE A temperature employed in heat treatment to refine structure, in particular, to refine the grain size.

REGENERATIVE QUENCHING Quenching carburized parts from two different temperatures to refine the case and core. (Often called "double quenching.")

REINFORCEMENT OF WELD Weld metal on the face or root of a groove weld in excess of the metal necessary for the specified weld size. See "Face reinforcement."

RESIDUAL STRESS Stress remaining in a structure or member as a result of thermal or mechanical treatment or both. Stress arises in fusion welding primarily because the weld metal contracts on cooling from the solidus to room temperature.

RESISTANCE WELDING A group of welding processes wherein coalescence is produced by the heat obtained from resistance of the work to the flow of electric current in a circuit of which the work is a part, and by the application of pressure.

REVERSE POLARITY   The arrangement of direct current arc welding leads with the work as the negative pole and the electrode as the positive pole of the welding arc. A synonym for "direct current electrode positive."

RIMMED STEEL   A steel that is poured containing enough oxygen to evolve appreciable gas during solidification. The gas evolution results in a finished product having a very pure surface with the impurities concentrated in the interior. The pure zone, which is readily shown by etching, is referred to as the "rim."

ROCKWELL HARDNESS TEST   Forcing a cone-shaped diamond or hardened steel ball into the specimen being tested under standard pressure. The depth of penetration is an indication of the Rockwell hardness.

ROOT BEAD   A weld deposit that extends into or includes part of or all the root of the joint.

ROOT CRACK   A crack in the weld or heat-affected zone occurring at the root of a weld.

ROOT OF JOINT   That portion of a joint to be welded where the members approach closest to each other. In cross section, the root of the joint may be a point, a line, or an area.

ROOT OF WELD   The points, as shown in cross section, at which the back of the weld intersects the base metal surfaces.

SCAB   A defect on the ingot caused by metal that splashes during teeming; on rolled or forged products it appears as a silverlike defect partially welded or mechanically bound to the parent metal surface.

SCALE   An iron oxide formed on the surface of hot steel, sometimes in the form of large sheets that fall off when the steel is rolled.

Scleroscope or Shore hardness test   This test consists of dropping a small diamond-tipped hammer from a standard height onto the surface of the specimen being tested. The height to which the hammer rebounds is a measure of the surface hardness of the specimen.

SEAL WELD   Any weld designed primarily to provide a specific degree of tightness against leakage.

SEAM   An elongated discontinuity in metal caused by a blowhole or other defect that has been closed by rolling or forging mechanically but not welded.

SEAM WELD   A continuous weld made between or upon overlapping members, in which coalescence may start and occur on the faying surfaces or may have proceeded from the surface of one member. The continuous weld may consist of a single weld bead or a series of overlapping spot welds.

SEGREGATION   Concentration of the components of steel with the lowest freezing point in parts of the ingot that solidify last.

SEMIAUTOMATIC ARC WELDING   Arc welding with equipment that controls only the filler-metal feed. The advance of the welding is manually controlled.

SEMIKILLED STEEL   Steel characterized by variable degrees of uniformity and compositions and having properties intermediate between those of killed and rimmed steels.

SHEAR STRENGTH   The stress required to produce fracture in the plane of a cross section, the conditions of loading being such that the directions of force and of resistance are parallel and opposite although their paths are offset a specified minimum amount.

SHIELDED-CARBON-ARC WELDING (CAW-S)   A carbon-arc welding process variation that produces coalescence of metals by heating them with an electric arc between a carbon electrode and the work. Shielding is obtained from the combustion of a solid material fed into the arc or from a blanket of flux on the work, or from both. Pressure may or may not be used, and filler metal may or may not be used.

SHIELDED-METAL-ARC CUTTING (SMAC)   A metal-arc cutting process in which metals are severed by melting them with the heat of an arc between a covered metal electrode and the base metal.

SHIELDED-METAL-ARC WELDING (SMAW)   An arc welding process that produces coalescence of metals by heating them with an arc between a covered metal electrode and the work. Shielding is obtained from decomposition of the electrode covering. Pressure is not used, and filler metal is obtained from the electrode.

SHIELDING GAS   Protective gas used to prevent atmospheric contamination.

SHORT-CIRCUITING TRANSFER (gas-metal-arc welding)   Metal transfer in which molten metal from a consumable electrode is deposited during repeated short circuits.

SHORTNESS   Brittleness.

SILKY FRACTURE   A steel fracture having a very smooth, fine-grain, or silky appearance.

SIZE OF WELD
   Groove weld   The joint penetration (depth of bevel plus the root penetration when specified). The size of a groove weld and its effective throat are one and the same.
   Fillet weld   *For equal leg fillet welds,* the leg lengths of the largest isosceles right triangle that can be inscribed within the fillet weld cross section. *For unequal leg fillet welds,* the leg lengths of the largest right triangle that can be inscribed within the fillet weld cross section.

SKELP   Steel or iron plate from which pipe or tubing is made.

SLAB   A thick, rectangular piece of steel for rolling down into plates.

SLAG   A result of the action of a flux on nonmetallic constituents of a processed ore, or on the oxidized metallic constituents that are undesirable. Usually consists of combinations of acid oxides and basic oxides with neutral oxides added to aid fusibility.

SLAG INCLUSION   Nonmetallic solid material entrapped in weld metal or between weld metal and base metal.

SLIP BANDS   A series of parallel lines running across a crystalline grain. Slip bands are formed when the elastic limit is passed by one layer or portion of the crystal slipping over another portion along a plane, known as the "slip plane."

SOAKING   Holding steel at a predetermined temperature for a sufficient time to assure heat penetration or to complete the solution of carbides, or both.

SOLIDIFICATION RANGE   The temperature range through which metal freezes or solidifies.

SOLID SOLUTION   A condition wherein one element is dissolved in another element while the dissolving element is in a solid and not liquid condition.

SOLIDUS   The highest temperature at which a metal or alloy is completely solid.

SONIMS   *So*lid *n*onmetallic *i*nclusions in *m*etal.

SPACER STRIP   A metal strip or bar prepared for a groove weld and inserted in the root of a joint to serve as a backing and to maintain root opening during welding. It can also bridge an exceptionally wide gap due to poor fit-up.

SPALLING   Cracking and flaking of a metal surface.

SPATTER   The metal particles expelled during welding and not forming a part of the weld.

SPATTER LOSS   Metal lost owing to spatter.

SPECIFIED GRAIN SIZE   Grain size can only be specified coarse (1 to 5) or fine (5 to 8) except in alloy steels, which allow a more restrictive gradient.

SPHEROIDIZING   Heating and cooling processes that make carbides spherical in shape. Steels are commonly spheroidized by prolonged heating at a temperature just below the lower limit of the transformation range, with subsequent slow cooling.

SPOT WELD   A weld made between or upon overlapping members in which coalescence may start and occur on the faying surfaces or may proceed from the surface of one member. The weld cross section (plan view) is approximately circular.

SQUARE GROOVE WELD   A type of groove weld.

STAGGERED INTERMITTENT WELDS   Intermittent welds on both sides of a joint in which the weld increments on one side are alternated with respect to those on the other side.

STICK ELECTRODE See "Covered electrode."

STICK ELECTRODE WELDING See the preferred term, "Shielded-metal-arc welding."

STITCH WELD See the preferred term, "Intermittent weld."

STRAIGHT POLARITY The arrangement of direct current arc welding leads in which the work is the positive pole and the electrode is the negative pole of the welding arc. A synonym for "direct current electrode negative."

STRANDED ELECTRODE A composite filler-metal electrode consisting of stranded wires that may mechanically enclose materials to improve properties, stabilize the arc, or provide shielding.

STRESS The load per unit area tending to deform a material.

STRESS-CORROSION CRACKING Failure of metals by cracking under combined action of corrosion and stress, residual or applied.

STRESS RELIEF CRACKING Intergranular cracking in the heat-affected zone or weld metal that occurs during the exposure of weldments to elevated temperatures during postweld heat treatment or high-temperature service.

STRESS RELIEF HEAT TREATMENT Uniform heating of a structure or a portion thereof to a sufficient temperature to relieve the major portion of the residual stresses, followed by uniform cooling.

STRESS RELIEVING Reducing residual stresses in a metal by heating to a suitable temperature for a certain time. This method relieves stresses caused by casting, quenching, normalizing, machining, cold working, or welding.

STRINGER BEAD A type of weld bead made without appreciable weaving motion.

STUD WELDING A general term for the joining of a metal stud or similar part to a workpiece. Welding may be accomplished by arc, resistance, friction, or other suitable process, with or without external gas shielding.

SUBMERGED-ARC WELDING (SAW) An arc welding process that produces coalescence of metals by heating them with an arc or arcs between a bare metal electrode or electrodes and the work. The arc and molten metal are shielded by a blanket of granular, fusible material on the work. Pressure is not used, and filler metal is obtained from the electrode and sometimes from supplementary sources (welding rod, flux, or metal granules).

SURFACING WELD A type of weld composed of one or more stringer or weave beads deposited on an unbroken surface to obtain desired properties or dimensions.

SWAGING Shaping metal by causing it to flow in a swage by pressing, rolling, or hammering (also called "swedging").

TACK WELD A weld made to hold parts of a weldment in proper alignment until the final welds are made.

TEMPER CARBON   A form of graphite in iron-base alloys produced by heating below the melting point.

TEMPERING   Reheating after hardening to a temperature below the critical and then cooling.

TEMPORARY WELD   A weld made to attach a piece or pieces to a weldment for temporary use in handling, shipping, or working on the weldment.

TENSILE STRENGTH   The maximum load per unit of original cross-sectional area obtained before rupture of a tensile specimen.

THEORETICAL THROAT   See "Throat of a fillet weld."

THERMAL STRESSES   Stresses in metal resulting from nonuniform temperature distributions.

THROAT DEPTH   In resistance welding, the distance from the centerline of the electrodes or platens to the nearest point of interference for flat sheets in a resistance welding machine. In the case of a resistance seam welding machine with a universal head, the throat depth is measured with the machine arranged for transverse welding.

THROAT HEIGHT   In resistance welding, the unobstructed dimension between arms throughout the throat depth in a resistance welding machine.

THROAT OF A FILLET WELD
   Theoretical throat   The distance from the beginning of the root of the joint perpendicular to the hypotenuse of the largest right triangle that can be inscribed within the fillet weld cross section. This dimension is based on the assumption that the root opening is equal to zero.
   Actual throat   The shortest distance from the root of weld to its face.
   Effective throat   The minimum distance minus any reinforcement from the root of weld to its face.

TIG WELDING   See the preferred term, "Gas-tungsten-arc welding."

T JOINT   A joint between two members located approximately at right angles to each other in the form of a T.

TOE CRACK   A crack in the base metal occurring at the toe of a weld.

TOE OF WELD   The junction between the face of a weld and the base metal.

TRANSFORMATION RANGE   The temperature range in which various changes occur in the structure of steel and above which it is necessary to heat steel to effect complete structural change. Normally, distinction should be made between the transformation range when heating and the range when cooling.

TUNGSTEN ELECTRODE   A non-filler-metal electrode used in arc welding or cutting, made principally of tungsten.

UNDERBEAD CRACK   A crack in the heat-affected zone generally not extending to the surface of the base metal.

UNDERCUT   A groove melted into the base metal adjacent to the toe or root of a weld and left unfilled by weld metal.

UNDERFILL   A depression on the face of the weld or root surface extending below the surface of the adjacent base metal.

UPSET   Bulk deformation resulting from the application of pressure in welding. The upset may be measured as a percent increase in interfacial area, a reduction in length, or a percent reduction in thickness (for lap joints).

UPSETTING   Deforming a heated bar by end-pounding.

VACUUM DEGASSING   A steelmaking process that permits cyclic degassing—plus the addition of alloying materials to molten steel in the absence of air before teeming. Undesirable oxide content is reduced.

VACUUM INDUCTION MELTING   Steel is melted in an induction electric furnace in vacuum chambers that also contain the ingot molds into which the melted steel is cast. Charging, melting, and casting are all performed under vacuum, thereby reducing undesirable oxide content.

VERTICAL POSITION   The position of welding in which the axis of the weld is approximately vertical.

VERTICAL POSITION, pipe welding   The position of a pipe joint in which welding is performed in the horizontal position, with the pipe rotated or kept stationary.

VICKERS HARDNESS TEST   See "Diamond pyramid hardness test."

WATER-QUENCH   In steel heat treatment, to cool steel from its quenching temperature with water.

WELD   A localized coalescence of metals produced either by heating the materials to suitable temperatures, with or without the application of pressure, or by applying pressure alone, with or without the use of filler material.

WELDABILITY   The capacity of a material to be welded under the fabrication conditions imposed into a specific, suitably designed structure and to perform satisfactorily in the intended service.

WELD BEAD   A weld deposit resulting from a pass. See "Stringer bead."

WELDBONDING   A joining method that combines resistance spot welding or resistance seam welding with adhesive bonding. The adhesive may be applied to a faying surface before welding or may be applied to the areas of sheet separation after welding.

WELD CRACK   A crack in weld metal.

WELDER   One who performs a manual or semiautomatic welding operation.

WELDER CERTIFICATION   Certification in writing that a welder has produced welds meeting prescribed standards.

WELDER PERFORMANCE QUALIFICATION   The demonstration of a welder's ability to produce welds meeting prescribed standards.

WELD GAGE A device designed for checking the shape and size of welds.

WELDING A process of joining materials, used in making welds.

WELDING MACHINE Equipment used to perform the welding operation, for example, spot welding machine, arc welding machine, seam welding machine.

WELDING OPERATOR One who operates machine or automatic welding equipment.

WELDING POSITION See "Flat position," "Horizontal position," "Horizontal fixed position," "Horizontal rolled position," "Inclined position," "Overhead position," and "Vertical position."

WELDING PROCEDURE The detailed methods and practices, including all welding procedure specifications, involved in the production of a weldment. See "Welding procedure specification."

WELDING PROCEDURE SPECIFICATION (WPS) A document providing in detail the required variables for a specific application to assure repeatability by properly trained welders and welding operators.

WELDING PROCESS A process of joining materials that produces coalescence of the materials by heating them to suitable temperatures, with or without the application of pressure, or by the application of pressure alone, with or without the use of filler metal.

WELDING ROD A form of filler metal used for welding or brazing that does not conduct the electrical current.

WELDING SEQUENCE The order of making the welds in a weldment.

WELDING TECHNIQUE The details of a welding procedure that are controlled by the welder or welding operator.

WELD INTERFACE The interface between weld metal and base metal in a fusion weld, between base metals in a solid-state weld without filler metal, or between filler metal and base metal in a solid-state weld with filler metal.

WELDMENT An assembly whose constituent parts are joined by welding, or parts that contain weld metal overlay.

WELD METAL That portion of a weld that has been melted during welding.

WELD-METAL AREA The area of the weld metal as measured on the cross section of a weld.

WELD METAL OVERLAY One or more layers of weld metal on the surface of a base metal to obtain desired properties, dimensions, or both.

WELD PASS A single progression of a welding or surfacing operation along a joint, weld deposit, or substrate.

WORK ANGLE The angle that the electrode makes with the referenced plane or surface of the base metal in a plane perpendicular to the axis of the weld. *Note:* This angle can be used to define the position of welding guns, welding torches, high-energy beams, welding rods, thermal cutting and thermal spraying torches, and thermal spraying guns.

WORK ANGLE, pipe   The angle that the electrode makes with the referenced plane extending from the center of the pipe through the molten weld pool. *Note:* This angle can be used to define the position of welding guns, welding torches, high-energy beams, welding rods, thermal cutting and thermal spraying torches, and thermal spraying guns.

WORK HARDNESS   Hardness resulting from mechanical working.

YIELD POINT   The load per unit of original cross-sectional area at which a marked increase in the deformation of the specimen occurs without increase in load. Usually calculated from the load determined by the drop of the beam of the testing machine or by use of dividers.

YIELD STRENGTH   The stress at which a material exhibits a specified deviation from proportionality of stress and strain. An offset of 0.2 percent is used for many metals.

ZYGLO INSPECTION   Metals are treated with a special dye containing water-washable oil that has the power to penetrate extremely small surface cracks. The part is then illuminated with short-wavelength light called "black light," causing the dye to glow with fluorescence and thus indicating the size and location of cracks or other defects.

B

Types of
Stainless Steels

1. Austenitics

			General Properties						
Composition or alloy content	Microstructure		Mechanical properties			Physical properties		Applications	
15–27% Cr; 8–35% Ni; 0–6% Mo, Cu, N	Austenite		Tensile strength: 490–860 MPa; Yield strength: 205–575 MPa; Elongation in 50 mm: 30–60%			Non-heat-treatable; nonmagnetic		Most widely used in general applications	
			Composition, %*						
Designation or type	UNS no.	C	Mn	Si	Cr	Ni†	P	S	Others
201	S20100	0.15	5.5–7.5	1.00	16.0–18.0	3.5–5.5	0.06	0.03	0.25 N
202	S20200	0.15	7.5–10.0	1.00	17.0–19.0	4.0–6.0	0.06	0.03	0.25 N
205	S20500	0.12–0.25	14.0–15.5	1.00	16.5–18.0	1.0–1.75	0.06	0.03	0.32–0.40 N
216 (XM-17)	S21600	0.08	7.5–9.0	1.00	17.5–22.0	5.0–7.0	0.045	0.03	2.0–3.0 Mo; 0.25–0.50 N
301	S30100	0.15	2.00	1.00	16.0–18.0	6.0–8.0	0.045	0.03	—
302	S30200	0.15	2.00	1.00	17.0–19.0	8.0–10.0	0.045	0.03	—
302B	S30215	0.15	2.00	2.0–3.0	17.0–19.0	8.0–10.0	0.045	0.03	—
303	S30300	0.15	2.00	1.00	17.0–19.0	8.0–10.0	0.20	0.15 min	0.06 Mo‡
303 Plus X (XM-5)	—	0.15	2.5–4.5	1.00	17.0–19.0	7.0–10.0	0.20	0.25 min	0.6 Mo
303 Se	S30323	0.15	2.00	1.00	17.0–19.0	8.0–10.0	0.20	0.06	0.15 min Se
304	S30400	0.08	2.00	1.00	18.0–20.0	8.0–10.5	0.045	0.03	—
304H	S30409	0.04–0.10	2.00	1.00	18.0–20.0	8.0–10.5	0.045	0.03	—
304HN	S30452	0.04–10.0	2.00	1.00	18.0–20.0	8.0–10.5	0.045	0.03	0.10–0.16 N
304L	S30403	0.03	2.00	1.00	18.0–20.0	8.0–12.0	0.045	0.03	—
304LN	S30453	0.03	2.00	1.00	18.0–20.0	8.0–10.5	0.045	0.03	0.10–0.15 N
304N	S30451	0.08	2.00	1.00	18.0–20.0	8.0–10.5	0.045	0.03	0.10–0.16 N
S30430	S30430	0.08	2.00	1.00	17.0–19.0	8.0–10.0	0.045	0.03	3.0–4.0 Cu
305	S30500	0.12	2.00	1.00	17.0–19.0	10.5–13.0	0.045	0.03	—
308	S30800	0.08	2.00	1.00	19.0–21.0	10.0–12.0	0.045	0.03	—

(continued)

1. Austenitics (continued)

Designation or type	UNS no.	Composition, %*							
		C	Mn	Si	Cr	Ni†	P	S	Others
308L	—	0.03	2.00	1.00	19.0–21.0	10.0–12.0	0.045	0.03	—
309	S30900	0.20	2.00	1.00	22.0–24.0	12.0–15.0	0.045	0.03	—
309 Cb	S30940	0.08	2.00	1.00	22.0–24.0	12.0–15.0	0.045	0.03	8 × %C min Cb
309 Cb + Ta	—	0.08	2.00	1.00	22.0–24.0	12.0–15.0	0.045	0.03	8 × %C min Cb+Ta
309S	S30908	0.08	2.00	1.00	22.0–24.0	12.0–15.0	0.045	0.03	—
310	S31000	0.25	2.00	1.50	24.0–26.0	19.0–22.0	0.045	0.03	—
310 Cb	S31040	0.08	2.00	1.50	24.0–26.0	19.0–22.0	0.045	0.030	Cb + Ta; 10 × C min 1.10 max
310S	S31008	0.08	2.00	1.50	24.0–26.0	19.0–22.0	0.045	0.03	—
312	—	0.15	2.00	1.00	30.0 nom	9.0 nom	0.045	0.03	—
316	S31600	0.08	2.00	1.00	16.0–18.0	10.0–14.0	0.045	0.03	2.0–2.0 Mo
316 Cb	S31640	0.08	2.00	1.00	16.0–18.0	10.0–14.0	0.045	0.030	Cb + Ta; 10 × C min 1.10 max; N 0.10 max; Mo 2.0–3.0
316F	S31620	0.08	2.00	1.00	16.0–18.0	10.0–14.0	0.20	0.10 min	1.75–2.5 Mo
316H	S31609	0.04–0.10	2.00	1.00	16.0–18.0	10.0–14.0	0.045	0.03	2.0–3.0 Mo
316L	S31603	0.03	2.00	1.00	16.0–18.0	10.0–14.0	0.045	0.03	2.0–3.0 Mo; 0.10–0.30 N
316LN	S31653	0.03	2.00	1.00	16.0–18.0	10.0–14.0	0.045	0.03	2.0–3.0 Mo; 0.10–0.30 N
316 Ti	S31635	0.08	2.00	1.00	16.0–18.0	10.0–14.0	0.045	0.030	Ti 5 × (C+N) min, 0.70 max; N 0.10 max; Mo 2.0–3.0

317	S31700	0.08	2.00	1.00	18.0–20.0	11.0–15.0	0.045	0.03	3.0–4.0 Mo
317L	S31703	0.03	2.00	1.00	18.0–20.0	11.0–15.0	0.045	0.03	3.0–4.0 Mo
317LM, 317LX, 317L Plus	—	0.03	2.00	1.00	18.0–20.0	12.0–16.0	0.045	0.03	4.0–5.0 Mo
321	S32100	0.08	2.00	1.00	17.0–19.0	9.0–12.0	0.045	0.03	5 × %C min Ti
321H	S32109	0.04–0.10	2.00	1.00	17.0–19.0	9.0–12.0	0.045	0.03	5 × %C min Ti
330HC	—	0.40	1.50	1.25	19.0 nom	35.0 nom	—	—	—
332	—	0.04	1.00	0.50	21.5 nom	32.0 nom	0.045	0.03	—
347	S34700	0.08	2.00	1.00	17.0–19.0	9.0–13.0	0.045	0.03	10 × %C min Cb + Ta‡
347 H	S34709	0.04–0.10	2.00	1.00	17.0–19.0	9.0–13.0	0.045	0.03	10 × %C min Cb + Ta
348	S34800	0.08	2.00	1.00	17.0–19.0	9.0–13.0	0.045	0.03	0.2 Cu; 10 × %C min Cb + Ta‡
348H	S34809	0.04–0.10	2.00	1.00	17.0–19.0	9.0–13.0	0.045	0.03	0.2 Cu; 10 × %C min Cb + Ta‡
384	S38400	0.08	2.00	1.00	15.0–17.0	17.0–19.0	0.045	0.03	—
385	—	0.08	2.00	1.00	11.5–13.5	14.0–16.0	0.045	0.03	—
18-18-2 (XM-15)	S38100	0.08	2.00	1.5–2.5	17.0–19.0	17.5–18.5	0.03	0.03	0.08–0.18 N
18-18 Plus	—	0.15	17.0–19.0	1.00	17.5–19.5	—	0.045	0.03	0.5–1.5 Mo; 0.5–1.5 Cu; 0.4–0.6 N
20Cb-3	N08020	0.07	2.00	1.00	19.0–21.0	32.0–38.0	0.045	0.035	2.0–3.0 Mo; 3.0–4.0 Cu; 8 × %C min Cb‡
20 Mo-6	—	0.03	1.00	0.50	22.0–26.0	33.0–37.0	0.03	0.03	5.0–6.7 Mo; 2.00–4.00 Cu
254 SMO	S31254	0.02	1.00	0.80	19.5–20.5	17.5–18.5	0.030	0.010	6.00–6.50 Mo; 0.50–1.00 Cu; 0.18–0.22 N

(continued)

1. Austenitics (continued)

Designation or type	UNS no.	Composition, %*							
		C	Mn	Si	Cr	Ni‡	P	S	Others
904L, JS904L, AL-4X, 2RK65	N08904	0.02	2.00	1.00	19.0–23.0	23.0–28.0	0.045	0.035	4.0–5.0 Mo; 1.0–2.0 Cu
AL-6X	N08366	0.03	2.00	0.75	20.0–22.0	23.5–25.5	0.030	0.003	6.0–7.0 Mo
Cronifer 2328	—	0.04	0.75	0.75	22.0–24.0	26.0–28.0	0.030	0.015	Cu 2.5–3.5; Ti 0.40–0.70; Mo 2.5–3.0
Cryogenic Tenelon (XM-14)	S21460	0.12	14.0–16.0	1.00	17.0–19.0	5.0–6.0	0.06	0.03	0.35–0.50 N
JS-700	N08700	0.04	2.00	1.00	19.0–23.0	24.0–26.0	0.04	0.03	4.3–5.0 Mo; 0.5 Cu; 8 × %C min Cb§; 0.005 Pb; 0.035 Sn
JS-777	—	0.04	2.00	1.00	19.0–23.0	24.0–26.0	0.045	0.035	4.0–5.0 Mo; 1.0–2.5 Cu
Nitronic 32¶	S24100	0.10	12.0	0.5	18.0	1.6	—	—	0.35 N
Nitronic 33	S24000	0.08	11.50–14.50	1.00	17.0–19.0	2.25–3.75	0.060	0.030	0.20–0.40 N
Nitronic 40	S21900	0.08	8.0–10.0	1.00	18.0–20.0	5.0–7.0	0.06	0.03	0.15–0.40 N
Nitronic 50	S21910	0.06	4.0–6.0	1.00	20.5–23.5	11.5–13.5	0.04	0.03	1.5–3.0 Mo; 0.2–0.4 N; 0.1–0.3 Cb; 0.1–0.3 V
Nitronic 60	S21800	0.10	7.0–9.0	3.5–4.5	16.0–18.0	8.9–9.0	0.04	0.03	—
Sanicro 28	N08028	0.020	2.0	1.0	26.0–28.0	29.5–32.5	0.020	0.015	Mo 3.0–4.0; Cu 0.6–1.4
Tenelon	S21400	0.12	14.5–16.0	0.3–1.0	17.0–18.5	0.75	0.045	0.03	0.35 N

* Single values are maximum values unless otherwise noted.
† For some tube-making processes, the Ni content of certain grades must be slightly higher than shown.
‡ Optional.
§ 0.50% maximum.
¶ Nominal composition; limits not available.

2. Ferritics

					General Properties				
Composition or alloy content	Microstructure		Mechanical properties		Physical properties			Applications	
11–30% Cr; 0–4% Ni; 0–4% Mo	Ferrite		Tensile strength: 415–650 MPa; Yield strength: 275–550 MPa; Elongation in 50 mm: 10–25%		Non-heat-treatable; magnetic; good resistance to chloride stress-corrosion cracking			Parts requiring combination of good general corrosion resistance with good stress-corrosion resistance; seawater applications	

Designation or type	UNS no.	C	Mn	Si	Cr	Ni	P	S	Others
					Composition, %*				
405	S40500	0.08	1.00	1.00	11.5–14.5	—	0.04	0.03	0.10–0.30 Al
409	S40900	0.08	1.00	1.00	10.5–11.75	—	0.045	0.045	6 × %C min Ti†
430	S43000	0.12	1.00	1.00	16.0–18.0	—	0.04	0.03	—
430F	S43020	0.12	1.25	1.00	16.0–18.0	—	0.06	0.15 min	0.6 Mo‡
430FSe	S43023	0.12	1.25	1.00	16.0–18.0	—	0.06	0.06	0.15 min Se
434	S43400	0.12	1.00	1.00	16.0–18.0	—	0.04	0.03	0.75–1.25 Mo
436	S43600	0.12	1.00	1.00	16.0–18.0	—	0.04	0.03	0.75–1.25 Mo; 5 × %C min Cb + Ta§

(continued)

2. Ferritics (continued)

Designation or type	UNS no.	Composition, %*							
		C	Mn	Si	Cr	Ni	P	S	Others
442	S44200	0.20	1.00	1.00	18.0–23.0	—	0.04	0.03	—
446	S44600	0.20	1.50	1.00	23.0–27.0	—	0.04	0.03	0.25 N
430Ti	S43036	0.10	1.00	1.00	16.0–19.5	0.75	0.04	0.03	5 × %C min Ti†
444 (18-2)	S44400	0.025	1.00	1.00	17.5–19.5	1.00	0.04	0.03	1.75–2.5 Mo; 0.035 max N; 0.2 + 4 (%C + %N) min (Ti + Cb)
18SR¶	—	0.04	0.30	1.00	18.0	—	—	—	2.0 Al; 0.4 Ti
182-FM	S18200	0.08	2.50	1.00	17.5–19.5	—	0.04	0.15 min	—
E-Brite 26-1 (XM-27)	S44627	0.01	0.40	0.40	25.0–27.5	0.50	0.02	0.02	9.75–1.5 Mo; 0.015 N; 0.2 Cu; 0.5 Ni + Cu; Cb 0.05–0.20
AL 29-4-2	S44800	0.010	0.30	0.20	28.0–30.0	2.0–2.5	0.025	0.02	3.5–4.2 Mo
Monit	S44635	0.025	1.00	0.75	24.5–26.0	3.5–4.5	0.04	0.03	3.5–4.5 Mo; 0.03–0.06 (Ti + Cb)
Sea-Cure/SC-1	S44660	0.025	1.00	0.75	25.0–27.0	1.5–3.5	0.04	0.03	2.5–3.5 Mo; 0.2 + 4 (%C + %N) min (Ti + Cb)
439 (XM-8)	S43035	0.07	1.00	1.00	17.00–19.00	0.50	0.040	0.03	Ti 0.20 + 4 (%C + %N) min 1.10 max; Al 0.15 max; N 0.04 max
AL 29-4C	S44735	0.025	1.00	0.75	28.00–30.00	0.50	0.040	0.03	3.50–4.50 Mo; Ti 0.2 + 4 (%C + %N) min

* Single values are maximum values unless otherwise noted.
† 0.75% max.
‡ Optional.
§ 0.70% max.
¶ Nominal composition; limits not available.

3. Martensitics

<table>
<tr><td colspan="10" align="center">General Properties</td></tr>
<tr>
<td>Composition or alloy content</td>
<td>Microstructure</td>
<td colspan="3">Mechanical properties</td>
<td colspan="2">Physical properties</td>
<td colspan="3">Applications</td>
</tr>
<tr>
<td>11–18% Cr; 0–6% Ni; 0–2% Mo</td>
<td>Martensite</td>
<td colspan="3">Tensile strength: 480–1000 MPa
Yield strength: 275–860 MPa
Elongation in 50 mm: 14–30%</td>
<td colspan="2">Hardenable by heat treatment; high strength</td>
<td colspan="3">High-strength parts; pumps; valves; paper machinery</td>
</tr>
</table>

Designation or type	UNS no.	C	Mn	Si	Cr	Ni†	P	S	Others
									Composition, %*
403	S40300	0.15	1.00	0.50	11.5–13.0	—	0.04	0.03	—
410	S41000	0.15	1.00	1.00	11.5–13.0	—	0.04	0.03	—
414	S41400	0.15	1.00	1.00	11.5–13.5	1.25–2.50	0.04	0.03	—
416	S41600	0.15	1.25	1.00	12.0–14.0	—	0.04	0.03	0.6 Mo‡
416Se	S41623	0.15	1.25	1.00	12.0–14.0	—	0.06	0.06	0.15 min Se
420	S42000	0.15 min	1.00	1.00	12.0–14.0	—	0.04	0.03	—
420F	S42020	0.15 min	1.25	1.00	12.0–14.0	—	0.06	0.15 min	0.6 Mo‡
422	S42200	0.20–0.25	1.00	0.75	11.0–13.0	0.5–1.0	0.025	0.025	0.75–1.25 Mo; 0.75–1.25W; 0.15–0.3 V
431	S43100	0.20	1.00	1.00	15.0–17.0	1.25–2.50	0.04	0.03	—
440A	S44002	0.60–0.75	1.00	1.00	16.0–18.0	—	0.04	0.03	0.75 Mo
440B	S44003	0.75–0.95	1.00	1.00	16.0–18.0	—	0.04	0.03	0.75 Mo
440C	S44004	0.95–1.20	1.00	1.00	16.0–18.0	—	0.04	0.03	0.75 Mo
Type 410Cb (XM-30)	S41040	0.18	1.00	1.00	11.5–13.5	—	0.04	0.03	0.05–0.30 Cb

(continued)

3. Martensitics (continued)

Designation or type	UNS no.	Composition, %*							
		C	Mn	Si	Cr	Ni†	P	S	Others
Type 410S	S41008	0.08	1.00	1.00	11.5–13.5	0.60	0.04	0.03	—
Type 414L	—	0.06	0.15	0.15	12.5–13.0	2.5–3.0	0.04	0.03	0.5 Mo; 0.03 Al
416 Plus X (XM-6)	S41610	0.15	1.00	1.00	12.0–14.0	—	0.06	0.15 min	0.6 Mo

* Single values are maximum unless otherwise indicated.
† For some tubemaking processes, the Ni content of certain grades must be slightly higher than shown.
‡ Optional.

4. Duplex

General Properties

Composition or alloy content	Microstructure	Mechanical properties	Physical properties	Applications
18–27% Cr; 4–7% Ni; 2–4% Mo, Cu, N	Austenite and ferrite	Tensile strength: 680–900 MPa; Yield strength: 410–900 MPa; Elongation in 50 mm: 10–48%	Non-heat-treatable	Shell-and-tube heat exchangers; wastewater treatment; cooling coils

Designation or type	UNS no.	Composition, %*								
		C	Mn	Si	Cr	Ni	Mo	P	S	Other
329, No. 7Mo	S32900	0.10	2.00	1.00	25.00–30.00	3.00–6.00	1.0–2.0	0.0045	0.03	—
3RE60	—	0.030	1.50	1.7	18.5	4.9	2.7	0.030	0.030	—
SAF2205/AF22	—	0.030	2.0	0.8	22	5.5	3.0	0.030	0.020	0.14 N
Ferralium 255	S32550	0.030	2.0	1.0	26	5.0	3.0	0.030	0.020	0.17 N, 2.0 Cu
DP-3	—	0.030	2.0	1.0	26	6.5	3.0	0.030	0.020	0.5 Cu, 0.15 N

* Single values are maximum unless otherwise indicated.

5. Precipitation Hardening

		General Properties		
Composition or alloy content	Microstructure	Mechanical properties	Physical properties	Applications
12–28% Cr; 4–25% Ni; 1–5% Mo, Al, Ti, Co	Austenite and martensite	Tensile strength: 895–1100 MPa; Yield strength: 276–1000 MPa; Elongation in 50 mm: 10–35%	Hardenable by heat treatment; very high strength	Parts requiring high strength, corrosion-resistance, and high-temperature resistance

Designation or type	UNS no.	Composition, % *							Others
		C	Mn	Si	Cr	Ni	P	S	
PH 13-8 Mo	S13800	0.05	0.10	0.10	12.25–13.25	7.5–8.5	0.01	0.008	2.0–2.5 Mo; 0.90–1.35 Al; 0.01 N
15-5 PH	S15500	0.07	1.00	1.00	14.0–15.5	3.5–5.5	0.04	0.03	2.4–4.5 Cu; 0.15–0.45 Cb + Ta
17-4 PH	S17400	0.07	1.00	1.00	15.5–17.5	3.0–5.0	0.04	0.03	3.0–5.0 Cu; 0.15–0.45 Cb + Ta
17-7 PH	S17700	0.09	1.00	1.00	16.0–18.0	6.5–7.75	0.04	0.03	0.75–1.5 Al
AM-350 (type 633)	S35000	0.07–0.11	0.5–1.25	0.50	16.0–17.0	4.0–5.0	0.04	0.03	2.5–3.25 Mo; 0.07–0.13 N
AM-355 (type 634)	S35500	0.10–0.15	0.5–1.25	0.50	15.0–16.0	4.0–5.0	0.04	0.03	2.5–3.25 Mo
AM-363		0.04	0.15	0.05	11.0	4.0			0.25 Ti
Custom 450 (XM-25)	S45000	0.05	1.00	1.00	14.0–16.0	5.0–7.0	0.03	0.03	1.25–1.75 Cu; 0.5–1.0 Mo; 8 × %C min Cb

* Single values are maximum values unless otherwise indicated.

C

Mechanical and
Physical Properties
of Some Stainless Steels

					Izod impact, ft·lb	

Representative Mechanical Properties at Room Temperature (Annealed)

AISI type number	Tensile strength, psi	Yield strength (0.2% offset)	Elongation percent in 2 in	Rockwell hardness	Izod impact, ft·lb	
302	90,000	40,000	50	B-85	+ 70F	110
					+ 32F	110
					− 40F	110
					− 80F	110
					− 320F	110
304	84,000	42,000	55	B-80	+ 70F	110
					+ 32F	110
					− 40F	110
					− 80F	110
					− 320F	110
304L	81,000	39,000	55	B-79		
309-309S	90,000	45,000	45	B-85	+ 70F	110
310-310S	95,000	45,000	45	B-85	+ 70F	90
316	84,000	42,000	50	B-79	+ 70F	110
					+ 32F	110
					− 40F	110
					− 80F	110
					− 320F	110
316L	81,000	42,000	50	B-79		
317-317L	90,000	40,000	45	B-85	+ 70F	110
321	90,000	35,000	45	B-80	+ 70F	110
347-348	95,000	40,000	45	B-85	+ 70F	110
					+ 32F	110
					− 40F	110

(continued)

Representative Mechanical Properties at Room Temperature (Annealed) *(continued)*

AISI type number	Tensile strength, psi	Yield strength (0.2% offset)	Elongation percent in 2 in	Rockwell hardness	Izod impact, ft-lb − 80F 110 − 320F 110
409	65,000	35,000	30	B-75	—

Physical Properties (Annealed)

Type	Density, lb/in³	Specific electrical resistance, μΩ/(cm)cm²	Specific heat, Btu/(lb)(°F)	Thermal conductivity, Btu/(h)(ft²)(°F)	Mean coefficient of expansion, °F		Modulus of elasticity tension, psi	Annealing temperature (suggested min), °F
					32-212	32-1200		
302	0.29	72	0.12	9.4	9.6×10^{-6}	10.4×10^{-6}	28×10^6	1900
304 304L	0.29	72	0.12	9.4	9.6×10^{-6}	10.4×10^{-6}	28×10^6	1900
309-309S	0.29	78	0.12	9.0	8.3×10^{-6}	10.0×10^{-6}	29×10^6	1900
310-310S	0.29	78	0.12	8.2	8.8×10^{-6}	9.7×10^{-6}	29×10^6	1900
316 316L	0.29	74	0.12	9.4	8.9×10^{-6}	10.3×10^{-6}	28×10^6	1900
317 317L	0.29	74	0.12	9.4	8.9×10^{-6}	10.3×10^{-6}	28×10^6	1900
321	0.29	72	0.12	9.3	9.3×10^{-6}	10.7×10^{-6}	28×10^6	1900
347 348	0.29	73	0.12	9.3	9.3×10^{-6}	10.6×10^{-6}	28×10^6	1900
409	0.28	60	0.11	14.4	5.5×10^{-6}	6.5×10^{-6}	29×10^6	1400/1600

Design Specifications
for Pressure Vessels

The ASME code on pressure vessel requirements is a minimum standard and was originally designed to protect the public from accidents involving the more common pressure vessels, such as air tanks, hot-water tanks, and similar static storage vessels. The code places the responsibility on the purchaser to calculate and specify many loadings and conditions that are not detailed in the code. Among these are corrosion, wind loads, cyclic stress, vibration, heat treatment, and even radiographic examination of welded joints.

In addition to being structurally sound, many process vessels must be built with due consideration to the process variables that the vessel will have to work with; information on such variables is supplied to the fabricator by the owner-operator. For heat exchangers, the list can be quite extensive, and the same is true of any vessel where some work is done in a process stream by the vessel. It is not realistic to expect vendors to submit an intelligent bid nor to produce a product to meet the buyer's vessel operating requirements unless the buyer clearly spells out stress and inspection requirements. It is not good practice to rely on code inspectors, since the manufacturer, not the code inspector, is held responsible for the vessel. The manufacturer should provide the necessary quality control, and the buyer should merely spot-check to see that the manufacturer does a good job.

It is not only reasonable but also absolutely necessary in most cases to ask for requirements that are more specific and more stringent than code requirements. Code requirements are minimum requirements.

Some guidelines to observe in ordering vessels are:

1. Buyer must provide specifications. The ASME code is just a minimum standard for design of common commercial vessels. The code does *not* provide for the various designs and operations of the unusual vessels and apparatuses required in advanced technology. Detailed study of the code will reveal that the designer is advised to include such variations in his or her stress analysis. One must ask for more than code requirements. For example, cyclic operations are not even provided for in code requirements, and the mechanical closures on pressure vessels and vacuum-pressure vessels require careful and, in the case of mechanical closures on pressure vessels, extensive stress analysis. Code formulas are generally shortcut formulas that normally (but not necessarily) adequately cover thinwalled vessels. However, on thick-walled vessels, the state of stress depends on the three principal stresses (radial, circumferential, and longitudinal) and *not* on the maximum one.

The buyer must spell out the requirements, including such critical details as charpy impact specifications (for vessels that may be pressurized at below 200°F), cyclic operating conditions, stress design, maximum surges of temperature and pressure, maximum steady-state operating pressures and temperatures, specifications for inspection techniques, and definitions of defects that will not be accepted *before* the job is placed with a supplier.

2. All vessels should be ordered for a maximum allowable working pressure (MAWP) of 25 percent, and never less than 15 percent, above the expected or required maximum operating pressure.

3. Requirements for heat treatment should be specified, where none may be required by the code.

4. Requirements for NDT and examination by other means, such as gas bubble tests on heat exchanger joints, should be specified.

5. Material known to provide good corrosion-resistant service should be specified.

6. The required corrosion allowance should be specified.

7. Provision for vendor inspection in the fabricator's shop should be made.

8. Design considerations must go beyond the vessel or equipment itself and provide for:

    a. Shipping, handling, lifting, installation containers, covers, bracing.

    b. Clearly identified and adequate lifting and handling lugs and material-handling equipment and instructions, to preclude damage to critical parts.

    c. Space for (1) installation of the apparatus or vessel and all its mechanisms, piping, wiring, furnaces, cooling, exhaust, pressurizing, vacuum, filter, and any other auxiliaries or components; (2) adequate access to all controls, valves, meters, and gages; (3) adequate access for lubrication and changing of belts, cleaning, testing, inspection, repair, entry into access holes, and replacement of internal tubes and other parts.

Two design specifications have been included in this appendix to illustrate the need for specifying details to a fabricator, beyond simply "ASME code construction."

1. Heat Exchanger Questionnaire[1]

Customer _____ Ref. No. _____

Address _____ Phone No. _____

Date _____

Service: ☐ Cooler ☐ Heater ☐ Condenser ☐ Reboiler ☐ Partial Condenser

Type: ☐ Shell & Tube ☐ Crossbore ☐ Cubic ☐ Falling Film ☐ Thimble ☐ Jacketed Pipe ☐ Bayonet ☐ Plate

Position: ☐ Horizontal ☐ Vertical ☐ Inclined @ _____ Degrees

Thermal Date

	Shellside	Tubeside
1. Name of Fluid (*)	_____	_____
2. Flow Rate (GPM or #/hr.)	_____	_____
3. Specific Gravity	_____	_____
4. Specific Heat (BTU/#°F)	_____	_____
5. Th. Cond. (BTU/hr. Ft. °F)	_____	_____
6. Viscosity (1)	cps @ _____ °F	cps @ _____ °F
(2)	cps @ _____ °F	cps @ _____ °F
7. Inlet Temperature (**) °F	_____	_____
8. Outlet Temperature °F	_____	_____
9. Operating Pressure PSIG	_____	_____
10. Allowable Pressure Drop PSI	_____	_____
11. Fouling Factors	_____	_____

For Condensing or Vaporizing

12. Fill in Items 1 to 11 above. Items 3, 4, 5, & 6 apply to the condensate, or liquid, portion.

	Shellside	Tubeside
13. Specific Heat of Vapor (BTU/#°F)	_____	_____
14. Latent Heat of Vapor (BTU/#)	_____	_____
15. Molecular Wt. of Vapor	_____	_____
16. Vapor Pressure Points (1)	(mmHg.) @ _____ °F	(mmHg.) @ _____ °F
(or condensing/vaporizing temp.) (2)	(mmHg.) @ _____ °F	(mmHg.) @ _____ °F
17. Noncondensables (#/hr.)	_____	_____
18. Specific Heat of Noncond. (BTU/#°F)	_____	_____
19. Molecular Wt. of Noncond.	_____	_____

Mechanical Data

(continued)

1. Heat Exchanger Questionnaire[1] (*continued*)

20.	Design Pressure	_____ PSI	_____ PSI
21.	Design Temperature	_____ °F	_____ °F
22.	Material of Construction	_____	
23.	Connections	_____	
24.	Gaskets	_____	
25.	Tubes	_____	
	O.D. _____ GAGE _____ NUMBER _____ MAX. LENGTH _____		
26.	Heads: ☐ Bonnet ☐ Channel		
27.	Code Stamp Required ☐ Yes ☐ No		

*If mixture, list quantity and name of each component.
**Maximum summer temperature if cooling water.
[1]Questionnaire courtesy of the Pfaudler Company, a Sohio Company, P.O. Box 1600, Rochester, NY 14692.

2. Specification Questionnaire for a Section VIII, Division 2, Pressure Vessel*

1.0 <u>PURPOSE</u>. The purpose of this specification is to set forth minimum requirements as to the intended operating conditions for a pressure vessel in such detail as to constitute an adequate basis for the company to select materials, design, fabricate, and inspect the vessel as required to comply with the rules of Division 2, Section VIII, *ASME Boiler and Pressure Vessel Code.*

2.0 <u>SCOPE</u>. The requirements of this specification apply to a vessel to be installed by _____
(user)
at a fixed location in _____
(city, state)
for the specific service described herein, where operation and maintenance control is retained during the useful life of the vessel by _____.
(user)

3.0 <u>CONFIGURATION</u>. The pressure vessel configuration shall be that of a right cylindrical shell of revolution having a removable upper closure.

4.0 <u>ORIENTATION and SUPPORT</u>. The vessel shall be installed with its axis oriented normal to the plane of the earth's surface, supported by a plane surface at its base.

5.0 <u>LOADINGS</u>. The loadings which shall be considered in the design of this vessel include the following:
  (a) Internal pressure, including the static head of the working fluid.
  (b) Vessel weight.
  (c) A normal contents weight of _____ pounds during operation, suspended from the upper closure, or supported from the base of the internal cavity.
  (d) Reaction of the supporting plane at the vessel base.
To the best of _____ knowledge, the following
(user)
loadings are either insignificant or non-existent:
  (a) External pressure.
  (b) Superimposed loads from equipment other than that to be furnished by this company as original equipment with the system of which this vessel is a part (e.g., insulation, piping).
  (c) Wind, snow, and earthquake loads.
  (d) Impact loads from sources external to the system of which this vessel is a part.

6.0 <u>WORKING FLUID</u>. The working fluid to be used in the routine operation of the pressure vessel is _____.

7.0 <u>CORROSION ALLOWANCE</u>. _____ has
(user)
significant experience with comparable equipment operating under similar conditions. Based on this experience, loss of vessel metal during its useful life due to corrosion, erosion, mechanical abrasion, or other environmental effects will either not occur or be of a superficial nature only. Accordingly, this company need not provide a corrosion allowance in the design of the vessel.

8.0 <u>VESSEL CAVITY</u>. The nominal configuration of the internal vessel cavity during operation shall be a right circular cylinder having a diameter of _____ inches and length of _____ inches.

9.0 <u>DESIGN BASIS</u>
9.1 <u>DESIGN PRESSURE:</u>   _____ psig
9.2 <u>DESIGN TEMPERATURE:</u>   _____ °F
9.3 <u>OPERATING PRESSURE:</u>   _____ psig
9.4 <u>TEST PRESSURE:</u>   _____ psig
9.5 <u>RUPTURE DISC</u>   _____ psig
10.0 <u>FATIGUE EVALUATION</u>. _____ *has/has not*
(user)

significant experience with comparable equipment operating under similar conditions. Based on this experience, a fatigue analysis of this vessel for cyclic service *is/is not* required (Ref: AD160). Parameters for fatigue evaluation are as follows:

(a) The expected (design) number of pressure cycles is _____ .

(b) The expected (design) range of pressure cycles is from 0 psig to the specified operating pressure.

(c) Stresses produced by other than the variation of the internal pressure during a normal operating cycle are not significant.

11.0 <u>LETHALITY.</u> _____ has determined that the
(user)

working fluid specified for use in this vessel is not a lethal substance as defined by the Code (Ref: A-301.1(c)).

Prepared by _____ _____
Date

Certification

The contents of this specification comply, to my belief and knowledge, with the applicable requirements of the "Alternative Rules for Pressure Vessels," *ASME Boiler & Pressure Vessel Code,* Section VIII, Division 2, 1980.

_____
(P. E.)

_____
(State)

_____
(Reg. No.)

*Questionnaire courtesy of National Forge Co.

Organizations Active in Pressure Vessel Rules and Codes

Organization	Information provided by organization
American Society of Mechanical Engineers c/o United Engineering Center 345 East 47th Street New York, New York 10017 TELEPHONE: (212) 705-7722	1. *ASME Boiler and Pressure Vessel Code* books, including *Welding and Brazing* (Section IX) and *Nondestructive Examination* (Section V) 2. Procedures and applications for ASME authorization to use the different stamps issued by the ASME, such as U for pressure vessels, S for power boilers
The National Board of Boiler and Pressure Vessel Inspectors 1055 Crupper Avenue Columbus, Ohio 43229 TELEPHONE: (614) 888-8320	1. National Board publications, including *National Board Inspection Code,* used by authorized inspectors in fieldwork 2. Procedures and applications for R, or repair, stamp issued by the National Board to qualified repair organizations
American Society for Nondestructive Testing 4153 Arlingate Plaza/Caller No. 28518 Columbus, Ohio 43228-0518 TELEPHONE: (614) 274-6003	1. Procedures in qualifying personnel for nondestructive testing, levels 1, 2, and 3
Uniform Boiler and Pressure Vessel Laws Society P.O. Box 512 Oceanside, New York 11572 TELEPHONE: (516) 356-5485	1. Publication outlining the requirements of the various jurisdictions in the United States and Canada concerning installation and reinspection of named boilers and pressure vessels that the jurisdiction has placed under legal requirements

Selected Bibliography

ASME Boiler and Pressure Vessel Code, American Society of Mechanical Engineers, New York, N.Y.

Fundamentals of Welding, American Welding Society, Miami, Fla.

Harvey, John F.: *Theory and Design of Modern Pressure Vessels,* 2d ed., Van Nostrand Reinhold, New York, 1974.

Hazard Survey of the Chemical and Allied Industries, American Insurance Association, New York, N.Y.

National Board Inspection Code, National Board of Boiler and Pressure Vessel Inspectors, Columbus, Ohio.

National Fire Protection Codes, National Fire Protection Association, Boston, Mass.

Recommended Practices for NDT Personnel Qualifications and Certifications, American Society for Nondestructive Testing, Evanston, Ill.

Standards for Closed Feedwater Heaters, Heat Exchange Institute, New York, N.Y.

Standards of Tubular Exchanger Manufacturers Association, New York, N.Y.

State, County, and City Synopsis of Boiler and Pressure Vessel Laws on Design, Installation, and Reinspection Requirements, Uniform Boiler and Pressure Vessel Laws Society, Oceanside, N.Y.

Index

Absolute pressure, 3
Absolute temperature, 4
AC (alternating-current) magnetization, 145
Acetylation, 34
Acoustic emission, 150
Actuator (control), 293
Aftercooler, 40
Agitator, 89
Air hardening, metallurgical, 388
Air tank, 33, 39–40
Alarms, 294–295
  change-of-state, 295
  deviation, 294
  limit, 294
  status, 295
  trend, 295
Alkylation, 34
Allowable pressure, 38, 207–252, 255
  braced surfaces, 232–233
  calculations for, 207–252
  dished heads, 226–230
  flatheads, 230–232
  openings, 237–240
  ordering vessels, 430
  shells, 220–225
  spheres, 225–226
  thick cylinders, 235–236
  thick spheres, 236–237
  tubes, 215–218
Allowable stress, 122, 124, 209–210
  alloy steel, 210
  carbon steel, 209
  cast iron, 111
Allowable working pressure, 38
  calculations for, 207–252
  ordering vessels, 430
  (See also Allowable pressure)
Alloy, definition of, 388
Alloy development, 200
Alloys of steel, 113–114
Alpha iron, 101–102
Alterations, NB rules for, 364–365
Alternating-current (ac) magnetization, 145

Aluminum, 114
American National Standards Institute (ANSI), 332
American Petroleum Institute (API), 11, 45, 79
American Society of Mechanical Engineers (see ASME; ASME code)
American Society for Nondestructive Testing (ASNDT), 152, 436
American Welding Society (AWS), 128, 331–332
Amidation, 34
Amination, 34
Ammonia, 199
Annealing, 115, 388, 406
  flame, 396
  full, 397
Anodic corrosion, 186–187, 204
ANSI (American National Standards Institute), 332
API (American Petroleum Institute), 11, 45, 79
Arc blow, 388
Arc cutting, 388
Arc gouging, 388
Arc strikes, 381, 389
Arc time, 389
Arc voltage, 389
Arc welding, 126–127, 389
Arc welding electrodes, 128–130, 389
  (See also Electrodes)
ASME (American Society of Mechanical Engineers), 11–13, 122–124, 208, 429, 436
ASME code, 11–13, 122–124
  certificate of authorization, 12
  chemical analysis (material), 123
  cryogenic service, 51–52
  hydrostatic tests, 346–347
  inspectors, 12, 15–16
  material, 122–124
  mill test, 123
  Section VIII, Divisions 1 and 2, 16–17
  sections of code, 122–123
  strength calculations, 207–252

ASNDT (American Society for Nonde-
structive Testing), 152, 436
Atmospheric corrosion, 187
Atmospheric pressure, 3
Austempering, 389
Austenite, 102, 389
Austenitic steels, 117–118, 161, 192, 389
Authorized inspector, 12, 15–16
insurance, 15–16, 131
jurisdiction, 15–16
owner-user, 15–16
Autogenous welding, 389
Automatic welding, 389
Avogadro's hypothesis, 7
AWS (American Welding Society), 128,
331–332
electrode classification system, 128–
129
Axis of weld, 389

Back-gouging, 381, 389
Backhand welding, 389
Backing ring, 389
Ball check valves, 287
Base metal, 389
Bayonet tube heat exchanger, 69
Bearing stress, 166
Bellows gage, 265
Benchmark readings, 324–325, 338
Bending stress, 165
tube-sheet, 218
Bending test, 390
Betz ring, 146
Billet, 390
Bimetallic gage, 267–268
Bleed valve, 282
Blister, 390
Blistering, 182
Block valve, 282
Blowback, 309–310, 317
Blowhole, 390
Blueprints, 327
Bonnet, 66
Borescope, 140
Boron, 113
Bourdon gage, 264–265, 318
Boyle's law, 5
Bracing surfaces, 232–233, 248
Brazing repairs, 381
Breakdown maintenance, 379
Breather valves, 315–317
Brinell hardness, 105, 390

Brittle, 50, 155
in cryogenic service, 50
in hydrogen service, 93
Brittle fracture, 180
Brittle materials, 105, 203
Brittleness, 155
Brix, 273
Bulges, 162
Buoyancy, 273
Bureau of Mines, 332
Burning (welding), 390
Bursting pressure, 213
Bursting proof test, 242–243, 252
Butt joint, 221–223, 390
efficiency of, 221–223
Butterfly valve, 285
Buttering (welding), 390
Bypasses in processes, 327

Carbide precipitation, 119, 373
Carbides, 390
Carbon-arc cutting (CAC), 390
Carbon-arc welding (CAW), 390
Carbon packing, 287
Carbon steel, 102–104, 112–117, 390
allowable stresses, 209
alloys, 113–114
heat treatment, 115–116
impurities in, 114–115
manufacture of, 112–113
types of, 116–117
(See also Steel)
Carbonitriding, 390
Carburizing, 390
Carburizing compounds, 391
Case hardening, 391
Cast iron, 101–102, 110–112
code restrictions, 111–112
gray, 111
lethal service, 30, 111–112
white, 111
Casting strains, 391
Catalyst damage, 254
Cathodic corrosion, 186–187, 204
Caustic embrittlement, 192
Caustic gouging, 195
Cementite, 101–103
Centigrade, 4
Central maintenance, 328
Change-of-state alarm, 295
Channel (heat exchanger), 66
Charles's law, 5

Charpy V-notch test, 51, 108, 391
Check analysis (metallurgical), 391
Check valve, 285–287
    ball, 287
    lift, 286
    spring-loaded, 287
    swing, 285
Checklist (maintenance), 327, 335
Chemical plant evaporators, 77–78
    calendria-type, 78
    forced-circulation, 78
    materials of construction, 78
    natural-circulation, 78
    rising-falling film, 78
Chemical plant exposures, 1–3, 9–11,
    254–255
Chemical plant pressure vessels, 84–91
    distillation columns, 84–86
    jacketed reactors, 86–88
    multiwall, 36, 84, 91
Chemical reactions (process), 9–10, 34–36
    acetylation, 34
    alkylation, 34
    amidation, 34
    amination, 34
    cracking, 35, 337–338
    depolymerization, 35
    diazotization, 35
    distillation, 35
    endothermic, 9–10
    esterification, 35
    exothermic, 9, 31, 254, 320
    extraction, 35
    Friedel-Crafts, 35
    halogenation, 35
    hydrogenation, 35, 96
    nitration, 36
    oxidation, 36, 194
    polymerization, 36
    reduction, 36
    runaway, 10, 36, 296–297
    sulfonation, 36
    vapor cloud, 9
Chemical reactors, 34
Chemical tests (metallurgical), 150–151
Chill ring (welding), 391
Chipping, 391
Chloride stress corrosion, 193, 205
Chromatograph, 335
Chromium, 113, 391
Chromium carbide, 119, 374
Circumferential stress (cylinder), 173–
    174, 220–225

Clad plate, 120–122, 161, 251
    allowable thickness, 121
Cladding, 391
Clearance for pressure vessels, 39
Cleavage plane, 391
Coalescence (welding), 391
Coating, 187, 191, 391
Cobalt, 141
Code, ASME (see ASME code)
Code construction, 15–18
Coefficient of discharge (safety valve),
    319
Coefficient of expansion, 174–175
    of stainless steels, 428
Cold finishing, 391
Cold heading, 392
Cold rolling, 392
Cold shut, 392
Cold working, 392
Columbium, 119
Combustible-gas meter, 274
Combustible liquids, 41
Complete fusion, 392
Component pressure vessel strength, 213
    calculations, 207–252
Compound gage, 318
Compression stress, 165
Compressor displacement, 40–41
Computer applications, 23–24, 26–29
    in hazard control, 26–29
    in process control, 259–263
Computer simulation, 23–24
Computerized maintenance, 327
Concentration cell corrosion, 188
Condensate in air tanks, 40
Confidentiality in specifications, 18
Conical head, 229
Consumable inserts, 392
Contract maintenance, 323–324
Control valves, 289–290
Controls, 38, 291–293
    discrete, 291
    distributed computer, 293–294
    pneumatic, 289
    proportioning, 291
    regulatory, 292–293
    sensors, 275–276
    transmitters, 291–292
Cookers, 34, 79–82
Corrodents, 191
Corrosion, 179, 184–190, 337
    anodic and cathodic, 186–187, 204
    atmospheric, 187

Corrosion (*Cont.*):
  chloride stress, 193, 205
  combating, 190
  concentration cell, 188
  crevice, 189, 195, 357
  electrochemical theory of, 186–187
  galvanic, 187–188, 194
  intergranular, 189, 400
  pits, 179, 405
  pitting, 188–189, 198, 368–369
  stainless steel, 192–194
  with stress, 190
  (*See also* Stress-corrosion cracking)
Corrosion allowance, 18, 248, 430, 433
Corrosion erosion, 356
Corrosion fatigue, 179, 189
Corrosion-fatigue testing, 192
Corrosive service, 22, 181
Counterflow, 59, 96
Covered electrode, 128–130, 392
Crack-arrest temperature, 180
Cracking, 35, 337–338
Cracks, 162, 178, 392
  growth rate, 178, 206
  propogation, 178–179
  repairs, 370–372
  in welding, 182–183
Crater (welding), 392
Creep, 109–110, 181, 183, 202, 392
  code requirements, 183–184
Creep strength, 392
Creep tests, 392
Crevice corrosion, 189, 195, 357
Critical range (metallurgical), 393
Critical temperature, 392
CRT (cathode-ray tube) screen, 147, 262, 327
Crude distillation column, 122
Cryogenic pressure vessels, 48–52, 96, 393
  brittle fracture, 50, 155, 180
  code requirements, 51–52
  material, 51, 121
  repairs, 52
  storage vessels, 49–51
Crystalline structure, 100
Cup fracture, 393
Curies, 41
Cyclic service, 429
Cylinder paper machine, 54
Cylindrical stress, 172–174, 220–225
  calculations for, 220–225

Cylindrical stress (*Cont.*):
  circumferential, 173–174, 248
  longitudinal, 172, 173
  thick cylinders, 235–237

Dalton's law, 5
Data reports, 124, 133, 249, 387
  use of computer, 263
DC (direct-current) electrodes, 393
DC (direct-current) magnetization, 145
Deaerator heaters, 73–74, 97
  inspection of, 354
Dealloying, 198
Decalescence, 393
Decarburization, 393
Deep drawing (metallurgical), 393–394
  drawing back, 394
  drawing quality, 394
Deep etching, 393
Defect analysis, 162
Defect removal, 366
Degrees API, 273
Degrees Baumé, 273
Delamination, 97
Delta ferrite, 101–102
Delta iron, 102
Density measurements, 272–274
  Brix, 273
  buoyancy, 273
  degrees API, 273
  degrees Baumé, 273
  fixed-volume weighers, 273
  radiation, 273
  refractometry, 273
  Richter, 273
  Sikes, 273
  specific gravity, 273
  torque, 274
  Tralles, 273
  twaddle, 273
  vibration, 274
Deoxidation, 393
Department of Labor, OSHA administration by, 41, 45–48
Depolymerization, 35
Deposited metal, 393
Deposition efficiency, 393
Deposition rate, 393
Deposition sequence, 393
Depth of fusion, 393

Design pressure, 247
  calculations for, 207–252
Design specifications, 429–433
Design temperature, 247–248
Destructive tests (material), 150–151
  chemical, 150
  hardness, 151
  mechanical, 151
  metallographic, 151
Deviation alarms, 294
Dezincification, 198
Diamond hardness test, 393
Diaphragm gage, 266
Diaphragm valve, 284–285
Diazotization, 35
Digesters, 79–82
  batch, 80
  continuous, 80
Direct-current (dc) electrodes, 393
  negative, 393
  positive, 393
Direct-current (dc) magnetization, 145
Discharge from safety valve, 305
Discontinuities, 145, 179, 366, 393
  notches, 179
Discrete controls, 291
Dished covers (heat exchangers), 240–241
  bolted type, 240–241
Dished heads, 226–230
  conical, 229
  ellipsoidal, 227–228
  external pressure on, 229–230
  hemispherical, 226–227
  torispherical, 228–229
Dissimilar metals repair, 375
Distillation, 35
Distillation columns, 84–86
  catalyst trays, 86
Distortion (metallurgical), 394
Distributed computer control, 293–294
Division 2 pressure vessels, 16–17, 124
  ordering of, 433–434
Double tube-sheet heat exchangers, 70–71, 97
Double-welded joint, 137, 221–224, 394
  efficiency of, 222
Drain valves, 282, 385
Drains from pressure vessels, 39
Drawing (metallurgical), 394
Drawing back, 394
Drawing quality, 394

Ductility, 155, 202, 393
Dump valve, 282, 385
Duplex stainless steel, 118, 194
  properties of, 424
Dye penetrant, 147–148, 345

Earthquake loads, 163, 243–244
EBC (electron-beam cutting), 395
EBW (electron-beam welding), 128, 395
Eccentricity of tubing, 338–339
Eddy-current test, 149–150, 346, 394
Effective length of weld, 394
  throat, 394
EGW (electrogas welding), 395
Elastic limit, 171, 200, 394
Electrical equipment, 290–292, 331, 385
  hazardous area class, 290–292
  hazardous area division, 292
  hazardous area group, 290–292
  hazardous areas inspection, 331, 385
Electrochemical theory of corrosion, 186–187
Electrode extension, 394
Electrode holder, 395
Electrode lead, 395
Electrodes, 128–130, 389, 394
  AWS classification, 129
  flux-coated, 129, 380, 392, 397
  low-hydrogen, 129
  nonconsumable, 157
Electrogas welding (EGW), 395
Electron-beam cutting (EBC), 395
Electron-beam welding (EBW), 128, 395
Electroslag welding (ESW), 395
Ellipsoidal head, 227–228
Elongation, 395
Embrittlement:
  caustic, 192
  cryogenic, 50
  hydrogen, 180–181
Emergencies, 256–257
  breakdowns, 257
  disaster drills, 256
  explosions, 256
  fire, 256
  leaks, 257
  simulated training, 256
  spill disposal, 257
Endothermic reactions, 9–10
Endurance limit, 176, 201, 395
Equipment inventory, 326–327, 377

Equipment reliability, 2
Erosion, 162, 196, 337
Erosion corrosion, 195
Essential variables (welding), 134
Esterification, 35
ESW (electroslag welding), 395
Etch test, 395
Etching, 395
Eutectic, 155
Eutectoid steel, 395
Evaluation (process), 24–29
    of computer-controlled processes, 26–29
    hazard analysis, 24
    indirect exposure, 28
    risk analysis, 24
    safety, 26
    team for, 26
Evaporators, 76–79
    chemical plant, 77–78
    power plant, 76–77
Exfoliation, 194
Existing installation equation, 220, 249
Exothermic reactions, 9, 31, 254, 320
Expansion, 174–175, 395
Explosimeter, 335
Explosion, 8–9
    chemical reaction, 9
    emergency training for, 256
    exothermic, 9, 31, 254, 320
    overpressure, 8
    vapor cloud, 9
Explosive limits, 31–32
    lower, 31–32
    upper, 31–32
Extensometer, 395
Extent of inspections, 338–339
External inspections, 339, 342
Extraction, 35
Extruding (metallurgical), 395

F numbers (welding), 134
Face of weld, 395
Face-bend test (weld), 135–136
Face reinforcement, 396
Factor of safety, 212–213, 220
Fahrenheit, 4
Fail-safe concept, 21, 262, 289
    control valves, 289
    design, 21
    emergencies, 262
Failure analysis, 162

Fatigue, 176, 380, 396
    corrosion, 189
Fatigue analysis, 124
Fatigue cracking, 206
    of tubes, 197
Fatigue limit, 176, 380, 396
Fatigue test, 396
Fault-tree analysis, 25, 32
Feedwater heater, 69, 72–73
Ferrite iron, 102–103
Ferrite materials, 110
    cast iron, 110–111
    pig iron, 110
    steel (see Steel)
Ferritic stainless steel, 118, 161, 194, 421–422
Ferroalloys, 396
Fiber (metallurgical), 396
Fiber stress, 396
Fiberglass-reinforced pressure vessels, 94–96
    glass attack, 94–95
    resin attack, 94
    thermoset polymers, 94
Fiberscope, 140
Field welding, 365–368
Filler metal, 125, 396
Fillet weld, 157, 396
Finished steel, 396
Finishing temperature, 396
Fire effects to pressure vessel, 8, 41
Fire hazard to pressure vessels, 9
Firebox quality steel, 396
Fixed-roof tanks, 315
Fixture in welding, 396
Flakes (metallurgical), 396
Flame annealing, 396
Flame cutting, 397
Flame hardening, 397
Flammable gases and liquids, 9, 41
    OSHA requirements, 45–48
    overpressure protection, 8, 41
    vapor cloud explosion, 9
Flange quality steel, 397
Flare stacks, 317
Flatheads, 230–232
Flaw inspection, 345–346
Floating head (tank), 66, 68
Floating-roof tank, 315
Flow-induced vibration, 197
Flowability (metallurgical), 397
Flowmeters, 270–272
    differential-pressure, 272–273

Flowmeters (*Cont.*):
  magnetic, 271
  point-velocity, 272
Fluorescent dye, 147
Fluoride, 89, 97
Flux, 129, 397
Flux-coated electrodes, 129
Flux-cored electrode, 397
Forehand welding, 397
Forging, 397
Foudrinier papermaking machine, 54
Fouling (heat exchangers), 59, 199–200,
    355–356
Fracture (metallurgical), 397
Fracture mechanics, 178
Fracture test, 397
Fracture toughness, 205
Fracture transition elastic (FTE), 180
Freeze-up protection, 259
  for instrumentation, 275
Frequency of inspection, 339–342
Fretting corrosion, 189
Friedel-Crafts reaction, 35
FTE (fracture transition elastic), 180
Full annealing, 397
Full-fillet weld, 398
Fume protection, 332, 336
Fusion (welding), 398
Fusion welding, 126
Fusion zone, 398

Gage pressure, 3, 39
Gages, 267–274
  density, 272–274
  extensometer, 395
  flow, 270–272
  level, 272
  pressure, 264–267
  temperature, 267–270
Galvanic corrosion, 187–188, 194
  in heat exchangers, 196
Gamma rays, 141
Gas constant, *R*, 5–6
Gas-metal-arc cutting (GMAC), 398
Gas-metal arc welding (GMAW), 127,
    398
Gas pockets, 143, 398
Gas-shielded-arc welding, 398
Gas-tungsten-arc cutting (GTAC), 398
Gas-tungsten-arc welding (GTAW), 127,
    398

Gases, 10
  flammable, 10
  molecular weight of, 308
  toxic, 10
Gasket material, 287
Gasket replacement, 383
Gasket seating, 338
Gate valve, 282–283
Glass-lined jacketed reactor, 88–89, 97
Globe valve, 283–284
Globular transfer (welding), 398
GMAC (gas-metal-arc cutting), 398
GMAW (gas-metal-arc welding), 127, 398
Gouging (welding), 398
Grain growth, 398
Grain refinement, 398
Grain size, 153
Grain structure, 153, 398
Graphitizing (metallurgical), 399
Groove (welding), 399
Groove angle, 399
Groove weld, 157, 399
GTAC (gas-tungsten-arc cutting), 398
GTAW (gas-tungsten-arc welding), 127,
    398
Guardrails, 334

Hairpin exchanger, 69
Half-pipe coil reactor, 87
Halogenation, 35
Hard restraints, 296–315
  chemical reactions, 296–298
  code safety valves, 298–302
  pressure relief valves, 298
  rupture disks, 301–304
Hardenability, 399
Hardening, 116, 399
Hardfacing, 399
Hardness, 105
  Brinell, 105, 390
  definition of, 155
  Rockwell, 105
  Shore scleroscope, 105
Hardness numbers, 106
Hardness tests, 151
HAZ (heat-affected zone), 82, 105, 181,
    358, 399
Hazardous areas (electrical equipment),
    290–292
  classes, 290
  divisions, 290
  electrical code for, 290

Hazardous areas (*Cont.*):
  explosion-proof equipment, 290
  groups, 290
  intrinsically safe equipment, 292
  safety considerations for electric con-
    trols, 292
  (*See also* Electrical equipment)
Hazards (process), 9, 24–29
  analysis, 24, 32, 376
  chemical reactions, 9
  evaluation, 24–29
  fire, 9
  pressure, 8, 10
  process streams, 21
  vapor cloud, 9
Head-to-shell attachments, 56–57
  rotating shells, 56–57
  stresses on, 230
Heads, 226–230
  dished, 226–230
  external pressure on, 229–230
  flat-, 230–232
Heat of steel, 399
Heat-affected zone (HAZ), 82, 105, 181,
    358, 399
Heat effects to pressure vessel, 8, 9, 41
Heat Exchange Institute (HEI), 79, 219
Heat exchanger types, 53–70
  bayonet tube, 69
  double pipe, 63, 69
  double tube-sheet, 70–71, 97
  hairpin, 69
  panel coil, 63
  plate fin, 63, 70
  reboiler, 65, 68
  shell-and-tube, 63, 65–69
  spiral, 63, 70
  TEMA shells, 64
  U tube, 66–67
Heat exchangers, 33–34, 57–73, 195–200
  codes and standards, 64, 79
  counterflow, 59
  fouling, 59, 199–200, 355–356
  inspection of, 354–358
  mean temperature difference, 59
  mechanical design, 60
  mechanical problems, 356–358
  ordering questionnaire, 431–432
  parallel flow, 59
  performance problems, 355–356
  repairs, 372–373
  service effects, 195–200
  TEMA classifications of (*see* TEMA)

Heat exchangers (*Cont.*):
  thermal performance, 59
  viscosity considerations, 61
Heat mass flow rate, 58
Heat-tracing lines and vessels, 259
Heat treatment, 399, 430
Heli-arc welding, 127
Hemispherical head, 226–227
High-alloy stainless steel, 118, 193–194
High-strength steel, 400
High-velocity flow, 214
Hold points, 133
Hooke's law, 170
Horizontal weld positions, 135, 400
  fillet weld, 400
  fixed position, 400
  groove weld, 400
  rolled position, 400
Hot-shortness (metallurgical), 400
Hot working, 400
Huddling chamber, 299, 317–318
Hydrogen, 115
Hydrogen embrittlement, 180–181
Hydrogen service, 93–94, 97
Hydrogenation, 35, 96
Hydrostatic testing, 108, 248, 346–347
  ASME, 346
  cast iron vessels, 112
  limits, 334
  precautions, 347–348
  safety, 334
  time duration of test, 347
  water temperature, 347

Impact loads, 163, 250
Impact test, 51, 400
  Charpy, 51, 400
Impingement of flow, 356
Inadequate joint penetration, 400
Inclined position (welding), 400
Inclusions, 138, 400
Incomplete fusion, 400
Incomplete penetration, 367
Indirect process exposure, 28
Induction hardening, 400
Inert gas blanketing, 48, 96
Inert gas welding, 400
Infrared gage, 269–270
Inspection agency, 12
Inspections, 334–359
  cleaning for, 341
  corrosion, 337

Inspections (*Cont.*):
  cracks, 338
  creep, 338
  dye-penetrant, 345
  eddy-current, 346
  erosion, 337
  extent of, 338–339
  external, 339, 342
  flaws with NDT, 345–346
  frequency of, 339–342
  heat exchangers, 354–358
  hydrostatic, 346–348
  interior illumination, 336–337
  internal, 335–341
  interpretation of results, 344
  jurisdictional, 339
  leak tests, 346
  lined vessels, 351–352
  low-pressure vessels, 354
  NDT, 342–346
  overlaid vessels, 351–352
  radiography, 345
  report of findings, 343
  rotating vessels, 335, 352–354
  safety precautions, 334–337
  stress-corrosion, 349–351
  thickness tests, 342–344
  ultrasonic tests, 346
  visual, 345
  wear rates, 344
  welding, 135–139
Inspectors, 12–16, 31, 131
  ASME, 12–15
  authorized, 15, 31
  code, 12–16
  insurance, 15, 31, 131
  jurisdictional, 15, 31, 131
  NDT, 151–152
  owner-user, 15, 31
  state, 15, 31, 131
Inspector's gage connection, 266
Instrumentation, 263–275
  ASME code, 263–264
  density, 272–273
  flowmeter, 270–272
  freeze protection, 275
  level, 272
  maintenance, 275
  pressure gages, 264–267
  temperature gages, 267–270
Interaction of operators, controls, and
    computers, 259–263
Intergranular corrosion, 189, 400

Interlocks, 295–296
Intermittent weld, 400
Interpass temperature, 401
Interpretation of tests, 344
Interruption of production, 2
Iridium, 141
Iron, 110–113, 401
  alloying, 113–114
  alpha, 101–102
  cast, 110–112
  delta, 101–102
  gamma, 101–102
  gray cast, 111
  white cast, 111
Isolating vessels, 333
Isothermal transformation diagrams, 103
Isotopes, 141
Izod test, 401
  for stainless steel, 426–428

Jacketed reactors, 86–87
  conventional, 87
  dimple, 87
  glass-lined, 88–89
  half-pipe coil, 87
  safety considerations, 89–91
Joining by welding, 124–140
  adverse weather, 130
  ASME code, 130–131
  electrode storage, 129
  electrodes, 128–129
  manufacturer's, and contractor's re-
    sponsibility in, 131
  qualification of welders, 134
  quality control, 131–133
  recertification of welders, 135
  repair of weld defects, 138
  tack welds, 137
  test plates, 135–136
  types of welding, 125–128
  welded-joint efficiencies, 137
  welder positions, 135
  welding defects, 138
  welding engineers, 125
  welding inspections, 138–139
  welding surfaces, 130
Joint (welding), 157, 221–224, 390, 396
Joint design, 401
Joint efficiency, 137, 221–224, 401
  butt, 221–224
  lap, 221–224, 401
Joint geometry, 401

Joint penetration, 142, 392, 401
Jominy end-quench test, 401
Jurisdictional inspections, 339, 385
Jurisdictional inspectors, 15, 31, 131

Kettle reboiler, 65, 68
Killed steel, 108, 116, 401
Kinetic reaction rate, 297
Kip, 401

Lack of fusion, 138, 143, 401
Lack of joint penetration, 401
Ladders, 334
Ladle analysis, 401
Laminations, 401
Lap joint, 221–223, 401
  cracks in, 372
Leak testing, 346, 349, 383
Leg of fillet weld, 401
Lethal service, 11, 18, 30
Lethal substance, 10–11, 18, 30
  ASME rules, 10–11
  weld efficiency required, 248
Level gages, 272
  diaphragms, 272
  displacement, 272
  inert seal, 272
Level 1,2,3 NDT, 151–152, 380
Liability suits, 2
Lift check valves, 286
Ligaments, 233–235
  cracking, 357
  efficiency of, 233–234
Lightly coated electrodes, 402
Limit alarms, 294–295
Limitations in NDT, 143–149
  liquid-penetrant inspection, 146
  magnetic-particle inspection, 145
  radiography, 143
  ultrasonic inspection, 149
Lined pressure vessels, 251
Liquefaction, 49
Liquid-penetrant inspection, 146–147
  fluorescent penetrant, 147
  limitations of, 146
Liquidus, 402
Loadings of pressure vessels (design), 29–30, 163
Loads on pressure vessels, 163, 243–244
  earthquakes, 243–244
  wind, 162, 243

Local preheating, 402
Locked-up stresses, 82, 158, 164, 402
Logarithmic mean temperature difference, 59
Longitudinal sequence in welding, 402
Longitudinal stress, 172–173, 220–225, 248
Low-alloy steel, 117
Low-hydrogen electrode, 129
Lower explosive limit, 31–32

MAC (metal-arc cutting), 403
Machine welding, 402
Macroetch test, 402
Magnaflux test, 402
Magnetic-analysis inspection, 144–146, 402
Magnetic-particle inspection, 144–146
  ac magnetization, 145
  dc magnetization, 145
  prod spacing, 160
Maintenance, 19–23, 275, 322–334
  benchmark readings, 324–325, 338
  breakdown, 379
  central organization, 328
  clearance for, 259
  computerized, 327
  instrumentation needs, 275
  management of, 377
  need for, 19–23, 324
  organizational structure for, 328–329, 379
  performance monitoring, 323
  planned, 322
  safety in, 329–334
  types of, 322–323
Malleability, 402
Malleable iron, 156
Management responsibility for safety, 340
Management safety support, 376–379
Manganese, 113
Manometer, 266
Manual reset control, 317
Manual valves, 282
Manual welding, 402
Maraging steel, 402
Martensite, 103, 402
Martensitic stainless steel, 118, 423–424
Masks, 332–333
  air-line, 333
  Bureau of Mines–approved, 333

Materials, ASME code, 122–124
  certification in, 124
Matrix (metallurgical), 402
Mean temperature difference, 59
Mechanical problems, heat exchangers,
    356–358
  corrosion erosion, 356
  crevice corrosion, 357
  HAZ, 358
  impingement, 356
  ligament cracking, 357
  stress-corrosion cracking, 357
  tube leakage, 356
  vibration damage, 357
  wire drawing, 357
Mechanical properties, material, 104,
    403
Mechanical tests, 151
Megahertz, 147
Meltback time, 403
Melting range, 403
Melt-thru, 403
Merchant bar–quality steel, 403
Mercury, 198–199
Metal-arc cutting (MAC), 403
Metal-arc welding, 403
Metallographic tests, 403
MIG welding (metal–inert gas welding),
    403
Mill scale, 156
Mill test report, 123, 156, 159, 382
Mode of control, 317
Modulus of elasticity, 170, 201, 403
  of stainless steels, 428
Molar calculations, 7
Mole, 7
Molecular weight of gases, 308
Molybdenum, 113
Multilayer construction, 91, 248
Multiwall construction, 36, 84, 91

National Board (NB) of Boiler and Pres-
    sure Vessel Inspectors, 12–15, 37,
    339, 436
  alterations, 364–365
  authorized inspectors, 15–16
  inspection code, 12–13
  R stamp, 359–364
  reinspection rules, 339
  repairs, 364–365
National Fire Protection Association
    (NFPA), 11, 322

NDT (nondestructive testing), 139–152
  acoustic emission, 150
  eddy current, 149–150
  inspections for flaws, 345–346
  level 1,2,3 inspectors, 151–152
  limitations in, 143–149
  liquid-penetrant, 146–147
  magnetic-particle, 144–146
  ordering the service, 430
  outside service, 152
  personnel, 151–152
  radiography, 151–152
  selection, 139, 159
  ultrasonic, 147–149, 346
  visual examination, 140–141
Needle valve, 284
Nelson curves, 181, 382
Network structure (metallurgical), 403
NFPA (National Fire Protection Associa-
    tion), 11, 322
Nickel, 113
Nickel steel, 403
Nil-ductility temperature, 108
Nitration, 36
Nitriding, 404
Nitrogen, 115
Nitrogen purge, 336
Nominal diameter pipe, 215
Nominal diameter tube, 215
Nominal dimension, 404
Nominal thickness, 251
Noncorrosive flux, 404
Nondestructive testing (see NDT)
Nonferrous metals, 215–218
  NDT of, 380
  properties of, 215–218
  use of, 251
Nonmetallic inclusions, 115
Nonsparking tools, 330
Normal stress, 164
Normalizing, 116, 154, 404
Notch sensitivity, 203
Notch toughness, 123, 181
Notched-bar impact test, 203
Notches, 170

OAC (oxygen-arc cutting), 404
Occupational Safety and Health Admin-
    istration (OSHA), 45–48
OFC (oxy-fuel gas cutting), 404
OFW (oxy-fuel gas welding), 127, 404
Oil quench, 404

Olsen test, 404
Open-circuit voltage (welding), 404
Openings, 237–240
Operating manuals, 256
Operating surveillance, 255
Operator training, 23, 255–256, 379
Ordering pressure vessels, 20, 255–256,
    379
Organizational charts, 132
  maintenance, 328–329, 379
  repairs, 363–364
Orsat analyzers, 335
OSHA (Occupational Safety and Health
    Administration), 45–48
Outside packed lantern ring, 68
Outside packed stuffing box, 68
Overhead position for welding, 404
Overlap, 367, 404
Overlay welding, 376, 381, 404
Overpressure protection, 41
  for flammable or combustible liquids,
    45–48
Overtemperature protection, 90
Owner-user responsibility, 17–19
Oxidation, 36, 194
Oxy-fuel gas cutting (OFC), 404
Oxy-fuel gas welding (OFW), 127, 404
Oxygen, 114
  deficiency of, 332–333, 336
Oxygen-arc cutting (OAC), 404
Oxygen gouging, 405
Oxygen welding, 405

P numbers in welding, 134
PAC (plasma-arc cutting), 405
Packed floating tube sheet, 66
Packing, 287, 384
  asbestos, 287
  carbon, 287
  polymer fiber, 287
  teflon, 287, 384
Parallel flow, 59
Parent metal, 405
Partial joint penetration, 142, 392, 401,
    405
Parting corrosion, 198
Passivation, 161, 405
Passive stainless steel, 161
Patch insert repairs, 369–370
PAW (plasma-arc welding), 405
Pearlite, 103, 405
Peening, 405

Penetrameter, 142, 160
Performance monitoring, 323
Permanent set, 405
Permits (welding and work), 331, 334,
    335
Phase diagrams for iron-carbon, 103
Phosphorus, 114
Photomicrograph, 405
Physical properties, hydrocarbons, 42–44
Physical properties, metals, 99, 104–115,
    405
  alloys, 113–115
  alpha iron, 102
  brittleness, 105
  creep, 109–110
  crystalline structure, 100
  delta iron, 102
  ductility, 105
  ferritic iron, 102
  formability, 103
  gamma iron, 102
  hardness, 105
  pig iron, 110–112
  reheating, 103
  slip planes, 102
  space lattice, 100–101
  stainless steel, 416–428
  steel, 112–117
  strength, 104
  toughness, 110
Pickling (metallurgical), 405
Piercing, 405
Pig iron, 110–112
Pilot-operated valves, 289, 312–313, 320
  concern on back pressure, 312
  types of, 312
Pipe, definition of, 405
Pits, 179, 405
Pitting, 188–189, 198, 368–369
  tubes, 198
Pitting repairs, 368–369
Plain steel, 116
Planned maintenance, 322
Plasma-arc cutting (PAC), 405
Plasma-arc welding (PAW), 127, 405
Plate heat exchangers, 70
Plug welds, 157, 223, 406
Plugged tubes, 372
Pneumatic controls, 289
Pneumatic-operated regulators, 289
Pneumatic tests, 249, 348–349, 380
Poisson's ratio, 201
Polymer fiber for packing, 287

Polymerization, 36
Porosity, 138, 367, 406
Porosity charts, 142
Postheating (welding), 406
Postweld heat treatment, 18, 29, 86, 138,
    325, 406
Precautions (see Safety)
Precipitation hardening, 406
Precipitation-hardening stainless steels,
    118, 425
Predictive maintenance, 323
Preheat, 139, 158, 181, 406
Preheat temperature, 406
Press rolls, 56
Pressure, 3–4
    absolute, 3
    atmospheric, 3
    gage, 3, 39
    vacuum, 3
Pressure gages, 264–267
    bellows, 265
    bourdon, 264–265, 318
    diaphragm, 266
    gas service, 267
    high temperature, 266
    inspector's connection, 266
    liquid service, 267
    manometer, 266
    positioning, 266–267
    siphon, 266, 318
Pressure switch, 321
Pressure-temperature-volume relations, 3
Pressure vessel laws, 12–14
    cities and counties, 14
    states, 13
Pressure vessels, 1–2, 9–18, 33–98
    air tanks, 33, 39–40
    ASME code, 12–14
    authorized inspector, 15–16
    chemical plant, 84–91
    chemical reactions in, 9–10
    chemical reactors, 34–36, 86–94
    clad plate, 161, 251
    codes and standards on, 11–12
    computer control of, 26–27
    cooker type, 34, 79–82
    cryogenic service (see Cryogenic pres-
        sure vessels)
    deaerators, 73–76
    digesters, 79–82
    distillation columns, 84–86
    Division 2 vessels, 16–17, 124
    effect of heat on, 8

Pressure vessels (Cont.):
    evaporators, 76–79
    feedwater heaters, 69, 72–73
    fiberglass-reinforced, 94–96
    fire hazard to, 9
    glass-lined, 88–89, 97
    heat exchangers, 33–34, 57–71
    hydrogen service, 93–94
    installation of, 38–39
    jacketed, 86–88
    lethal service, 10–11, 30
    lined, 251
    loadings, 29–30, 163
    maintenance, 19–23, 322–334
    multiwall, 91–93
    ordering, 20, 255–256, 379
    owner-user responsibility for, 17–18
    questionnaire for ordering, 20, 352,
        429–434
    quick-opening doors, 82–84
    risk analysis on, 24–25
    rotating (see Rotating pressure vessels)
    storage type (see Storage pressure ves-
        sels)
    supports for, 38, 244
    systems of vessels, 1–3
    thick-wall, 91
    vibration in, 39
Preventive maintenance, 322, 379
    high-pressure plant, 329
Primary flaws, 140
Priority alarms, 295
Procedure welding qualification, 134, 406
Process annealing, 406
Prod magnetic-particle testing, 160
Prod spacing, 160
Product impairment, 2, 253
Proof tests, 242–244
    bursting type, 242–243, 252
    yielding type, 242
Proportional limit, 169, 406
Proportioning controls, 291
Protective atmosphere (welding), 406
Protective clothing, 330–332
Pull-through floating head, 66, 69
Pulse-echo testing, 148
Pulsed-power welding, 406
Purging, 255, 333, 335–336, 380

Qualification tests (welder), 131–136
Qualified code inspector, 12–16
Qualified inspection agency, 12–16

Qualified NDT inspector, 151–152
Qualified welders, 130, 134–139
Quality control:
  in fabrication, 131–133, 386
  in repairs, 361–364
Quality control manual, 363–364
Quenching, 116, 154, 407
  regenerative, 407
Quenching medium, 407
Quenching temperature, 407
Questionnaire for pressure vessel ordering, 20, 325, 429–434
Quick-opening doors, 82–84, 245–247
  bar locking, 83
  expanding ring, 83
  inspection of, 245–247
  interlock, 245–247
  interlocking lug, 82–83

R stamp, 359–364
Radiation effect on materials, 184
Radiography, 124, 141–144, 345, 407
  degree of examination, 221–224
  fully radiographed, 221–224
  gamma rays, 141
  limitations, 143–144
  penetrameter, 142
  spot-examined, 221–224, 251
  x-rays, 141
Reactions, 9–10, 34–36
  chemical, 9–10, 34–36
  exothermic, 9, 31, 254, 320
  runaway, 10, 36, 296–297
Reactors in processes, 34–36, 87–91
  closed vessels, 87
  glass-lined, 88–89
  jacketed, 86
  safety considerations, 89–91
Recalescence, 407
Red-shortness, 407
Reduced-section tension test, 135–136
Reducing atmosphere (metallurgical),
  407
Reduction (process), 36
Reduction of area (material testing),
  407
Refinement of grain structure, 407
Refining temperature (heat treatment),
  407
Refractometry, 273
Regenerative quenching, 407
Regulators, 289
  pilot-operated, 289

Regulators (Cont.):
  pneumatic, 289
  spring-loaded, 289
Regulatory controls, 292–293
Reheating, 103
Reinforcement:
  of openings, 237–240, 250
  of weld, 407
Reinspection laws, 12–14
Relief valves, 298–315
  capacity, 304–305
  code requirements, 298–302
  discharge lines, 305
  installation, 305
  pressure setting, 304
  safety, 8, 39, 298
  seals, 306
  stamping, 298–301
  steam safety valves, 309–311
  test levers, 306
  vacuum, 313–315, 320
Repair organization, 360–364
Repairs, 364–376, 386
  cracks, 370–372
  determining reason for failure,
    365
  dissimilar metals, 375
  heat exchangers, 372–373
  NB rules, 364
  patches, 369–370
  pitting, 368–369
  R stamp, 359–364
  seal welding, 373
  sleeving, 373
  stainless steel, 373–375
  wasted areas, 369
  weld defect, 138
  weld problems in field, 365–368
Repetitive stress, 179
Residual stress, 366–367
Resilient materials, 203
Resistance of materials to stress, 164–
  169
Resistance welding, 407
Resonant testing, 148
Reverse polarity, 408
Richter scale, 273
Rimmed steel, 116, 408
Risk analysis, 24
Risk evaluation, 26
Rockwell hardness, 105
  of stainless steels, 427–428
Rockwell hardness test, 408
Root of joint, 408

Root of weld, 408
Root bead, 408
Root-bend test, 135–136
Root crack, 408
Rotating pressure vessels, 33, 52–57, 335, 352–354
  cracking in, 53
  cylinder type, 54
  Foudrinier paper machine, 54
  inspection of, 352
  paper machine, 54
  press rolls, 56
  safety precautions, 335
  slashers, 52
  suction rolls, 56
  Yankee roll, 54–56
Rupture, 161
Rupture disks, 87, 301–304
  burst pressure, 302
  cyclic service, 303
  discharge, 304
  materials, 303, 320
  setting, 302
  stamping, 301
  thickness, 303

Safety, 329–337
  electrical equipment in process, 331
  hydrostatic tests, 334, 346–348
  in illumination, 336–337
  internal inspections, 335–336
  leak tests, 349
  lifelines, 333
  maintenance, 329–334
  management responsibility for, 340
  management support, 376–379
  masks, 332–333
  pneumatic tests, 348–349
  protective clothing, 330–332
  in reactor vessels, 89–91
  rotating vessels, 335
  skin contact, 336
  supervision, 333–334
  in welding, 329, 331–332
Safety audits, 377
Safety factor, 212–213, 220
Safety relief valves, 8, 39, 298
  (See also Relief valves; Safety valves)
Safety valves, 298–302, 304–312
  capacity, 304
  code requirement, 298–302
  conversion from steam rating, 306–309
  discharge from, 305

Safety valves (Cont.):
  installation of, 305
  with rupture disks, 300–302
  seals, 306
  setting, 304
  spring, 309
  stamping, 298–300
  steam type, 309–311
  test levers, 306
  vacuum, 313–315
SAW (submerged-arc welding), 127, 411
Scab (metallurgical), 408
Scale, 408
  (See also Fouling)
SCAW (shielded-carbon-arc welding), 409
SCC (See Stress-corrosion cracking)
Scleroscope hardness test, 105, 408
Seal welding, 372, 408
Seals (safety relief valves), 306
Seam, 408
Seam weld, 408
Search gas test, 349
Secondary flaw, 140
Segregation (metallurgical), 408
Selective leaching, 204
Semiautomatic arc welding, 409
Semikilled steel, 116, 409
Sensitization, 119, 192–193, 374
Sensors, 275–276
Service fluid test, 349
Setting, safety and relief valves, 304
Shear stress, 166, 409
Shell-to-head juncture stresses, 230
Shielded-carbon-arc welding (SCAW), 409
Shielded-metal-arc cutting (SMAC), 409
Shielded-metal-arc welding (SMAW), 127, 157, 386, 409, 411
Shielding gas (welding), 409
Shipping of pressure vessels, 430
Shore hardness test, 105, 408
Short-circuiting transfer (welding), 409
Sigma phase, 119–120, 374
Sikes scale, 273
Silicon, 114
Single-train process, 22, 37, 378
Siphon, 266, 318
Size of weld, 409
Skelp, 409
Skin contact in inspections, 336
Slab, 409
Slag, 410
Slag inclusions, 365, 381, 410

Sleeving, 373
Slip bands, 410
Slip planes, 102
SMAC (shielded-metal-arc cutting), 409
SMAW (shielded-metal-arc welding), 127,
    157, 386, 409, 411
S-N diagrams, 176
Soaking (metallurgical), 410
Solidification range, 410
Solidus, 410
Solution anneal, 205, 385
Sonims, 410
Space lattice, 100–101, 154
Spacer strip (welding), 410
Spalling, 410
Spare equipment, 327
Spatter loss in welding, 410
Specific gravity, 273
Specific heat, 6
    of stainless steels, 428
Specific volume, 6
Specified grain size, 410
Sphere, 225–226
    thick, 236–237
Spheroidizing, 410
Spiral heat exchanger, 63, 70
Spot radiography, 221–224, 251
Spot weld, 410
Spring, safety-valve, 309
Spring-loaded check valve, 287
Square groove weld, 410
Stabilized stainless steel, 119
Staggered welds, 410
Stainless-steel liners, 36, 86
Stainless steels, 117–120, 416–425
    alloy content, 416–425
    application, 416–425
    austenitic, 117, 161, 417–420
    carbide precipitation, 119
    cladding, 120
    composition, 416–425
    corrosion, 192–194
    duplex, 118, 194, 424
    ferritic, 118, 161, 194, 421–422
    handling of, 118–119
    high-alloy, 118, 193–194
    martensitic, 118, 423–424
    mechanical properties, 416–428
    microstructure, 416–425
    passive, 161
    physical properties, 416–428
    precipitation-hardening, 118, 425
    repairs, 373–375

Stainless steels (*Cont.*):
    sensitization, 119, 374
    sigma phase, 119–120, 374
    types of, 416–425
    vessel inspection, 351–352
    welding of, 119, 380
Stamping (ASME or NB code):
    pressure vessels, 124
    safety valves, 298–300
Start-up procedures, 257–259
    for heat exchangers, 258–259
Status alarm, 295
Staybolt, 232
Staying, 232–233, 248
Steam safety valve, 309–311
    blowback, 309–310
    capacity of, 311
    escape pipe, 310
    number of, 311
    setting, 311
Steel, 112–118
    alloys, 113–114
    annealing, 115
    carbon (*see* Carbon steel)
    distillation columns, 122
    hardening, 116
    heat treatment, 115
    killed, 108, 116, 401
    low-alloy, 117
    low-temperature service, 121–122
    manufacture of, 112–113
    normalizing, 116
    plain, 116
    quenching, 116
    rimmed, 116, 408
    semikilled, 116, 409
    stainless (*see* Stainless steels)
    stress relief, 115
    stress table of, 209–211
Stick electrode, 411
Stick electrode welding, 157, 411
Stitch weld, 411
Storage pressure vessels, 33, 37, 41–48,
    315–317
    air tank, 39–40
    breather valves, 315–317
    combustible storage, 41
    cryogenic service, 49–50
    fixed-roof, 315
    flammable storage, 41
    floating-roof, 315
    variable-vapor-space, 315
    volatile-substance, 315

Straight polarity, 411
Strain, 167
Stranded electrode, 411
Strength calculations, 207–252
  bracing and staying, 232–233
  conical head, 229
  dished heads, 226–230
  ellipsoidal head, 227–228
  external pressure, 229–230
  flatheads, 230–232
  hemispherical head, 226–227
  openings, 237–240
  shells, 220–225
  spheres, 225–226
  thick shells, 235–236
  thick spheres, 236–237
  torispherical head, 228–229, 249
  tube-sheet thickness, 218–219
  tubes, 213–218
Stress, 164–167, 172–174, 220–225, 411
  bearing, 166
  bending, 165
  circumferential, 173–174, 220–225
  compression, 165
  longitudinal, 172–173, 220–225, 248
  normal, 164
  shear, 166, 409
  tension, 165
  thermal, 163, 174–175, 412
  torsion, 167
  tube-sheet joint, 219–220
Stress concentration, 164
Stress-concentration factors, 175–176, 203
Stress-corrosion cracking (SCC), 109, 190–191, 349–350, 411
  causes of, 349–350
  chloride effect, 193
  heat exchangers, 196–197, 357
  inspection for, 350–357
  stainless steels, 192–194
Stress relief, 114, 154, 411
Stress-strain diagram, 167–169
Stress table, 209–211
  for cast iron, 111
Stringer bead, 411
Strip chart, 274–275
Stud welding, 411
Submerged-arc welding (SAW), 127, 411
Suction rolls, 56
Sulfonation, 36
Sulfur, 114
Supports for pressure vessels, 38, 244

Surfacing weld, 411
Swaging (metallurgical), 411

T joint, 412
Tack welds, 137, 411
Tantalum plug, 89
Teflon packing, 287, 384
Telltale holes, 86, 97–98, 121
TEMA (Tubular Exchange Manufacturers Association), 11, 64, 207
  class B, C, R designation, 67
  shell-and-tube type classification, 64–69
Temper carbon, 412
Temperature, 4–5
  absolute, 4
  centigrade, 4
  cryogenic, 49, 96
  Fahrenheit, 4
Temperature gages, 267–270
  bimetallic, 267–268
  electronic, 267–270
  filled systems, 268–270
  infrared, 269–270
  thermistors, 269
  thermocouples, 269
Temperature gradient, 163, 174–175
Tempering, 412
Temporary weld, 412
Tensile strength, 200, 412
  of stainless steels, 427–428
Tension stress, 165
Test lever, 306
Test plates (welding), 135
Theoretical throat (fillet weld), 412
Thermal stress, 163, 174–175, 412
Thermistors, 269
Thermit welding, 126
Thermocouples, 269
Thick cylinders, 235–236
Thick spheres, 236–237
Thick-wall pressure vessels, 91
Thickness tests, 342–344
Throat of a fillet weld, 412
Throat depth, 412
Throat height, 412
TIG (tungsten-inert-gas welding), 127, 412
Titanium, 119
Toe weld crack, 412
Torispherical head, 228–229, 249
Torsion, 167

Tough materials, 203
Toxic gases, 10
Training, 377
  for inspection, 377
  for maintenance, 377
  for operation, 377
Tralles scale, 273
Transducer, 147–148
Transformation range, 412
Transformation temperature, 155
Transition temperature, 50, 154
Transmitters, 291–292
  electric, 291–292
  electronic, 291–292
  pneumatic, 291–292
Trend alarms, 294
Trepanning, 384
Tube-sheet thickness, 218–219
  bending, 219
  shear, 218
Tube-to-tube-sheet joint, 71–72
  stresses on, 219–220
Tubes, 213–218, 356–358
  allowable pressure, 215–218
  effect of flow, 214
  erosion, 196
  fatigue, 197
  fouling, 199–200
  leakage, 356
  manufacture, 214–215
  pitting, 198
  plugging, 358
  removal, 74–75
  turbining, 385
  vibration on, 197–198, 201, 357
Tubular Exchange Manufacturers Association (see TEMA)
Tungsten, 114
Tungsten electrode, 412
Tungsten-inert-gas welding (TIG), 127, 412
Turbining tubes, 385
Twaddle scale, 273

U stamp, 15
U tube, 66–67, 250
U-tube manometer, 266
Ultimate strength, 170, 213
Ultimate stress, 213
Ultrasonic testing, 147–149, 346
  A and B scan, 149
  pulse-echo, 148

Ultrasonic testing (Cont.):
  resonant, 148
Underbead cracking, 412
Undercut, 138, 367, 413
Underfill, 413
Uniform Boiler and Pressure Vessel Laws Society, 436
Upper explosion limit, 31–32
Upset, 413
  in forging, 413
  in welding, 413

Vacuum degassing, 413
Vacuum induction melting, 413
Vacuum pressure, 3, 89
Vacuum relief valves, 313–315, 320
  diaphragm pallet, 313–315
  liquid-seal, 313–315
  solid pallet, 313–315
Vacuum test, 349
Valves, 276–290
  ball check, 287
  bleed, 282
  block, 282
  butt-weld connections, 288
  butterfly, 285
  check, 285–287
  code requirements, 276–277
  connections, 287–288
  control, 289–290
  drain, 282, 385
  dump, 282, 385
  flanged end connections, 288
  gate, 282–283
  globe, 283–284
  lift check, 286
  maintenance, 287
  manual, 282
  material, 277–281
  needle, 284
  relief (see Relief valves)
  safety (see Safety valves)
  screwed connection, 288
  selection of, 277
  socket-weld connection, 288
  spring-loaded check, 287
  swing check, 285
  trim, 321
  types of, 281–287
  vent, 315–316, 321
Vendor surveillance, 326

Vents, 45–46, 72, 199, 315–316, 321
  capacity, 47
  combustible liquids, 45–46
  feedwater heater, 72
  flammable liquids, 45–46
  heat exchangers, 199
  valves, 315–316, 321
Verification inspections, 340
Vertical position (welding), 413
Vibration, 39–40, 72, 163, 244
  air tanks, 39–40, 244
  feedwater heaters, 72
  on tubes, 197–198, 201, 357
Vickers hardness test, 413
Visual inspection, 140–141, 345
  borescope, 140
Volatile substance storage, 315
Volume of gas, 6–7
Vortex shedding, 201
Vulcanizers, 79–80

Wasted area repair, 369
Water quenching, 413
Wear rates, 339, 342, 344
Weight of contents, 29, 38, 163, 347
Weight-loss corrosion test, 204
Weld, 413
  butt, 137, 157, 222
  classification, 157
  cracks, 182, 367, 413
  failures, 180
  heat treatment, 18, 29, 86, 138, 325, 406
  metal-arc, 403
  metal overlay, 376, 381, 404, 414
  overlay repairs, 376
  reinforcement, 142, 158
  repairs, 365–368
Weld bead, 413
Weld gage, 414
Weld interface, 414
Weld metal, 414
Weld-metal area, 414
Weld pass, 414
Weldability, 413
Weldbonding, 413
Welded joint categories, 221
  butt, 221–223, 390
  efficiency, 137, 221–224
  fillet, 157, 396
  for lethal substances, 248
  locations on vessel, 221

Welder qualification, 130, 134–139
  positions for, 135, 414
  recertification in, 135
  test plate for, 135–136
Welding, 103, 108, 124–139, 325, 414
  arc, 126–127, 389
  ASME code, 130–139
  autogenous, 389
  automatic, 389
  clad plate, 121
  contractor's responsibility, 131
  defects, 138
  definition of, 125
  electrodes, 128
  electrogas, 395
  electron-beam, 128, 395
  filler metal, 125, 396
  gas-metal-arc, 127, 398
  gas-tungsten-arc, 127, 398
  heli-arc, 127
  incomplete penetration, 367
  inspection, 135–139
  joints, 157, 221–224, 390, 396
  operator position, 135–136
  operator qualification, 130–134, 414
  oxy-fuel gas, 127, 404
  plasma-arc, 127, 405
  positions, 135, 414
  postheating, 18, 29, 86, 138, 325, 406
  procedure specifications, 414
  procedures, 130, 134, 386, 414
  safety, 329, 331–332
  stainless steel, 119, 380
  surfaces, 138
  types of, 125–128
  vessel welding ordering, 325
  (See also Joining by welding; Weld)
Welding engineers, 125
Welding in field, 365–368
  cracks, 367
  defect removal, 366
  defects, 138
  discontinuities, 366
  environmental problems, 365
  incomplete penetration, 367
  overlap, 367
  porosity, 367
  residual stress, 366–367
  slag inclusion, 367
  undercut, 367
Welding rod, 414
Welding sequence, 414
Welding technique, 414

White cast iron, 111
Wickel-type multilayer pressure vessels,
    91
Wind loads, 162, 243
Wire drawing, 357
Work hardening, 415
Work order, 327
    (*See also* Permits)

X-rays, 141
    degree of examination, 221–224
    (*See also* Radiography)

Yankee dryer roll, 54–56, 97
    cracking hazard, 56, 97
Yield point, 169, 415
Yield proof tests, 242
Yield strength, 415
    for stainless steel, 427–428
Yoke-type magnetizing units, 146

Zyglo inspection, 415

ABOUT THE AUTHOR

Anthony L. Kohan has more than 35 years' experience operating, testing, and inspecting boilers and pressure vessels
in power generating plants as well as vastly diverse industrial and chemical plant complexes. During his employment
with the engineering departments of various insurance
companies, he has investigated and reviewed accident reports on pressure vessel failures, prepared material for loss
prevention inspections, and supervised inspectors and
trained them in ASME pressure vessel code requirements.
Mr. Kohan is the coauthor of three other McGraw-Hill
books, including *Standard Heating and Power Boiler Plant
Questions & Answers*, 2d ed., and *Standard Boiler Operators' Questions and Answers*.